Yams: Botany, Production and Uses

Yams: Botany, Production and Uses

Anthony Keith Thompson and Ibok Oduro

CABI

CABI is a trading name of CAB International

CABI
Nosworthy Way
Wallingford
Oxfordshire OX10 8DE
UK

Tel: +44 (0)1491 832111
Fax: +44 (0)1491 833508
E-mail: info@cabi.org
Website: www.cabi.org

CABI
WeWork
One Lincoln St
24th Floor
Boston, MA 02111
USA

Tel: +1 (617)682-9015
E-mail: cabi-nao@cabi.org

A catalogue record for this book is available from the British Library, London, UK.

Library of Congress Cataloging-in-Publication Data

Names: Thompson, Anthony Keith (Agriculture), author. | Oduro, Ibok, author.
Title: Yams : botany, production and uses / Anthony Keith Thompson, Ibok Oduro.
Other titles: Botany, production and uses.
Description: Boston, MA : CAB International, [2021] | Series: Botany, production and uses | Includes bibliographical references and index. | Summary: "Dioscorea species, commonly known as yam, are tuberous plants which constitute a major staple food in many parts of Africa, Southeast Asia and the South Pacific. This book covers botany, taxonomy, composition and uses, cultivation, production, diseases and pests, and storage"-- Provided by publisher.
Identifiers: LCCN 2021006789 (print) | LCCN 2021006790 (ebook) | ISBN 9781789249279 (hardback) | ISBN 9781789249286 (ebook) | ISBN 9781789249293 (epub)
Subjects: LCSH: Yams.
Classification: LCC SB211.Y3 T48 2021 (print) | LCC SB211.Y3 (ebook) | DDC 635/.23--dc23
LC record available at https://lccn.loc.gov/2021006789
LC ebook record available at https://lccn.loc.gov/2021006790

References to Internet websites (URLs) were accurate at the time of writing.

ISBN-13: 9781789249279 (hardback)
9781789249286 (ePDF)
9781789249293 (ePub)

DOI: 10.1079/ 9781789249279.0000

Commissioning Editor: Rebecca Stubbs
Editorial Assistant: Emma McCann
Production Editor: James Bishop

Typeset by SPi, Pondicherry, India
Printed and bound in the UK by Severn, Gloucester

Contents

About the authors

Anthony Keith Thompson is Visiting Professor at King Mongkut's Institute of Technology Ladkrabang in Thailand and was formerly Professor of Plant Science, University of Asmara, Eritrea; Professor of Postharvest Technology and head of department, Cranfield University, UK; Principal Scientific Officer, Tropical Products Institute, London; Team Leader for EU funded projects in the Windwards Islands and Jamaica; Postharvest Expert for the UN in Sudan, Yemen and Korea for the Food and Agriculture Organization, Ghana and Sri Lanka for the International Trade Centre and Gambia for the World Bank; Advisor to the British, Jamaican and Colombian governments in postharvest technology of fruit and vegetables; Research Fellow in Crop Science, University of the West Indies, Trinidad; Demonstrator in Biometrics, University of Leeds. He has also carried out almost 100 consultancies for various commercial and government organizations in many countries throughout the world. He has published over 100 scientific papers and reports (including eight on yams) and many scientific textbooks.

Ibok Oduro is a Professor at the Department of Food Science and Technology, Kwame Nkrumah University of Science and Technology, Kumasi, Ghana. She is dedicated to promoting indigenous and underutilized food crops in West Africa and her work aims at converting scientific findings into practical solutions. Her interest is in development of competitive products and students' empowerment on commercialization of research output leading to the establishment of start-ups or spin-off businesses. She has over 20 years of research, theory, practice and policy knowledge, teaching and entrepreneurial experience in development and product diversification of underutilized crops, perishable crops processing, food compositional studies and traditional foods and systems. Professor Oduro teaches both undergraduate and postgraduate students and has supervised three PhD and five masters theses. She has authored over ten research paper publications on yams and more than 100 other publications published in international journals along with five book chapters. Professor Oduro is the Immediate Past first female Provost of the College of Science and the Kwame Nkrumah University of Science and Technology. She was the coordinator for IITA New and Diverse Food Products developed from the *D. alata* project; Collaborating Scientist for the IFAD/WECARD/IITA project on Poverty Alleviation and Enhanced Food Availability in West Africa through improved yam technologies; and Collaborating Scientist for the Root and Tuber Improvement Programme concerned with on-farm storage of yam tubers.

Acknowledgements

Thanks are due to the following for advice, critical help and allowing the publication of photographs: Sarinya Kumpila; Professor Dr Vincent Lebot; Professor Dr Graham Seymour; Associate Professor Dr Praphan Pinsirodom; Assistant Professor Dr Chirdsak Thapyai; Assistant Professor Jiraporn Sirison; Associate Professor Dr Ratiporn Haruenkit; Ing. Joseph O. Akowuah; Dr Nana Pepra-Ameyaw; Dr (Mrs) Irene A. Idun; Dr Dewayne Stennett; Dr Roger Bancroft; Dr Marcella Corcoran; Pikunthong Nukthamna; Shamina Maccum; and Life Sciences, Springer, Holland.

Preface

When thinking about writing this book we considered primarily what its function should be and, effectively, the purpose of any scientific textbook. Our conclusions were that it should evaluate the current state of knowledge and how this knowledge affects current practices. This would give indications of why this particular industry of yam production and utilization is where it is now and what has influenced it getting there. This, in turn, would influence where it might go in the future. Therefore, our approach to the yam industry in this book has been to review the appropriate literature and reach a consensus on the findings of scientists carrying out research into all aspects of yams. Also, to look at commercial yam production and marketing and the situation in terms of producer countries, in order to give some indication as to why they are at the point they are. This gives an indication of where they could usefully develop in the future. *Dioscorea* is such a diverse genus in terms of its morphology and chemistry as well as its geographical origin, domestication and breeding, that there is a need to examine *Dioscorea* species that currently have little or no commercial or domestic value, but might have value in future or could, in some way, contribute to the future production and utilization of this crop.

A challenge in preparing the book is the number of specific and common names applied to yams. In published papers and books, it is often not clear which yam is being referred to. It is not claimed that this book has solved or even usefully contributed to the debate of taxonomists on this issue, but hopefully it makes the situation a little clearer. An example of this challenge is the species *D. oppositifolia* and *D. opposite*. It is clear that most authorities accept that they are the *same* species, but publications may refer to one or the other and it may be a varietal or cultivar difference or they may have even got the species completely wrong. So, we have largely reviewed the work under the specific name given by the authors of the publication or that used by farmers.

A further challenge in producing this book is the number of publications. S.M. Lawani and M.O. Odubanjo produced a bibliography in 1976 which listed 1,562 publications related to *Dioscorea*, which was, even then, nowhere near complete. So, producing such a book as this will result in justifiable criticism in terms of relevant research that unfortunately has been omitted, but hopefully this criticism will be minimal and positive.

1

Introduction

In this book the name 'yam' is applied only to members of the *Dioscorea* genus, although many other root and tuber crops are sometimes referred to as yams. The name yam derives from older names, including jugnamis, iniamea, yamme and yame. In French the word ignome derives from the earlier igniame. In Spanish they are called niame and nyame. In Portuguese they are called inhame, derived from ynhame. In Dutch they are called iniame. In Italian they are called gname and ignamo. In German they are called ignamkolle and yamswurtzel. In Arabic they are called ighnam. The Mandé groups in West Africa (Benin, Burkina Faso, Côte d'Ivoire, Gambia, Ghana, Guinea, Guinea-Bissau, Liberia, Mali, Mauritania, Niger, Nigeria, Senegal and Sierra Leone) often use the word 'niam' for yam (Cumo, 2013).

The order Dioscoreales Hook. f. contains three families: Dioscoreaceae, Burmanniaceae and Nartheciaceae. The highest diversity of the order is in the tropics of South America, the Caribbean islands, Africa, Madagascar, China and Indochina with Dioscoreaceae and Burmanniaceae occurring mainly in the wet tropics and Nartheciaceae mainly in mountainous habitats. Dioscoreaceae was reported to contain four genera: *Dioscorea* Plum ex L. (653 species); *Stenomeris* Planch. (2 species); *Tacca* J.R. Forst and G. 75 Forst (17 species); and *Trichopus* Gaertn. (1 or 2 species) (Caddick *et al.*, 2002; Govaerts *et al.*, 2007). *Tacca* was previously in the family Taccaceae. However, it has also been reported that Dioscoreaceae contains six genera: *Dioscorea*, *Epipetrum*, *Rajania*, *Stenomeris*, *Tacca* and *Trichopus* (Burkill, 1960; Coursey, 1967; The Plant List, 2018) although Bornstein (1989) reported that R. Knuth recognized ten genera in Dioscoreaceae in 1930. Wikiwand (2018) reported that there were 613 accepted species and varieties in Dioscoreaceae and the *Encyclopedia of Life* (EOL, 2019) listed 626 *Dioscorea* species.

Dioscorea was named after a Greek physician and naturalist called Pedanios Dioscoride. The history of *Dioscorea* species has been linked to humans for thousands of years by a slow and gradual process of domestication (Ayensu and Coursey, 1972). The occurrence of *Dioscorea* species in southern Asia, Africa and South America and their domestication in these areas appears directly linked to indigenous populations. Rubatzky and Yamaguchi (1997) reported that artefacts dated about 50,000 BCE from West Africa indicate that wild yams were being used for food at that time and cultivation of yams appears to have developed independently in West Africa, Tropical America and South East Asia from about 3,000 BCE. Since that time some

DOI: 10.1079/9781789249279.0001

species have been introduced by man from different areas, for example *D. alata* was introduced at an uncertain date into West Africa, and *D. alata* and *D. esculenta* into Tropical America in or after the 16th century (Coursey, 1967).

It was estimated (FAO Stats, 2019) that the area under yam cultivation in 2017 was about 8 million hectares yielding about 73 million tonnes of tubers. Different species had their origin in different parts of the world, but the majority that are used for food are of tropical origin, mainly in West Africa (Table 1.1). However, yams are important in other regions, for example in parts of India (Table 1.2) and South America. Belachew Garedew *et al.* (2017) reported that in the Sheko district of Ethiopia, yams were a main staple food and farmers were growing *D. abyssinica*, *D. alata* and *D. bulbifera*, with *D. alata* preferred because of its taste and high yield. Hahn (1995) reported that only six species were of major economic importance as food plants: *D. rotundata*, *D. alata*, *D. cayenensis*, *D. bulbifera*, *D. dumetorum* and *D. esculenta*.

Tubers of *Dioscorea* species produce a high value food and the plants are easily grown, mature quickly in the right soil conditions and their tubers can generally be stored after harvest for protracted periods. In addition, some species are economically important due to secondary metabolites present in the tubers, particularly diosgenin, which has a molecular structure similar to steroidal hormones and has been used to synthesize steroids for the production of birth control pills. Most *Dioscorea* species grow best in rainfall greater than 1,500 mm year^{-1} and require a minimum 6-month growing season with well distributed rainfall, but do not tolerate poorly drained soils or waterlogging. They are mildly drought tolerant but do not compete well with weeds for soil nutrients. Most species should be staked to improve yield, reduce weed competition and the incidence of diseases. Yams exhibit early shade tolerance during establishment but require full sun for optimum yield.

As well as being a starch staple in many areas, some wild *Dioscorea* species are famine foods and, as mentioned above, others are sources of drugs both in traditional

Table 1.1. Geographic origin of what was referred to as the ten most important cultivated *Dioscorea* species. Modified from Lebot (2019).

Species	Geographical origin
D. alata	South East Asia, Melanesia
D. bulbifera	South America, Africa, Asia,
D. cayenensis	West Africa
D. esculenta	South East Asia, Melanesia
D. opposite, *D. japonica*	Japan, China
D. nummularia	Melanesia
D. pentaphylla	South East Asia, Melanesia
D. rotundata	West Africa
D. transversa	Australia, Melanesia
D. trifida	South America

Table 1.2. Wild edible tubers, rhizome, corm, roots and stems consumed by the tribal Valaiyans in India. Mohan and Kalidass (2010).

Botanical name	Plant parts used
Aponogeton natans (L.) Engler and k.	Tuber
Boerhavia chinensis (L.) Asch and Schweinf.	Root
Caralluma adscendens (Roxburgh) Haw. var. *attenuata* (Wight) Grav. and Mayuranathan	Stem
Caralluma pauciflora (Wight) N.B.Br.	Stem
Canna indica L.	Stem
Cissus quadrangularis L.	Rhizome
Cissus vitiginea L.	Tuber
Colocasia esculenta (L.) Shott	Corm
Cycas circinalis L.	Tuber
Cyphostemma setosum (Roxb) Alston	Tuber
Decalepis hamiltonii Wight and Arn.	Tuber
Dioscorea pentaphylla L. var. *pentaphylla*	Tuber
Dioscorea oppositifolia L. var. *oppositifolia*	Tuber
Dioscorea spicata Roth.	Tuber
Dioscorea tomentosa Koen. Ex. Spreng.	Tuber
Hemidesmus indicus (L.) R. Br. var. *indicus*	Root
Ipomoea staphylina Roem and Schultes	Root
Kedrostis foetidissima (Jacq.) Cogn.	Tuber
Maerua oblongifolia (Forsk.) A. Rich	Tuber
Momordica diocia Roxb ex Willd	Tuber
Nymphaea pubescens Willd	Tuber
Nymphaea rubra Roxb ex Andrews	Tuber
Parthenocissus neilgherriensis (Wight) Planch	Tuber

Chinese and Western medicine, including *D. nipponica* and *D. zingiberensis* that are major sources of steroid precursors. Abhyankar and Upadhyay (2011) commented that some ethnic groups in India are dependent on plant extracts for curing various ailments, including *D. bulbifera*, *D. aculeata*, *D. sativa* and *D. arachnida*. This use has been recently decreasing, perhaps because of the lack of confidence of younger generations in traditional medicine and the increasing availability of modern medicines. *D. oppositifolia* tubers are used in Chinese medicine and as herbal tonics (Tu, 2002). Dried *D. japonica* tubers were said to be cut into shavings and used in traditional Chinese medicine and as a tonic (Birkill, 1935). In Korea, Kim *et al.* (2011a) reported that *D. japonica* tubers have been used as a medicine known as San Yak and they also reported that *D. quinqueloba* tubers must be correctly prepared for traditional medicine, otherwise their use may cause life-threatening acute kidney injury. *D. villosa* tubers were reported to be used in traditional medicine by Native Americans in the USA (Moerman, 1998). In parts of India, Choudhary *et al.* (2009) reported that *D. bulbifera* was used as a medicinal plant and Swarnkar and Katewa (2008) reported its use as a contraceptive. In summary, *Dioscorea* species occur throughout the world and have many uses, as well as being a basic staple food.

2

Botany

Dioscorea species are mostly vines (Fig. 2.1) that scramble and twine over other vegetation. The direction of twining of the stems is species-specific (Prain and Burkill, 1938), for example *D. zingiberensis*, *D. sansibarensis* and *D. tentaculigera* twine to the left and *D. glabra*, *D. transversa* and *D. polystachya* twine to the right. *Dioscorea* species generally have simple cordate leaves, borne on long narrow stems that vary in shape even on the same plant (Fig. 2.2). *Dioscorea* species bear racemes of inconspicuous dioecious flowers – that is, the male and female flowers are separate and are usually borne on separate plants, although some species may be monoecious. Ding *et al.* (2000) reported that *D. zingiberensis* is the only Chinese species recorded as being monoecious, Hawley (1956) reported that the Brazilian species, *D. lagoa-santa* was monoecious and Hammel (2000) reported that the Costa Rican species, *D. lepida* was monoecious. Female flowers are borne in axillary spikes (unbranched, indeterminate inflorescences bearing sessile flowers) and male flowers in panicles (inflorescences with many branches) that are up to about 30 cm long (Fig. 2.3). There are many more male than female flowers. The fruit is a dry dehiscent, trilocular capsule varying in size from about 1 to 3 cm long and longer than wide or equally wide as long, depending on species. The capsules contain winged seeds that are dispersed by wind (Figs 2.4, 2.5) (Coursey, 1967; Ding and Gilbert, 2010).

Two types of tuber are produced by *Dioscorea* species, those above ground, commonly called bulbils and sometimes referred to as aerial tubers, and those produced below ground that are normally just referred to as tubers or sometimes rhizomes.

Bulbils are produced in leaf axils by some species (Fig. 2.6) including: *D. alata*, *D. bemarivensis*, *D. bulbifera*, *D. filiformis*, *D. kamoonensis*, *D. oppositifolia*, *D. pentaphylla*, *D. persimilis*, *D. polystachya*, *D. prazeri*, *D. puber*, *D. pyrenaica*, *D. sansibarensis*, *D. schimperiana*, *D. smilacifolia*, *D. togoensis* and *D. transversa*. Structurally bulbils are stems and have evolved as a method of reproduction, where they may have advantages over underground tubers since they are less dense and can float in water, which may assist distribution in some cases. The limited number of studies conducted on bulbil germination, all performed under laboratory conditions, suggest that bulbils are not dormant and may germinate shortly after dispersal (Okagami and Tanno, 1991). However, Walck *et al.* (2010) reported that germination of bulbils of *D. polystachya* in China generally occurs in spring at the beginning of a favourable environment for growth since they are mostly dormant when dispersed in summer

DOI: 10.1079/9781789249279.0002

Fig. 2.1. *D. alata* (a) and *D. esculenta* (b) growing in a woodland area in Thailand, at Sarinya's yam farm in Chachoengsao province, eastern region, 2018.

Fig. 2.2. *D. pentaphyla* (a, b and c) showing different leaves from the same plant and *D. oryzetorum* (d), all growing at Sarinya's yam farm, 2018.

or autumn and dormancy can be overcome with cold stratification during winter.

In general, there are two types of underground tubers produced by plants: tuberous roots that are lateral roots, modified and enlarged to function as a storage organ – e.g. cassava – and stem tubers that are stems modified and enlarged to function as a storage organ – e.g. potatoes and yams. Stem tubers develop from swellings of stolons arising from the crown of the plant. Stolons generally grow horizontally in the soil as do tubers, but tubers of some species grow vertically downwards in the soil, including *D. acuminata*, *D. daunea*, *D. filiformis*, *D. hamiltonii and D. japonica*. Tubers can produce new shoots and roots at points along their length to form new plants. Stolons of *Dioscorea* species vary in length from less than 5 cm to about 50 cm and form from thickened rhizomes where the top parts of the tuber generally produce shoots and the lower parts generally produce roots, but if the top part is removed shoots can be produced from the lower part (Fig. 2.7). At the end of the growing season the tops of the plants die down and after a period of dormancy the tubers begin to regrow. There may be several tubers produced from each plant or there may be only a single tuber.

Rubatzky and Yamaguchi (1997) commented that there were some anomalous features of *Dioscorea* tubers that make them difficult to categorize. These include the lack of remnants of scale leaves or vestigial nodes on the surface, and the fact that they

Fig. 2.3. (a) Male inflorescences of diploid accessions of *D. alata* (first two on the left), tetraploid *D. alata* (third inflorescence) and inflorescences of *D. nummularia* cultivars; (b) male inflorescences of *D. nummularia*; (c) and (d) inflorescences of *D. nummularia* growing in the Pacific islands. Lebot *et al.* (2016).

Fig. 2.4. Fruit of *D. nummularia* growing in the Pacific islands: (a) female inflorescence; (b) female inflorescence with developing capsules; (c) capsules with developing seeds. Lebot *et al.* (2016).

produce no preformed buds or eyes on, or near, the surface as there are in potato tubers. Their growing point may have no terminal bud and most species exhibit strong geotropic growth. Rubatzky and Yamaguchi also pointed out that there was evidence that the *Dioscorea* storage organ originates from hypocotyl tissue as the first meristematic activity in tuber formation occurs in that region. In tuber growth, the primary meristem is at the distal end. As growth continues, thin layers of meristematic cells are produced beneath the cortex, which, with continued activity, results in the increase in girth. Some of the parenchyma cells just below the epidermis can become meristematic and produce cork cells protecting the tuber (see also 'Tuber development', p. 19). Most tubers have a dark brown skin and usually white flesh, although some have other colours. Yams are perennial but the tops die back and there is a period of dormancy before the plant regrows from the tubers. The appearance of small protuberances under the skin layer is an indication of the end of dormancy. The length of tuber dormancy is endogenously controlled but temperature and soil moisture can affect when they regrow. Postharvest, the physiological age of tubers as well as light, temperature, humidity and partial pressure of the surrounding gases can affect when they will regrow (see section on 'Sprouting', p. 70).

Fig. 2.5. *D. glabra* – mature fruit in natural forest in Chaiyaphum province in northeastern Thailand. Reproduced with permission of Dr Chirdsak Thapyai.

Taxonomy

Girma *et al.* (2016) commented that distinguishing *Dioscorea* species, based on morphological traits, is extremely difficult and unreliable. The number of species reported in *Dioscorea* varies with different authors but generally it is commonly claimed that it contains around 650 species (Caddick *et al.*, 2002; Acevedo-Rodríguez and Strong, 2005; Govaerts *et al.*, 2007; Lebot, 2009). A total of 1,600 names have been attributed to *Dioscorea*, among species, varieties and subspecies, with 77 considered as synonyms

Fig. 2.6. Bulbils from two *Dioscorea* species: (a) *D. bulbifera* bulbil in leaf axil; (b) *D. bulbifera* bulbil sliced longitudinally, both from a wild race in Siritkit Dam, Uttaradit Province of northern Thailand (reproduced with permission of Dr Chirdsak Thapyai); and (c) *D. alata* var *purpurea* from Sarinya's yam farm in Chachoengsao province in the eastern region of Thailand.

Fig. 2.7. Yam for sale at a supermarket in Huddersfield in January 2018.

(The Plant List, 2013). The genus *Dioscorea* is divided into five clades (sections) within which the species are grouped:

> **Enantiophyllum**, where the vines twine in a clockwise direction, including *D. polystachya*, *D. rotundata*, *D. alata*, *D. cayenensis*, *D. japonica*, *D. opposita* and *D. transversa*.
> **Lasiophyton**, where the vines twine anti-clockwise , including *D. pentaphylla*, *D. dumetorum* and *D. hispida*.
> **Opsophyton**, where the vines twine anti-clockwise, including *D. bulbifera*.
> **Combilium**, where the vines twine anti-clockwise, including *D. esculenta*.
> **Macrogynodium** (Macroura), where the vines twine anti-clockwise, including *D. trifida* and *D. sansibarensis*.

IPGRI/IITA (1997) published descriptors of *Dioscorea* as an aid to identification. This comprehensive list of *Dioscorea* species is as follows:

D. abysmophila Maguire & Steyerm.
D. abyssinica Hochst. ex Kunth
D. acanthogene Rusby
D. acerifolia Phil.
D. aculeata L
D. acuminata Baker
D. adenantha Uline
D. aesculifolia R. Knuth
D. aguilarii Standl. & Steyerm.
D. alata L.
D. alatipes Burkill & H. Perrier
D. althaeoides R. Knuth
D. altissima Lamarck
D. amaranthoides C. Presl
D. amazonum Mart. ex Griseb.
D. amoena R. Knuth
D. analalavensis Jum. & H. Perrier
D. ancachsensis R. Knuth
D. andina Phil.
D. andromedusae O. Tellez
D. anomala Griseb.
D. antaly Jum. & H. Perrier
D. antucoana Uline ex R. Knuth
D. arachidna Prain & Burkill
D. araucana Phil.
D. arcuatinervis Hochr.
D. argyrogyna Uline ex R. Knuth
D. arifolia C. Presl
D. aristolochiifolia Poepp.
D. asclepiadea Prain & Burkill
D. aspera Humb. & Bonpl. ex Willd.
D. aspersa Prain & Burkill
D. asperula Pedralli
D. asteriscus Burkill
D. atrescens R. Knuth
D. auriculata Poepp.
D. bahiensis R. Knuth
D. bako Wilkin
D. balcanica Kosanin
D. bancana Prain & Burkill
D. banzhuana S.J. Pei & C.T. Ting
D. bartlettii C.V. Morton
D. basiclavicaulis Rizzini & A. Mattos
D. baya De Wild.
D. beecheyi R. Knuth
D. belophylla (Prain) Voigt ex Haines
D. belizensis Lundell
D. bemandry Jum. & H. Perrier
D. bemarivensis Jum. & H. Perrier
D. benthamii Prain & Burkill
D. berenicea McVaugh
D. bermejensis R. Knuth
D. bernoulliana Prain & Burkill
D. besseriana Kunth
D. beyrichii R. Knuth
D. bicolor Prain & Burkill
D. biformifolia S.J. Pei & C.T. Ting
D. biloba (Phil.) Caddick & Wilkin
D. biplicata R. Knuth
D. birmanica Prain & Burkill
D. birschelii Harms ex R. Knuth
D. blumei Prain & Burkill
D. bolivarensis Steyerm.
D. bonii Prain & Burkill
D. bosseri Haigh & Wilkin

D. brachybotrya Poepp.
D. brachystachya Phil.
D. bradei R. Knuth
D. brandisii Prain & Burkill
D. brevipetiolata Prain & Burkill
D. bridgesii Griseb. ex Kunth
D. brownii Schinz
D. bryoniifolia Poepp.
D. buchananii Benth.
D. buckleyana Wilkin
D. bulbifera L.
D. bulbotricha Hand.-Mazz.
D. burchellii Baker
D. burkilliana J. Miège
D. cachipuertensis Ayala
D. calcicola Prain & Burkill
D. caldasensis R. Knuth
D. calderillensis R. Knuth
D. callacatensis R. Knuth
D. cambodiana Prain & Burkill
D. campanulata Uline ex R. Knuth
D. campestris Griseb.
D. campos-portoi R. Knuth
D. carionis Prain & Burkill
D. carpomaculata O. Téllez & B.G. Schub.
D. castilloniana Hauman
D. catharinensis R. Knuth
D. caucasica Lipsky
D. cayennensis Lamarck
D. ceratandra Uline ex R. Knuth
D. chacoensis R. Knuth
D. chagllaensis R. Knuth
D. chancayensis R. Knuth
D. chaponensis R. Knuth
D. chiapasensis Matuda
D. chimborazensis R. Knuth
D. chingii Prain & Burkill
D. choriandra Uline ex R. Knuth
D. chouardii Gaussen
D. cienegensis R. Knuth
D. cinnamomifolia Hook.
D. cirrhosa Lour.
D. cissophylla Phil.
D. claessensii De Wild.
D. claussenii Uline ex R. Knuth
D. claytonii Ayala
D. cochleariapiculata De Wild.
D. collettii Hook. f.
D. communis (L.) Caddick & Wilkin
D. commutata R. Knuth
D. comorensis R. Knuth
D. composita Hemsl.
D. contracta R. Knuth
D. convolvulacea Cham. & Schltdl.
D. conzattii R. Knuth
D. cordifolia Laness.
D. coreana (Prain & Burkill) R. Knuth
D. coriacea Humb. & Bonpl. ex Willd.
D. coripatenis J.F. Macbr.
D. coronata Hauman
D. cotinifolia Kunth
D. craibiana Prain & Burkill
D. crateriflora R. Knuth
D. crotalariifolia Uline
D. cruzensis R. Knuth
D. cubensis R. Knuth
D. cumingii Prain & Burkill
D. curitybensis R. Knuth
D. cuspidata Humb. & Bonpl. ex Willd.
D. cuyabensis R. Knuth
D. cyanisticta J.D Sm.
D. cymosula Hemsl.
D. cyphocarpa C.B. Rob. ex Knuth
D. daunea Prain & Burkill
D. davidsei O. Tellez
D. demourae Uline ex R. Knuth
D. debilis Uline ex R. Knuth
D. decaryana H. Perrier
D. decipiens Hook. f.
D. decorticans C. Presl
D. deflexa Griseb.
D. delavayi Franch.
D. delicata R. Knuth
D. deltoidea Wall. ex Griseb.
D. dendrotricha Uline
D. densiflora Hemsl.
D. depauperata Prain & Burkill
D. diamantinensis R. Knuth
D. dicranandra Donn. Sm.
D. dielsii R. Knuth
D. dissimulans Prain & Burkill
D. divaricata Blanco
D. diversifolia Griseb.
D. dodecaneura Vell.
D. dregeana (Kunth) T. Durand & Schinz
D. duchassaingii R. Knuth
D. dugesii C.B. Rob.
D. dumetorum (Kunth) Pax
D. dumetosa Uline ex R. Knuth
D. ekmanii R. Knuth
D. elegans Ridl. ex Prain & Burkill
D. elephantipes (L'Hér.) Engler
D. entomophila Hauman
D. epistephioides Taub.

D. escuintlensis Matuda
D. esculenta (Lour.) Burkill
D. esquirolii Prain & Burkill
D. exalata C.T. Ting & M.C. Chang
D. fandra H. Perrier
D. fasciculocongesta (Sosa & B.G. Schub.) O. Téllez
D. fastigiata Gay
D. fendleri R. Knuth
D. ferreyrae Ayala
D. filiformis Blume
D. flabellifolia Prain & Burkill
D. flaccida R. Knuth
D. floribunda M. Martens & Galeotti
D. floridana Bartlett
D. fodinarum Kunth
D. fordii Prain & Burkill
D. formosana R. Knuth
D. fractiflexa R. Knuth
D. fuliginosa R. Knuth
D. furcata Griseb.
D. futschauensis Uline ex R. Knuth
D. galeottiana Kunth
D. galiiflora R. Knuth
D. gallegosi Matuda
D. garrettii Prain & Burkill
D. gaumeri R. Knuth
D. gentryi O. Tellez
D. gillettii Milne-Redh.
D. glabra Roxburgh
D. glandulosa (Griseb.) Klotzsch ex Kunth
D. glomerulata Hauman
D. gomez-pompae O. Tellez
D. gracilicaulis R. Knuth
D. gracilipes Prain & Burkill
D. gracilis Hook. ex Poepp.
D. gracillima Miq.
D. grandiflora Mart. ex Griseb.
D. grandis R. Knuth
D. grata Prain & Burkill
D. gribinguiensis Baudon
D. grisebachii Kunth
D. guerrerensis R. Knuth
D. guianensis R. Knuth
D. haenkeana C. Presl
D. hamiltonii Hook.f.
D. hassleriana Chodat
D. hastata Mill.
D. hastatissima Rusby
D. hastifolia Nees
D. hastiformis R. Knuth
D. haumanii Xifreda
D. havilandii Prain & Burkill
D. hebridensis R. Knuth
D. hemicrypta Burkill
D. hemsleyi Prain & Burkill
D. heptaneura Vell.
D. herbert-smithii Rusby
D. herzogii R. Knuth
D. heteropoda Baker
D. hexagona Baker
D. hieronymi Uline ex R. Knuth
D. hintonii R. Knuth
D. hirtiflora Benth.
D. hispida Dennst.
D. holmioidea Maury
D. hombuka H. Perrier
D. hondurensis R. Knuth
D. howardiana O. Téllez B.G. Schub. & Geeta
D. humifusa Poepp.
D. humilis Bertero ex Colla
D. hunzikeri Xifreda
D. igualamontana Matuda
D. incayensis R. Knuth
D. inopinata Prain & Burkill
D. insignis C.V. Morton & B.G. Schub.
D. intermedia Thwaites
D. ionophylla Uline ex R. Knuth
D. iquitosensis R. Knuth
D. irupanensis R. Knuth
D. itapirensis R. Knuth
D. itatiensis R. Knuth
D. jaliscana S. Watson
D. jamesonii R. Knuth
D. japonica Thunb.
D. javariensis Ayala
D. juxtlahuacensis (O. Téllez & Dávila) Caddick & Wilkin
D. kalkapershadii Prain & Burkill
D. kamoonensis Kunth
D. keduensis Burkill ex Backer
D. kerrii Prain & Burkill
D. killipii R. Knuth
D. kimiae Wilkin
D. kingii R. Knuth
D. kituiensis Wilkin & Muasya
D. kjellbergii R. Knuth
D. knuthiana De Wild.
D. koepperi Standl.
D. koyamae Jayas.
D. kratica Prain & Burkill
D. kuntzei Uline ex Kuntze
D. lacerdaei Griseb.
D. laevis Uline

D. lamprocaula Prain & Burkill
D. lanata Bail
D. larecajensis Uline ex R. Knuth
D. laurifolia Wall. ex Hook. f.
D. lawrancei R. Knuth
D. laxiflora Mart. ex Griseb.
D. lehmannii Uline
D. lepcharum Prain & Burkill
D. lepida C.V. Morton
D. leptobotrys Uline ex R. Knuth
D. liebmannii Uline
D. lijiangensis C.L. Long & H. Li
D. linearicordata Prain & Burkill
D. lisae Dorr & Stergios
D. listeri Prain & Burkill
D. litoralis Phil.
D. loefgrenii R. Knuth
D. loheri Prain & Burkill
D. longicuspis R. Knuth
D. longipes Phil.
D. longirhiza Caddick & Wilkin
D. longituba Uline
D. lundii Uline ex R. Knuth
D. luzonensis Schauer
D. macbrideana R. Knuth
D. maciba Jum. & H. Perrier
D. macrantha Uline ex R. Knuth
D. macrothyrsa Uline
D. macvaughii B.G. Schub.
D. madecassa H. Perrier
D. madiunensis Prain & Burkill
D. maianthemoides Uline ex R. Knuth
D. mamillata Jum. & H. Perrier
D. mandonii Rusby
D. mangenotiana J. Miège
D. mantigueirensis R. Knuth
D. margarethia G.M. Barroso
E.F. Guim. & Sucre
D. marginata Griseb.
D. martensis R. Knuth
D. martiana Griseb.
D. martini Prain & Burkill
D. matagalpensis Uline
D. matudae O. Téllez & B.G. Schub.
D. mayottensis Wilkin
D. megacarpa Gleason
D. megalantha Griseb.
D. melanophyma Prain & Burkill
D. melastomatifolia Uline ex Prain
D. membranacea Pierre ex Prain & Burkill
D. menglaensis H. Li
D. meridensis Kunth
D. merrillii Prain & Burkill
D. mesoamericana O. Téllez & Mart.-Rodr.
D. mexicana Scheidw.
D. microbotrya Griseb.
D. microcephala Uline
D. microura R. Knuth
D. mindanaensis R. Knuth
D. minima C.B. Rob. & Seaton
D. minutiflora Engler
D. mitis C.V. Morton
D. mitoensis R. Knuth
D. modesta Phil.
D. mollis Kunth
D. monadelpha (Kunth) Griseb.
D. monandra Hauman
D. morelosana (Uline) Matuda
D. moritziana (Kunth) R. Knuth
D. mosqueirensis R. Knuth
D. moultonii Prain & Burkill
D. moyobambensis R. Knuth
D. mucronata Uline ex R. Knuth
D. multiflora Mart. ex Griseb.
D. multiloba Kunth
D. multinervis Benth.
D. mundii Baker
D. nako H. Perrier
D. namorokensis Wilkin
D. nana Poepp.
D. nanlaensis H. Li
D. natalensis R. Knuth
D. natalia Hammel
D. neblinensis Maguire & Steyerm.
D. nelsonii Uline ex R. Knuth
D. nematodes Uline ex R. Knuth
D. nervata R. Knuth
D. nervosa Phil.
D. nicolasensis R. Knuth
D. nieuwenhuisii Prain & Burkill
D. nipensis R.A. Howard
D. nipponica Makino
D. nitens Prain & Burkill
D. nuda R. Knuth
D. nummularia Lamarck
D. nutans R. Knuth
D. oaxacensis Uline
D. obcuneata Hook.f.
D. oblonga Gleason
D. oblongifolia Rusby
D. obtusifolia Hook. & Arn.
D. olfersiana Klotzsch ex Griseb.
D. oligophylla Phil.
D. omiltemensis O. Tellez

D. opaca R. Knuth
D. opposita Thunb.
D. oppositiflora Griseb.
D. oppositifolia L.
D. orangeana Wilkin
D. orbiculata Hook.f.
D. oreodoxa B.G. Schub.
D. organensis R. Knuth
D. orientalis (J. Thiébaut) Caddick & Wilkin
D. orizabensis Uline
D. orthogoneura Uline ex Hochr.
D. oryzetorum Prain & Burkill
D. ovalifolia R. Knuth
D. ovata Vell.
D. ovinala Baker
D. palawana Prain & Burkill
D. paleata Burkill
D. pallens Schltdl.
D. pallidinervia R. Knuth
D. palmeri R. Knuth
D. panamensis R. Knuth
D. panthaica Prain & Burkill
D. pantojensis R. Knuth
D. paradoxa Prain & Burkill
D. pavonii Uline ex R. Knuth
D. pedicellata Phil.
D. pencana Phil.
D. pendula Poepp. ex Kunth
D. pentaphylla L.
D. peperoides Prain & Burkill
D. perdicum Taub.
D. perenensis R. Knuth
D. perpilosa H. Perrier
D. petelotii Prain & Burkill
D. philippiana Uline ex R. Knuth
D. piauhyensis R. Knuth
D. pierrei Prain & Burkill
D. pilcomayensis Hauman
D. pilgeriana R. Knuth
D. pilosiuscula Bertero ex Spreng.
D. pinedensis R. Knuth
D. piperifolia Humb. & Bonpl. ex Willd.
D. piscatorum Prain & Burkill
D. pittieri R. Knuth
D. planistipulosa Uline ex R. Knuth
D. plantaginifolia R. Knuth
D. platycarpa Prain & Burkill
D. platycolpota Uline ex B.L. Rob.
D. plumifera C.B. Rob.
D. pohlii Griseb.
D. poilanei Prain & Burkill
D. polyclados Hook. f.
D. polygonoides Humb. & Bonpl. ex Willd.
D. polystachya Turcz.
D. pomeroonensis R. Knuth
D. potarensis R. Knuth
D. praehensilis Benth.
D. prainiana R. Knuth
D. prazeri Prain & Burkill
D. preslii Steud.
D. preussii Pax
D. pringlei C.B. Rob.
D. proteiformis H. Perrier
D. psammophila R. Knuth
D. pseudomacrocapsa G.M. Barroso E.F. Guim. & Sucre
D. pseudonitens Prain & Burkill
D. pseudorajanioides R. Knuth
D. pseudotomentosa Prain & Burkill
D. pteropoda Boivin ex H. Perrier
D. puber Blume Enum
D. pubera Blume
D. pubescens Poir.
D. pumicicola Uline
D. pumilio Griseb.
D. puncticulata R. Knuth
D. purdiei R. Knuth
D. putisensis R. Knuth
D. putumayensis R. Knuth
D. pynaertii De Wild.
D. pyrenaica Bubani & Bordère ex Gren.
D. pyrifolia Kunth
D. quartiniana A. Rich.
D. quinquelobata Thunb.
D. quispicanchensis R. Knuth
D. racemosa (Klotzsch) Uline
D. regnellii Uline ex R. Knuth
D. remota C.V. Morton
D. remotiflora Kunth
D. reticulata Gay
D. retusa Mast.
D. reversiflora Uline
D. ridleyi Prain & Burkill
D. riedelii R. Knuth
D. rigida R. Knuth
D. rimbachii R. Knuth
D. rockii Prain & Burkill
D. rosei R. Knuth
D. rumicoides Griseb.
D. rupicola Kunth
D. rusbyi Uline
D. sabarensis R. Knuth
D. sagittata Poir.
D. sagittifolia Pax

D. salicifolia Blume
D. salvadorensis Standl.
D. sambiranensis R. Knuth
D. sanchez-colini Matuda
D. sandiensis R. Knuth
D. sandwithii B. G. Schub.
D. sanpaulensis R. Knuth
D. sansibarensis Pax
D. santanderensis R. Knuth
D. santosensis R. Knuth
D. sarasinii Uline ex R. Knuth
D. saxatilis Poepp.
D. scabra Humb. & Bonpl. ex Willd.
D. schimperiana Hochst. ex Kunth
D. schubertiae Ayala
D. schunkei Ayala & T. Clayton
D. schwackei Uline ex R. Knuth
D. scortechinii Prain & Burkill
D. secunda R. Knuth
D. sellowiana Uline ex R. Knuth
D. semperflorens Uline
D. septemloba Thunb.
D. septemnervis Vell.
D. sericea R. Knuth
D. seriflora Jum. & H. Perrier
D. serpenticola Hoque & P.K. Mukh.
D. sessiliflora McVaugh
D. sexrimata Burkill
D. simulans Prain & Burkill
D. sincorensis R. Knuth
D. sinoparviflora C.T. Ting M.G. Gilbert & Turland
D. sinuata Vell.
D. sitamiana Prain & Burkill
D. skottsbergii R. Knuth
D. smilacifolia De Wild. & T. Durand
D. sonlaensis R. Knuth
D. sororopana Steyerm.
D. soso Jum. & H. Perrier
D. spectabilis R. Knuth
D. spicata Roth
D. spiculiflora Hemsl.
D. spiculoides Matuda
D. spongiosa J.Q. Xi M. Mizuno & W.L. Zhao
D. sprucei Uline ex R. Knuth
D. standleyi C.V. Morton
D. stegelmanniana R. Knuth
D. stellaris R. Knuth
D. stemonoides Prain & Burkill
D. stenocolpus Phil.
D. stenomeriflora Prain & Burkill
D. stenopetala Hauman
D. stenophylla Uline
D. sterilis O. Weber & Wilkin
D. stipulosa Uline ex R. Knuth
D. subcalva Prain & Burkill
D. subhastata Vell.
D. sublignosa R. Knuth
D. submigra R. Knuth
D. subtomentosa Miranda
D. sumatrana Prain & Burkill
D. sumiderensis B.G. Schub. & O. Téllez
D. suratensis R. Knuth
D. sylvatica Eckl.
D. synandra Uline
D. syringifolia (Kunth) Kunth & R.H. Schomb. ex R. Knuth
D. tabatae Hatus. ex Yamashita & M.N. Tamura
D. tacanensis Lundell
D. tamarisciflora Prain & Burkill
D. tamoidea Griseb.
D. tamshiyacuensis Ayala
D. tancitarensis Matuda
D. tarijensis R. Knuth
D. tarmensis R. Knuth
D. tauriglossum R. Knuth
D. tayacajensis R. Knuth
D. temascaltepecensis R. Knuth
D. tenebrosa C.V. Morton
D. tenella Phil.
D. tentaculigera Prain & Burkill
D. tenuipes Franch. & Sav.
D. tenuiphyllum R. Knuth
D. tenuis R. Knuth
D. tequendamensis R. Knuth
D. ternata Griseb.
D. therezopolensis Uline ex R. Knuth
D. togoensis R. Knuth
D. tokoro Makino ex Miyabe
D. toldosensis R. Knuth
D. tomentosa J. König ex Spreng.
D. torticaulis R. Knuth
D. trachyandra Griseb.
D. trachycarpa Kunth
D. traillii R. Knuth
D. transversa R. Br.
D. triandria Sessé & Moc.
D. trichantha Baker
D. trichanthera Gleason
D. trifida L. f.
D. trifoliata Kunth
D. trifurcata Hauman
D. trilinguis Griseb.

D. trimenii Prain & Burkill
D. trinervia Roxburgh ex Prain & Burkill
D. trisecta Griseb.
D. trollii R. Knuth
D. truncata Miq.
D. tsaratananensis H. Perrier
D. tubiperianthia Matuda
D. tubuliflora Uline ex R. Knuth
D. tubulosa Griseb.
D. uliginosa Phil.
D. ulinei Greenm. ex R. Knuth
D. undatiloba Baker
D. urceolata Uline
D. urophylla Hemsl.
D. uruapanensis Matuda
D. valdiviensis R. Knuth
D. vanvuurenii Prain & Burkill
D. variifolia Bertero
D. velutipes Prain & Burkill
D. vexans Prain & Burkill
D. vilis Kunth
D. villosa L.
D. volckmannii Phil.
D. wallichii Hook. f.
D. warburgiana Uline ex Prain & Burkill
D. warmingii R. Knuth
D. wattii Prain & Burkill
D. weberbaueri R. Knuth
D. widgrenii R. Knuth
D. wightii Hook.f.
D. wittiana R. Knuth
D. wrightii Uline ex R. Knuth
D. xizangensis C.T. Ting
D. yunnanensis Prain & Burkill
D. zingiberensis C.H. Wright.

Of the *Dioscorea* species listed above, several are used for food including *D. alata* L., *D. altissima* Lamarck, *D. bulbifera* L., *D. cayenensis* Lamarck, *D. composita* Hemsl., *D. convolvulacea* Cham. and Schltdl., *D. dumetorum* (Kunth) Pax, *D. elephantipes* (L'Hér.) Engler, *D. esculenta* (Lour.) Burkill, *D. esculenta* (Lour.) Burkill var. *esculenta*, *D. esculenta* (Lour.) Burkill var. *tiliaefolia* (Kunth) Fosberg and Sachet, *D. floribunda* M. Martens and Galeotti, *D. floridana* Bartlett, *D. hirticaulis* Bartlett, *D. hispida* Dennst., *D. japonica* Thunb., *D. macrostachya* Benth., *D. nummularia* Lamarck, *D. oppositifolia* L., *D. batatas* Decne., *D. pentaphylla* L., *D. pilosiuscula* Bertero ex Spreng., *D. polygonoides* Humb. and Bonpl. ex Willd., *D. praehensilis* Benth., *D. preussii* Pax, *D. quaternata* J.F. Gmel., *D. glauca* Muhl. ex Bartlett, *D. quaternata* J.F. Gmel. var. *glauca* (Muhl. ex Bartlett) Fernald, *D. villosa* L. var. *glabrifolia* (Bartlett) Fernald, *D. rotundata* Poir., *D. sansibarensis* Pax, *D. spiculiflora* Hemsl., *D. trifida* L. f., *D. villosa* L., *D. villosa* L. var. *floridana* (Bartlett) H.E. Ahles and *D. villosa* L. var. *hirticaulis* (Bartlett) H.E. Ahles.

Many of these *Dioscorea* species are listed as threatened to some degree (Red List, 2019). The 'Red List' which includes these species was prepared and is monitored by the International Union for Conservation of Nature (IUCN), which established the list in 1964. It has become the most comprehensive information source on the global conservation status of animal, fungi and plant species.

As would be expected with so many species of *Dioscorea*, their occurrence in the wild and their collection, introduction and cultivation over several thousand years has resulted in much confusion over specific and varietal names. The International Institute for Tropical Agriculture in Nigeria has a gene bank that holds more than 3,200 diverse germplasm of several species of *Dioscorea*. Since *Dioscorea* species have been vegetatively propagated over many generations there could be a lack of genetic diversity among individual species that are grown commercially. Wu *et al.* (2014) showed this lack of diversity by evaluating 21 landraces of *D. opposita*, *D. alata*, *D. persimilis* and *D. fordii*, and observed high levels of polymorphism: 95.3% for inter saMigmple sequence repeat and 93.5% for sequence related amplified polymorphism.

Molecular Taxonomy

Genomics has been used successfully to classify *Dioscorea* at the species and even variety and cultivar levels. There are many examples of genomics being used in the taxonomy of *Dioscorea*. For example, Zheng *et al.* (2007) tested the molecular authentication of *D. polystachya* and other *Dioscorea* species and found that chloroplast DNA (cpDNA) sequencing was a rapid, efficient, credible way to distinguish *D. polystachya* and related species. They concluded that their results strongly supported the view that the horticulture classification of *D. polystachya* using leaf and tuber characteristics was related to the molecular level.

Wu *et al.* (2015) reported that the combination of chemical characterization and multivariate data analysis provided a feasible classification for *Dioscorea*. Techen *et al.* (2017) reported that various molecular analyses have been published to help with the identification of species and varieties within *Dioscorea*. They analysed a single nuclear genomic region, biparentally inherited, and found it to be useful as a molecular marker for identifying and discriminating between *D. bulbifera*, *D. villosa*, *D. nipponica*, *D. alata*, *D. caucasica* and *D. deltoidea*. They found that the *LFY* genomic region can be useful as a molecular marker to distinguish between these six species. Malapa *et al.* (2005) used amplified fragment length polymorphism (AFLP) markers to assess the genetic relationship between *D. alata*, *D. abyssinica*, *D. bulbifera*, *D. rotundata-cayenensis*, *D. esculenta*, *D. nummularia*, *D. pentaphylla*, *D. persimilis*, *D. transversa* and *D. trifida*. They found that each species could be 'fingerprinted' uniquely with AFLP. Unweighted pair groups with AFLP of *D. alata* accessions from distant geographical origins and different ploidy levels showed the existence of three major groups of genotypes.

Sun *et al.* (2012) tested DNA 'barcoding' and found it applicable using three candidate DNA barcodes (*rbc*L, *mat*K, and *psb*A-*trn*H) to identify *Dioscorea* species. They found that *mat*K gave the best results but a combination of two or three loci achieved a higher success rate than one locus alone. Girma *et al.* (2016) reported that the development of more robust DNA barcoding systems was an ongoing challenge. Their results also indicated that the *rbc*L plus *mat*K combination can be utilized as multi-locus DNA barcode regions for identifying *Dioscorea* species. However, DNA barcoding was also reported to be relatively unsuccessful in discriminating between *Dioscorea* species, with the highest discrimination rate of only just over 23%, which was derived from *mat*K sequences.

Xia *et al.* (2019) compared genic and intergenic regions of three *Dioscorea* chloroplast genomes and found that the density of single nucleotide polymorphism and indels in intergenic sites was about twice and seven times higher, respectively, than that of single nucleotide polymorphism and indels in the genic regions. However, using this technique, they successfully identified the majority of the species they tested, except for species in the Enantiophyllum section of *Dioscorea*. They suggest that the variable loci derived from comparative analysis of plastid genome sequences was a possible DNA barcode candidate for taxonomic analysis and species delimitation in some *Dioscorea* species. Another approach to taxonomy of *Dioscorea* had previously been reported by Su (1987) and Schols *et al.* (2001) who both showed good correlation between pollen size and tuber species.

Genomics

Mignouna *et al.* (2007) commented that the genetics of *Dioscorea* species was least understood among the major staple food crops and developed an initial c-DNA library for *D. rotundata* and *D. alata* in order to produce expressed sequence tags for gene discovery and sources of additional molecular markers. Sun *et al.* (2012) subsequently commented that the classification and identification of *Dioscorea* species was controversial. Peng *et al.* (2016) examined the genetic relationships among 14 *D. polystachya* cultivars and found a high level of polymorphism, but intron sequence amplified polymorphism (ISAP) markers were consistent with their morphological characteristics and with the classification of these cultivars by leaf and tuber shapes. Based on leaf and tuber shapes and molecular data they concluded that *D. doryphora* might be a single species and a progenitor of *D. polystachya* cultivars. Using ISAP they were able to discriminate the popular local *D. polystachya* cultivar 'Tiegun' from the other 13. *Dioscorea* species were obligate outcrossing plants that displayed highly heterozygous genomes. This made the methods used in genetic analysis applied in inbreeding species inapplicable to *Dioscorea* species (Tamiru *et al.*, 2017). However, a reference genetic map for *D. alata* (Cormier *et al.*, 2019) and genome sequence assemblies for *D. rotundata* were prepared by Tamiru *et al.* (2017) who performed genome analysis and assembled a 594-Mb genome, 76.4% of which was distributed among 21 linkage groups. They predicted a total of 26,198 genes and used phylogenetic analyses with 2,381 conserved genes, which showed that *Dioscorea* was of a unique lineage among monocotyledons. Scarcelli *et al.* (2019) investigated the domestication of *D. rotundata* using whole-genome resequencing and statistical models, and found that the cultivated yam in West Africa was domesticated from a forest species and they inferred that its expansion started in the Niger river basin.

The chromosome number of many *Dioscorea* species has been counted and showed high variability. For example Baquar (1980) found that in *D. bulbifera* $2n$ varied from 36 to 80 depending on the origin of the variety. Dansi *et al.* (2001a) also showed that *Dioscorea* species exhibited considerable interspecific and intraspecific variations in ploidy level, which ranged from 2× to 16× based on basic chromosome numbers of either 9 or 10. For example, three ploidy levels (4×, 6×, 8×) were determined in accessions of *D. rotundata-cayenensis* complex from Benin. *D. nummularia* and natural interspecific hybrids with *D. alata* had chromosome numbers ranging from $2n = 3x = 60$ to $2n = 6x = 120$ (Lebot *et al.*, 2016). Ploidy has been given for some *Dioscorea* species by various workers as: *D. alata* $2n$ = 20, 30, 40, 50, 60, 70, 80, *D. rotundata* $2n$ = 40, 80 and *D. cayenensis* $2n$ = 36, 54, 60, 63, 66, 80, 120, 140 (Dansi *et al.*, 2001a, 2001b); *D. oppositae* $2n$ = 40, *D. japonica* $2n$ = 40, *D. dumetorum* $2n$ = 36, 40, 45, 54, *D. hispida* $2n$ = 40, 60, *D. bulbifera* $2n$ = 30, 40, 50, 60, 70, 80, 100, *D. esculenta* $2n$ = 30, 40, 60, 90, 100 and *D. trifida* $2n$ = 54, 72, 81 (Coursey, 1967; Purseglove, 1972; Degras, 1993; Onwueme and Charles, 1994; Asiedu *et al.*, 1997). In Cameroon the level of ploidy varied from diploid in *D. dumetorum* to octoploid in *D. cayenensis* and some other species were also shown to have varying ploidy levels (Ngo-Ngwe, 2014). Both tetraploid and hexaploid *D. rotundata* were also found. Arnau *et al.* (2017) found a wide genetic diversity and structuring in *D. alata* associated with geographical origin, ploidy levels, morphological and agronomic characteristics. They identified 17 major groups of genetically close varieties, including 11 groups of diploids, four groups of triploids and two groups of tetraploids. In China considerable variations in terms of major tuber nutritional composition and bioactive constituents were found among 25 landraces from *D. alata*, *D. oppositae*, *D. persimilis* and *D. fordii* by Wu *et al.* (2015). Loko *et al.* (2016) assessed the genetic diversity and relationships of 64 *D. rotundata-cayenensis* complex landraces in Benin. Thirteen were found to be polymorphic, giving 113 polymorphic alleles. The number of alleles per locus averaged 8.69.

Breeding

IPGRI/IITA (1997) commented that plant breeding programmes start with a comprehensive characterization of the germplasm collected from farmers' fields, which is done using internationally standardized agro-morphological descriptors. In root crops a critical step also involves the identification of genotypes with good tuber quality and equivalent ploidy levels showing traits relevant to the objectives of the breeding programme. These are then selected for hybridization. Yam breeding schemes have mainly concentrated on yield, quality and pest and disease susceptibility. However, in order to make yams more economic, much breeding work has highlighted selection of plants that can grow and yield well without staking and also that produce tubers close to the surface to facilitate harvesting. Improvement and conservation of *Dioscorea* germplasm is being carried out under the framework of the Consultative Group for International Agricultural Research (CGIAR) with the following objectives:

a) optimize *ex situ* and *in situ* yam conservation methodologies

b) increase coverage of yam gene pools
c) evaluate, genotype and phenotype yam collections for important traits
d) enrich databases with information on yam collections and make it freely accessible to users
e) improve procedures for safe exchange of roots, tubers and banana genetic resources.

The International Institute of Tropical Agriculture (IITA) established a core collection of 391 accessions of *Dioscorea* in 2006. Subsequently this gene bank was developed to includes eight species of more than 3,000 accessions, mainly of West African origin, which was reported to consist of *D. rotundata* (67%), *D. alata* (25%), *D. dumetorum* (1.6%), *D. cayenensis* (2%), *D. bulbifera* (2%), *D. esculenta* (0.7%), *D. praehensilis* (0.3%) and *D. mangenotiana* (0.25%). They classified genotypes using 18 DNA-based markers mainly focused on clonal selection from landraces and hybridization of elite clones of *D. alata* and *D. rotundata*. These included the evaluation of IITA-derived breeding lines with the National Root Crop Research Institute at Umudike in Nigeria and the Crops Research Institute in Ghana. The IITA released ten cultivars of *D. rotundata* and five cultivars of *D. alata* that had multiple pest and disease resistance, including resistance to fungi and viruses and nematodes in *D. dumetorum*. Some work was also carried out on interspecific hybridization, but problems of incompatibility and synchronization of flowering were encountered. For example, *D. rotundata* can be crossed to *D. cayenensis*, but crossing either of the two with *D. alata* has not been successful. In the 1960s in Guadeloupe, the INRA (Institut National de Recherches Agronomiques) bred new cultivars of *D. trifida* with yields of approximately 30 tonnes ha^{-1} not staked (Degras, 1993).

Subsequently research into cryopreservation of the *Dioscorea* collection has been carried out by IITA in collaboration with the Centre de Coopération Internationale en Recherche Agronomique pour le Développement (CIRAD) using liquid nitrogen. CIRAD also characterized physico-chemical attributes of *D. alata* in terms of food quality, sensory evaluation and pasting characteristics of yam flour for use in major food products. Some 300 *D. alata* genotypes were collected and evaluated by the United States Department of Agriculture (USDA) in Puerto Rico in the 1970s, resulting in the selection and distribution of the cultivars 'Florido', 'Forastero', 'Veeven', 'Gemelos' and 'Leone Globe' (Martin *et al.*, 1975). Collections of *Dioscorea*, mainly *D. cayenensis*/*D. rotundata* complex, have been assembled in several other countries including Côte d'Ivoire (Hamon and Touré, 1990a, 1990b) and Benin (Dansi *et al.*, 1999). A collaborative evaluation of breeding lines conducted by the IITA and the National Root Crop Research Institute of Nigeria (NRCRI) permitted the release of new cultivars (Mignouna *et al.*, 2007). Breeders at CIRAD primarily work on *D. alata*, *D. rotundata* and *D. cayenensis*. The genotypes of *D. alata* that they tested varied significantly in the number of chromosomes, flesh colour, tuber shape and response to diseases (including anthracnose) (CIRAD, 2018).

The Central Tuber Crops Research Institute (CTCRI) at Kerala, India, released several improved cultivars of *D. alata* (Abraham, 2002). *Dioscorea* collections are also maintained at the Vietnam Association of Supporting Industries (VASI) in Vietnam, the Philippine Root Crop Research and Training Center (PhilRootcrops) in the Philippines and the Vanuatu Agricultural Research and Technical Centre (VARTC) in the Pacific region. Lebot (2019) reported that a *D. alata* germplasm collection in Vanuatu comprised more than 300 accessions. He showed that the average yield was related to ploidy with plants with 40 chromosomes (diploids) yielding, on average, 2 kg fresh weight tubers plant^{-1}, plants with 60 chromosomes (triploids) yielding 2.5 kg fresh weight tubers plant^{-1} and plants with 80 chromosomes (tetraploids) more than 3 kg fresh weight tubers plant^{-1}. This was the case for two *D. alata* cultivars in New Caledonia, 'Nouméa Blanc' and 'Nouméa Rouge', both of which have 80 chromosomes and were shown to be very high yielding. Small collections of *D. opposita* and *D. japonica* are maintained in China and Japan. Agbaje *et al.* (2003) compared seven cultivars of *D. rotundata* (three currently grown and four locally bred) and found that their tuber yields were highest for two of the locally bred cultivars ('89/02665' and '95/01924') with a mean of 16 and 14 tonnes ha^{-1} respectively and were lowest for '89/01438' (6.6 tonnes ha^{-1}) and 'Danacha' (5 tonnes ha^{-1}). Nematode infestation of tubers was mild in all cultivars tested and there was no correlation between yield and nematode level, but there were positive correlations between tuber yield and virus infection.

In *D. rotundata*, rapid genetic advances can be achieved for yields under uniform environments by single plant selections followed by clonal propagation (Akoroda, 1983). For *D. cayenensis*, various phenotypical variances, coefficients of variation, expected genetic advances and correlation coefficients have been estimated for seven agronomic traits. Plant leafiness, virus infection, number of tubers per plant and tuber yield showed high expected genetic advances, while highly significant positive correlations were observed between tuber yield, plant leafiness and vine dry weight. It was also suggested that resistance to leaf virus infection, as expressed by foliage vigour, was the most important trait for selecting high-yielding genotypes (Akoroda, 1984). In Nigeria, a study aimed at understanding genetic control of yam mosaic virus resistance in the *D. rotundata* showed that it can be expressed by the action of a single dominant gene in simplex condition or a major recessive gene in duplex condition (Mignouna *et al.*, 2001a). Resistance of *D. alata* to a moderately virulent strain of the fungus *Colletotrichum gloeosporioides*, which causes anthracnose, was found to be strain-specific and appeared to be controlled by a single major dominant locus (Mignouna *et al.*, 2001b). Mignouna *et al.* concluded that breeding for anthracnose resistance, cultivars carrying many different genes for resistance need to be developed in order to provide a sustainable advantage against a fungus since *C. gloeosporioides* has many different strains. Siadjeu *et al.* (2020) approached breeding of *D. dumetorum* for resistance to postharvest hardening of tubers and other traits by sequencing the genome. They found that the genome was highly heterozygous, with two haplophases that were largely separated. They reported a total of 35,269 protein-encoding gene models and 9,941 non-coding RNA genes.

Physiology

Photoperiod plays an essential role in both the formation and growth of yam tubers. Different yam species differ in their reaction to photoperiod. Marcos *et al.* (2009) reported that the period from emergence to harvest in *D. alata* in Guadeloupe varied from three to six months, with the longer cycles corresponding to early planting. On average, the time from emergence to tuber initiation represented about one-third of the time from emergence to harvest, and was shown to be mainly affected by photoperiod and to a lesser extent by temperature. They found that long days promoted development of the foliage and stems, while short days initiate senescence of the foliage and tuber bulking. Yoshida *et al.* (1999, 2000, 2001) reported that in *D. opposita*, flower induction and tuber bulking were highly influenced by photoperiod and that more aerial bulbils were produced in short days (10–11 hours) than in days of more than 12 hours in both *D. opposita* and *D. bulbifera*. In Guadeloupe, Degras (1993) found that there was a significant effect of the planting date on the final yield of *D. alata* indicating the influence of photoperiod. Lacointe and Zinsou (1987) reported that the duration of the vegetative growth phase and of tuber dormancy varied between planting dates, but tuber sprouting always occurred in April which made them conclude that the problems encountered when attempting to cultivate this species out-of-season are related mostly to sensitivity to photoperiod.

Photosynthesis

Cornet *et al.* (2007, 2014, 2016) tested 23 genotypes belonging to seven *Dioscorea* species from either Benin or Guadeloupe and found that they were all C3 photosynthetic types. There are three metabolic pathways for carbon fixation in photosynthesis C3, C4 and CAM (crassulacean acid metabolism). About 85% of plant species are C3 plants including cassava, potatoes, cereal grains and most trees. C3 photosynthesis converts carbon dioxide, water and ribulose bisphosphate into 3-phosphoglycerate. In certain circumstances C3 photosynthesis may be disadvantageous, and is never advantageous, compared to C4 and CAM because of photo-respiration, which can reduce the utilization efficiency of light energy in the photosynthetic process (Hogan, 2011).

The photosynthetic efficiency of plants, including *Dioscorea* species, can be affected by several factors (Rodriguez-Montero, 1997). Padhan and Panda (2018) found that *D. oppositifolia*, *D. hamiltonii* and *D. pubera* showed superior photosynthetic efficiency compared to *D. alata*. Aighewi and Ekanayake (2004) tested the photochemical efficiency of two cultivars of *D. rotundata*, 'Pepa' and 'Abi', and showed that cultivar had a significant ($p = 0.05$) effect on photosynthetic efficiency and that the peak period for both cultivars was at 18 weeks after transplanting. Hgaza *et al.* (2009), working on *D. alata*, showed that the photosynthetic capacity of both young leaves (below fourth position) and older leaves (above twentieth position) was lower than the intermediate mature leaves with the maximum photosynthetic capacity at between the fourth to the seventh positions. They also found that the photosynthetic performance of mature leaves was related to the plant growth stage and its nutritional status, being higher with optimum use of fertilizers and during the tuber bulking phase.

Tuber development

Yam tubers are stem tubers that were described as 'developing by the plagiotrophic lobing of a degenerate rhizome' (Burkill, 1960; Martin and Ortiz, 1962). A rhizome is a modified subterranean stem that can produce roots and shoots. Tubers are enlarged structures used by plants as storage organs and for perennation as a means of asexual reproduction. In yams the part of the tuber nearest to the vine is more physiologically mature than the distal region, therefore it tends to sprout earlier. Roots, stems and tubers all appear to develop from common tissue arising on the planting sett, termed by Ferguson (1972) as the 'primary nodal complex'. In *D. esculenta*, where stolon-like structures precede tuber development, the time of tuber initiation is likely to be at the formation of these structures, which are formed at about 2 months after planting. However, in *D. alata* the time of tuber initiation has been suggested to be at, or soon after, the formation of the primary nodal complex (Ferguson, 1973). The first external signs of tuber development occurred about 3 months after planting in *D. alata* (Campbell *et al.*, 1962) and about 4.5 months after planting in *D. trifida* (Ferguson, 1969). The growth of the tubers in yams follows a similar pattern where at first there is a period of slow, more or less exponential growth and this is followed by a period of rapid tuber bulking and finally a period of reduced rates of increase (sometimes a decrease) and maturation. This pattern has been observed on *D. alata*, *D. trifida* and *D. esculenta* by Enyi (1972a, 1972b) and Ferguson (1969).

Melteras (2005, quoted by Lebot, 2009) reported that in Vanuatu, the first tubers of *D. esculenta* appeared 21 weeks after planting and their number increased over the following months and could reach a maximum of 27 tubers per plant between 37 and 41 weeks after planting. Yellowing of leaves became evident between 33 and 41 weeks after planting and by 50 weeks the vines were dead. Tuber dry weight production increased rapidly from 24 to 33 weeks after planting, when vine growth had ceased, and 45 weeks after planting. Tuber dry matter declined, probably due to the level of tuber respiration during dormancy.

In a detailed study of tuber development in *D. alata*, Ferguson (1973) described four distinct phases of growth, which represented periods of different physiological activity in the tubers. The first phase extended from emergence to about 13 weeks, and consisted mostly of the development of the primary nodal complex. However, he found that it was difficult to clearly separate the tuber from the primary nodal complex in *D. alata* because the primary nodal complex and the tuber seem to grow as a single organ, and he therefore regarded the primary nodal complex as an integral part of the tuber. The dry matter content of tubers remained at about 13% during this period. The second phase extended from 13 to 19 weeks. Active tuber development, beginning at 13 weeks, was associated with a reduction in per cent dry matter accumulation and began, during this period, with the fresh weight increasing relatively faster than dry weight. This period was interpreted as one of rapid cell division and some cell expansion in which many of the cells involved in subsequent tuber bulking were laid down. Growth up to the end of the second phase was considered to be exponential. The third phase extended from the 19th to the 32nd week and was a period of rapid accumulation of dry matter in tubers. The dry matter percentage of tubers also increased during this phase and hence tuber dry matter showed a relatively greater increase than tuber fresh weight. Dry weight could be considered to increase more or less linearly with time during this phase. During the final phase from week 32 to week 36, dry matter accumulation slowed and ceased altogether as maturity approached.

Also, in studies of *D. alata*, Enyi (1972a, 1972b) found that the size of tubers increased steadily between 13 and 28 weeks after planting, but then began to decrease. This decrease suggests that possibly a number of small tubers are resorbed during the final weeks of growth. Many of the tubers produced at maturity were, in any case, small and unmarketable. Enyi also showed that the average number of tubers per plant of marketable size at harvest were probably developed early, whereas tubers that were initiated and began bulking later may be poor competitors for available assimilates and hence are likely to remain small. *D. esculenta* produces many more tubers than *D. alata*, and tuber number tended to increase up to about six months after planting. However, when many tubers were initiated the size of individual tubers was small and the proportion of initiated tubers that never reach harvestable size was greater. Ferguson (1973) also reported similar observations on *D. alata*. It appears that factors which promote the initiation of an excessive number of tubers could therefore result in low yield in yams. Maturation of tubers is also related to foliage, as it is in other root crops including potatoes (Thompson, 1967). In yams, Lebot (2009) reported that the senescence of foliage starts 5 to 7 months after planting for early maturing genotypes, which is usually synchronized with the onset of suberization of the tubers. This senescence and suberization is probably initiated by photoperiod, but there is a genetic component as well.

Bulbil development

Ayensu (1972) reported that bulbils are produced from axillary buds in leaf axils and Burkill (1960) categorized bulbils as modified axillary branches. Ferguson (1972) and Wickham *et al.* (1982) described the histological development of bulbils and showed the significance of the primary nodal complex in bulbil ontogeny of *D. alata* and *D. bulbifera*. The control of initiation of bulbils in *D. bulbifera* was reported by Ubuebo (1971) to be under hormonal control. In *D. alata*, bulbils were reported to be formed just after the period of maximum bulking of the underground tubers. The build-up of assimilates, especially sugars in the phloem, may be a factor involved in the formation of bulbils. Bulbils in *D. alata* were also regarded as secondary sinks and seem to develop only after the primary sinks, the tubers, have passed their maximum period of bulking (Ferguson, 1977). Bulbils are used in propagation of yams and Lebot (2009) commented that bulbils can contribute to propagation of yams in natural situations and as noted, having a lower dry matter content than tubers, they can float on water, thus increasing dispersion potential.

Tuber dormancy

Hamadina (2011), who published an excellent review of tuber dormancy in *Dioscorea*, commented that there was considerable confusion in the definition of tuber dormancy and also with some of the terms associated with the start, duration and end of dormancy in tubers of *Dioscorea* species. In their review, Craufurd *et al.* (2001) reported that hormones and chemicals have been tested, with different degrees of success, to either prolong dormancy and storage life or to break dormancy of yam tubers. These chemicals include abscisic acid, ethylene chlorhydrin, gibberellic acid, clormequat chloride, indole acetic acid, 2,4-dichlorophenoxyacetic acid, 4-chlorophenoxyacetic acid, indole acetic acid, maleic hydrazide, kinetin, tetrachloronitrobenzene, pentachlkoro nitrobenzene and isopropyl n-chlorophenylcarbamate. Wickham *et al.* (1981) investigated dormancy and sprouting of *D. alata*, *D. esculenta*, *D. rotundata* and *D. trifida* tubers and found that the first sign of the breaking of dormancy was active cell division and differentiation that occurred in the meristematic region of the tubers. They found that this cell division and differentiation can lead to the formation of a localized mass of cells, called the 'primary thickening meristem' or the 'tuber germinating meristem'. The cells of the primary thickening meristem are small, either irregular or oblong in shape and arranged in a horizontal array. However, Wickham *et al.* commented that it was unclear how the site for renewed growth is determined. The tuber germinating meristem, which is distinguished from the primary thickening meristem by the widespread nature of cell activity in the meristematic layer and the change in the shape of the cells from a horizontal to a more vertical array, is the first event leading to the formation of shoot buds. The tuber germinating meristem is typically 10 to 40 cell layers thick, depending on the level of development and the area with the most activity. A developing apical shoot meristem is seen as an organized group of cells at the apex of the tuber germinating meristem. This event marks the progression into advanced stages of apical shoot bud formation, with the shoot apical meristem developing tangentially to the tuber germinating meristem. Leaf primordia are initiated from the peripheral cells of the shoot apical meristem marking the development of complete apical shoot buds. Up to this point, all of these events occur within the tuber with no external indications. Further development of the complete shoot bud leads to the appearance of the shoot bud on the surface of the tuber, which is commonly referred to as 'sprouting'. Thin, short-lived roots may be formed on tuber surfaces during storage that may or may not precede sprouting. Such roots can also be formed when tubers are exposed to high humidity (Wickham *et al.* 1981; Wilson *et al.* 1998; Hamadina, 2011).

3

Composition and Uses

Yam tubers are cultivated mainly as a staple crop and serve as a major food staple for many people throughout the world. Besides carbohydrates, yam tubers contain protein, fats, vitamins and minerals. The common species of *Dioscorea*, mostly produced and consumed by millions of people in West Africa, include *D. rotundata*, *D. alata* and *D. cayenensis*. In West Africa, yam tubers are probably available in every local market (Fig. 3.1) and constitute a significant portion of West African diets. In fact, some foods prepared from yams have cultural significance for indigenous people in West Africa, for instance, '*eto*' which is a food made from mashing cooked yam with palm oil, served with boiled eggs, is used to celebrate the birth of twins by the Ga tribe in Ghana.

Yam tubers can be prepared as boiled, roasted, fried or baked into different foods. There are some cases where the tubers are eaten raw, for example *D. polystachya* in Japan. Kay (1987) also reported that some succulent *D. bulbifera* varieties are also eaten raw. Yam tubers can be transformed into dry products, either by sun drying, often at the village level, or in an industrial drier where the dried product is often milled to produce flour or starch that can have both food and non-food applications. New applications for yam tubers are being explored with the aid of new and emerging technology. For example, many years ago Abe (1973) described freeze-drying of whole fresh yam tubers, which influences their functionality as a dry product and presented opportunities for applications in novel products.

There is a lot of potential for yams to become an increasingly valuable raw material for industry, particularly for underutilized varieties and non-edible species that may not be suitable for food use. Although non-edible *Dioscorea* species contain toxins that may be harmful, they still have functional properties that could be useful for non-food applications. An interesting fact was reported by Botany VM (2018) (on *D. kamoonensis*): that edible species have opposite leaves while toxic species have alternate leaves. Perhaps such a fact needs to be interpreted with caution as levels of toxins can vary considerably between yam species (see 'Non-food Uses', p. 40).

Currently, industrial processing of yams is underdeveloped, compared to such crops as cassava, despite the potential of yam as a food source and an economically viable raw material. There are a number of challenges that affect yam tuber processing. Among them is the high cost of some popular yam tuber species particularly when they are out of season. *D. rotundata* are especially expensive

DOI: 10.1079/9781789249279.0003

as a raw material for yam flour. Also, in some West African areas an inadequate value chain system limits the ability of processors to obtain quality yams from farmers on a consistent basis. In many developing countries, the high cost of specialized yam processing equipment and access to reliable power is an additional problem for small-scale processing industries. The irregular shape of most yam tubers also presents a challenge for processing, and specialized equipment is required to remove the yam peel efficiently prior to milling (Boateng *et al.*, 2019). Ayodeji *et al.* (2014) described the design and fabrication of yam peeling and slicing machines that could be adopted by local small-scale processing industries.

Fig. 3.1. Yam tubers displayed at a West African market (Dassa Zoumè market in Côte d'Ivoire).

Nutrients

A general figure for composition of an unspecified *Dioscorea* sp. is given in Table 3.1. Also, other data on the chemical composition of some species is given under the headings for the specific species in Chapter 8. However, the data from analyses varies considerably, not only between species, varieties and cultivars, but even when the same species of tuber has been grown in the same area of land. This clearly shows how difficult it is to be precise on chemical composition since this will vary from year to year, region to region, germplasm source and soil conditions, especially available nutrients that clearly affect so many constituents, especially the mineral content of the tubers. As would be expected there is considerable variation from different venues where the crop has been produced, which may be due to both variation in the growing environment and

Table 3.1. The composition of unspecified species of yam tubers for 100 g^{-1} fresh weight. Modified from USDA-NRCS (2012).

Component	Level		Component	Level	
Water	69.60	g	Thiamin	0.112	mg
Energy	118	kcal	Riboflavin	0.032	mg
Protein	1.53	g	Niacin	0.552	mg
Total lipid (fat)	0.17	g	Vitamin B-6	0.293	mg
Carbohydrate, by difference	27.88	g	Folate	23	μg DFE
Total dietary fibre	4.1	g	Vitamin B-12	0.00	μg
Sugars, total	0.50	g	Vitamin A	7	μg RAE
Calcium	17	mg	Vitamin A	138	IU
Iron	0.54	mg	Vitamin E (α-tocopherol)	0.35	mg
Magnesium	21	mg	Vitamin D (d2 + d3)	0.0	μg
Phosphorus	55	mg	Vitamin D	0.0	IU
Potassium	816	mg	Vitamin K (phylloquinone)	2.3	μg
Sodium	9	mg	Fatty acids, total saturated	0.037	g
Zinc	0.24	mg	Fatty acids, total monounsaturated	0.006	g
Total ascorbic acid	17.1	mg	Fatty acids, total polyunsaturated	0.076	g

the particular genotype grown in that area (Table 3.2). The mineral content, of course, reflects the mineral content of the soil in which they were grown (Table 3.3), but can also vary between genotypes (Tables 3.4, 3.5).

Osagie (1992) gave the nutritional value of yams in Benin in 100g of edible portion as: calories 71–135 kcal, moisture 65–81%, protein 1.4–3.5 g, fat 0.2–0.4 g, carbohydrate 16.4–31.8 g, fibre 0.40–10.0 g, ash 0.6–1.7 g, calcium 12–69mg, phosphorus 17–61 mg, iron 0.7–5.2 mg, sodium 8–12 mg, potassium 294–397 mg, β-carotene 0–10 mg, thiamine 0.01–0.11 mg, riboflavin 0.01–0.04 mg, niacin 0.3–0.8 mg, ascorbic acid 4–18 mg. Ukom *et al.* (2014) gave the chemical analysis of *D. cayenensis*, *D. dumetorum* and *D. bulbifera* tubers grown in Nigeria as: moisture 665 g kg^{-1}, crude protein 52.4 g kg^{-1}, crude fibre 52.5 g kg^{-1}, crude fat 3.4g kg^{-1}, ash 31.5g kg^{-1} and carbohydrate 195 g kg^{-1}, and the mean major carotenoids as lutein

Table 3.2. Fresh weight composition of seven *Dioscorea* species (values are means of different samples). Modified from Bradbury and Holloway (1988).

Dioscorea species	*alata*	*esculenta*	*bulbifera*	*nummularia*	*pentaphylla*	*rotundata*	*trifida*
Sample number (*n*)	16	99	25	12	9	3	3
Moisture%	77.3	74.2	71.9	81.7	82.5	65.7	80.7
Energy kj/100g	406	443	258	266	550	284	580
Protein%	2.06	2.04	1.94	1.65	1.42	1.52	0.53
Starch%	16.7	19.3	23.2	11.7	13.9	30.2	14.2
Sugars%	1.03	0.55	0.22	0.20	0.12	0.32	0.23
Dietary fibre%	1.88	1.15	1.84	1.42	0.66	0.63	1.02
Fat%	0.08	0.06	0.06	0.05	0.03	0.09	0.04
Ash%	0.81	0.82	0.95	0.69	0.76	0.73	0.70

Table 3.3. Fresh weight mineral composition of seven *Dioscorea* species (values are means of different samples in mg 100 g^{-1}). Modified from Bradbury and Holloway (1988).

Dioscorea species	*alata*	*esculenta*	*bulbifera*	*nummularia*	*pentaphylla*	*rotundata*	*trifida*
Mineral content							
Ca	8.2	7.5	6.5	8.4	13	4.6	8
P	38	39	40	27	26	28	38
Mg	17	26	30	19	23	17	15
Na	3.3	3.1	8.6	2.7	6.1	4.7	2.9
K	318	303	448	346	374	361	350
S	12	16	15	9.0	13	12	8.2
Fe	0.60	0.75	0.38	0.56	0.44	0.60	0.54
Cu	0.15	0.17	0.34	0.21	0.25	0.12	0.13
Zn	0.39	0.46	0.50	0.31	0.36	0.30	0.35
Mn	0.04	0.24	0.04	0.13	0.05	0.03	0.03
Al	0.64	0.51	0.29	0.49	0.62	0.63	0.41
B	0.09	0.07	0.05	0.10	0.17	0.08	0.11

Table 3.4. Mean variation in major components for 110 varieties of *D. alata* (% of dry weight) from different Pacific islands. Modified from SPYN (2003).

	varieties *n*	dry matter%	starch	amylose	protein	minerals	sugars
Papua New Guinea	43	23.5	67.5	17.5	12.0	5.1	3.3
Vanuatu	48	23.4	73.1	17.2	11.9	3.3	1.85
Fiji	19	25.2	68.5	18.6	8.03	4.25	2.46

1.03 µg g^{-1}, α-carotene 1.24 µg g^{-1}, all-trans-β-carotene 1.89 µg g^{-1}, 9-cis-β-carotene 1.05 µg g^{-1} and13-cis-β-carotene 0.32 µg g^{-1}. In Sri Lanka Wanasundera and Ravindran (1994) gave the crude protein content of *D. alata* tubers as 7.4%, starch as 75.6–84.3%, vitamin C from 13.0 to 24.7 mg 100 g^{-1} fresh weight, phytic acid 58.6 to 198.0 mg 100 g^{-1} dry matter and total oxalate as 486–781 mg 100 g^{-1} dry matter. In China 25 landraces from *D. alata, D. opposita, D. persimilis* and *D. fordii* were analysed on a dry matter basis and had starch 60.7–80.6%, protein 6.3–12.2%, Mg 326.8–544.7 mg kg^{-1}, allantoin 0.62–1.49% and dioscin 0.032–0.092% (Wu *et al.*, 2015).

In Ghana, Coursey and Aidoo (1966) commented that ascorbic acid was the only vitamin present in yam tubers in any significant quantity. They measured the ascorbic acid levels of 40 *D. alata, D. cayenensis* and *D. rotundata* varieties and found that they varied greatly between species and varieties with levels of 5.0–8.2 mg 100^{-1} in *D. alata,* 4.5–8.2 mg 100^{-1} in *D. cayenensis* and 6.5–11.6 mg 100^{-1} in *D. rotundata*, but one variety of *D. alata* had as much as 21.5 mg 100g^{-1}. There was little variation in levels in relation to the part of the country where the crop was grown even though the soil and climactic conditions varied. Losses of ascorbic acid during storage of *D. rotundata* for up to 4 months in ambient conditions were about 20% in sound tubers, but much higher if the tubers were damaged or bruised before storage. The retention of ascorbic acid during cooking of the tubers varied with different cooking methods with 95% retention if they were boiled without peeling and 65% if they were boiled after peeling. Frying in palm oil resulted in 93% retention of ascorbic acid and roasting or baking resulted in 85% retention. FAO (1965) compared the nutritional content of raw tubers and flour (Table 3.6) and also for their levels of some vitamins and mineral (Table 3.7).

Agbo-Egbe and Trèche (1995) measured the chemical composition of 98 cultivars of eight *Dioscorea* species that were extensively grown in Cameroon. On the basis of dry matter content, the yams were divided into three groups: low dry matter (23–25 g 100 g^{-1}, *D. alata, D. dumetorum* and *D. schimperiana*); intermediate dry matter (28–30 g 100 g^{-1}, *D. esculenta* and *D. bulbifera*); and high dry matter (32–37 g 100 g^{-1}, *D. cayenensis/rotundata* complex and *D. liebrechtsiana*). Mean fat levels were all in the range of 0.1–0.9 g 100 g^{-1}

Table 3.5. Mean proximate composition of different species and cultivars of yam approximately 2 weeks after harvest and shelf life in storage in a barn 30°C and 95% RH in Nigeria. Modified from Ogbo Frank and Agu Kingsley (2014b).

Species and cultivar	Shelf life	Moisture	Carbohydrate	Protein	Crude fibre
D. dumentorum 'Ona'	31 weeks	64%	29%	0.19%	3.8%
D. alata 'Abana Mmee'	27 weeks	70%	24%	1.05%	3.5%
D. alata 'Abana Ocha'	32 weeks	69%	25%	0.79%	3.0%
D. rotundata 'Adaka'	36 weeks	61%	32%	0.67%	4.3%
D. rotundata 'Abi'	40 weeks	63%	30%	0.48%	2.8%

Table 3.6. Composition of *Dioscorea* species tubers g 100 g^{-1} or mg 100 g^{-1} edible portion. Modified from FAO (1965).

Energy kcal	Moisture	Protein	Fat	Carbohydrate	Fibre	Ash
			Raw tuber			
119	69.0 56.3–78.6	1.9 1.5–4.2	0.2 trace-0.6	27.8	0.8 0.1–1.6	1.1 0.3–1.6
			Flour			
335	14.2 6.6–18.7	3.4 1.9–5.2	0.4 0.2–0.7	80.0	1.6 1.4–1.8	2.0 1.6–3.2

Table 3.7. Composition of *Dioscorea* species tubers g or mg 100 g^{-1} edible portion. Modified from FAO (1965).

Calcium	Phosphorus	Iron	β-carotene equivalent	Thiamine	Riboflavin	Niacin	Tryptophan	Ascorbic acid
				Raw tuber				
52	61	0.8	10	0.11	0.02	0.3	26	6
8–157	28–157	0.2–2.4	–	0.10–.11	0.02–0.03	0.3–0.4	–	5–10
				Flour				
20	110	1.1	tr.	0.10	0.08	1.1	35	–
20–22	–	–	–	0.05–.21	0.05–.14	0.7–2.0	–	–

Table 3.8. Selected yam species and their relative usefulness for food and medicine where a score of 1 is low and 5 is high. Modified from PFAF *D. japonica* (nd).

Species	Edibility rating	Medical rating
D. alata	4	1
D. batatas	5	5
D. bulbifera	4	2
D. cayennensis	4	0
D. communis	1	2
D. deltoidea	2	2
D. esculenta	4	0
D. japonica	4	2
D. tokoro	2	2
D. trifida	4	0
D. villosa	2	4

dry matter and mean starch content was within the range of 70.4–72.9 g 100 g^{-1}, except for *D. cayenensis/rotundata* complex and *D. liebrechtsiana*, which had levels higher than 80 g 100 g^{-1}. PFAF *D. japonica* (nd) listed several *Dioscorea* species and rated their edibility and medical uses (Table 3.8).

Dioscorin (not to be confused with dioscorine, which is a toxin also found in *Dioscorea* tubers) is an alkaloid with high angiotensin converting enzyme-inhibitory capacity. Dioscorin is the major soluble storage protein in yams and was reported to constitute 80–85% of the total tuber proteins in some *Dioscorea* species (Sharma *et al.*, 2017). Pérez *et al.* (2014), working in the Venezuelan Amazonas region, found dioscorin content varied between species and could be used to differentiate between species, in this case *D. japonica*, from other species growing in the region. Nagai and Nagashima (2006), working with *D. opposita* tubers, reported that dioscorin has a wide spectrum of strong antioxidative and antihypertensive activities and it could be utilized as a source of natural antioxidant. Hsu *et al.* (2002) also reported that dioscorin can inhibit angiotensin-converting enzyme activity and be used in the treatment of hypertension. Kumoro and Hartati (2015) described a successful technique for dioscorin extraction from *D. hispida* using microwaves.

Processing

Drying

Processing yam tubers into dry products is important, particularly to extend their shelf life. Yam tubers contain high amounts of moisture which make them susceptible to spoilage. Drying is one of the methods frequently used for their preservation. Reducing the moisture content of yams helps to prevent microbial spoilage and other deteriorative reactions. Traditionally, drying in the sun is mostly preferred due to its low cost. In some communities, yam tubers are sliced and laid out in the open to dry, which exposes them to the elements. Solar dryers or solar air heaters can, however, be built with wooden frames to protect the yam from contamination from the ground and from dust and rain (Fig. 3.2). Some solar dryers are fitted with solar panels to aid drying by collecting energy during the day for use during the night (Fig. 3.3). This allows drying for extended periods of time

Fig. 3.2. Solar dryers at the Kwame Nkrumah University of Science and Technology in Ghana.

Fig. 3.3. Solar dryer fitted with a solar panel at the Kwame Nkrumah University of Science and Technology in Ghana.

and continuous processing especially for commercial purposes.

Other methods of drying have been investigated. For example, Falade *et al.* (2007) studied hot air drying of *D. alata* and *D. rotundata* slices, Lin *et al.* (2007) investigated far-infrared-radiation-assisted freeze drying and Xiao *et al.* (2012) used superheated steam blanching of *D. alata* slices. Abano and Amoah (2015) demonstrated that microwave-assisted drying could be used to enhance heat and mass transfer processes to achieve better quality dried yam tubers.

Milling

Milling involves the process of size reduction to produce a powdered product. Yam tubers can be milled into a paste, flour or starch that have both food and non-food applications. The milling process and the type of mill can influence the quality of yam flour. Badmus *et al.* (2013) investigated the use of an attrition mill in producing yam flour and reported that the functional characteristics of the yam flour, which included its bulk density, water absorption and swelling power, was influenced by milling parameters such as the speed, flow rate and the moisture content of the yam tubers. They found that an optimum moisture content of 12% gave a flour yield of 54.2%, 19.1% grit and 23.6% meal. Kim *et al.* (2016) reported that the most starch damage occurred during milling using a jet mill to produce yam flour, at levels of up to 8.83%, compared to a roller mill which resulted in 7.47% starch damage. In communities where public commercial mills are used, Somorin *et al.* (2011) reported that these mills can introduce microorganisms into the yam flour that could affect its quality. They found that the yam flour samples from a public commercial mill in a market contained *Bacillus megaterium* and *Staphylococcus saprophyticus*. They also isolated the fungi *Fusarium oxysporum*, *Aspergillus niger* and *Rhizopus nigricans* from both *D. rotundata* flour and *D. alata* flour. They reported that regular cleaning and maintenance of the milling equipment was enough to limit contamination of the yam flour by these bacteria and fungi.

Flour

Flour is produced from yam tubers by washing, peeling, slicing, parboiling, drying and milling. The flour can then be reconstituted in boiling water and used in various dishes. The following method for the preparation of flour or starch from *D. hispida* tubers was suggested by Kay (1987):

1. The tubers are thoroughly washed in clean water, either by hand or mechanically, to remove adhering soil, etc.
2. The tubers are mashed with water; a potato rasping machine is suitable for the preparation of flour, but for the production of starch, the tubers must be ground very finely in order to rupture the cell walls and liberate the starch granules.
3. In order to detoxify the material, the pulp is treated with lime water containing potassium permanganate; usually lime water equivalent to five times the weight of tubers and containing 0.005% of potassium permanganate, is used. Any excess potassium permanganate is removed by treating the starch milk with sulphur dioxide.
4. The starch is allowed to settle out and is then washed and centrifuged as in the manufacture of sweet potato starch.

Jarmai and Montford (1968) also described flour extraction in which the tubers were sliced to a thickness of about 1 cm, peeled and dried in the sun. When dry, the hard slices are ground to give a coarse flour. In a slightly more sophisticated process, washed tubers are cut into slices about 5 cm thick, cooked until soft, then peeled and mashed into a pulp, which is spread out to a depth of about 2 cm and dried for 6 to 8 hours at a temperature of 50–70°C, until the moisture content is reduced to about 10%. The dried material is finely ground and passed through a sieve before being packed into polyethylene bags for storage or marketing.

Dried yams can also be directly packaged as flakes without finely grinding them. Yam flakes have been prepared in some of the Caribbean islands from *D. alata* and in Nigeria and Ghana from *D. rotundata*. Yam flakes can also be packaged conveniently in polyethylene bags. In the Sudan, wild detoxified *D. dumentorum* tubers have been ground into a flour that has been used as a base for the preparation of beer (Rašper and Coursey, 1967a). It was reported that in Japan about half of the *D. polystachya* were used as raw material in the preparation of various food products, including pastry, beanjam bun, fish paste and yam flour (Kawakam, 1970). The average composition of the flour extracted from *D. hispida* tubers was given as protein 5.28%, fat 0.23%, starch 88.34%, fibre 5.33% and ash 0.66% (Sulit, 1967).

Varieties of *D. rotundata* producing numerous smaller tubers are preferred for flour (e.g. 'Kokoro' in Benin). The sun-dried tuber pieces are pulverized into flour using hammer mills. The yam flour is stirred over boiling water and cooked for a few minutes in order to obtain a thick viscous fufu, which resembles the one obtained with pounded boiled yam (Lebot, 2019). In the Philippines it has recently been suggested that starch or flour could be produced on a commercial scale by extracting the tubers with 95% alcohol followed by treatment with 5% sodium chloride or acidified water. Also, in the Philippines, Salda *et al.* (1998) described using yam flour for noodles, snacks and baby food products. In Taiwan, Hsu *et al.* (2004) incorporated *D. alata* flour in bread at a ratio of 25% yam flour to 75% wheat flour. They found that the antioxidant capacity of the bread was increased and the bread was acceptable to consumers. Adebowale *et al.* (2018) pre-treated peeled and sliced yam tubers with 0.28% potassium metabisulphite for 15 minutes or blanched them at 70°C for 15 minutes. All the slices were then dried either at 60°C for 48 hours or in the sun for 3 days, then milled into flour. The species used were *D. rotundata*, *D. alata*, *D. dumetorum* and *D. cayenesis*. The antinutritional factors in the flour were significantly reduced by blanching compared with the use of potassium metabisulphite. Antinutritional values ranged from 165–979 mg 100 g^{-1} for total phenol, 123–440 mg 100 g^{-1} for tannin, 13.4–79.2 mg 100 g^{-1} for phytate, 1.61–13.0 mg 100 g^{-1} for saponin, 0.02–0.11 mg 100 g^{-1} for alkaloid and 9.02–49.3 mg 100 g^{-1} for oxalate. The main and combined effects of species, pre-treatment and drying method were significantly different ($p < 0.05$) except for saponin which showed no significant effect. The range of vitamins in the flour were in mg 100g^{-1} 0.18–1.05 for B_1, 0.44–4.55 for B_2, 2.00–4.34 for B_6 and 20.9–30.91 for C. Generally, samples treated with potassium metabisulphite had higher vitamin B content, while the vitamin C content from blanching pre-treatment was higher. The sun-dried samples, irrespective of species and pre-treatment had the lowest ascorbic acid content.

Starch

Yam tubers have a high starch content. Changes in starch content of *D. rotundata*, *D. alata* and *D. esculenta* tubers during dormancy and sprouting in ambient conditions in India showed that starch decreased on average by 35% in the three species over a period of 90 days storage (Hariprakash and Nambisan, 2007). For starch characteristics, Amani *et al.* (2004) reported that yam starch gels are characterized by their weak stability to 'technological stress' and they evaluated starches from *D. alata* and *D. cayenensis-rotundata*, characterized by a large diameter grain (approximately 25 µm), a high amylose content (around 25% dry weight), a high intrinsic viscosity (mean of 190 cm^3 g^{-1}) and a high apparent viscosity and clarity of the paste. *D. esculenta* had smaller granules (approximately 6 µm), a low intrinsic viscosity (121 cm^3 g^{-1}), a high gelatinization enthalpy change (19 J g^{-1}) and a low paste viscosity. *D. dumetorum* starch differed from *D. esculenta* starch by having a pure A-type crystalline form and an opaque paste. Amani *et al.* (2004) also tested the stability of starch from different species and cultivars during thermal processing and found that the gel of 'Kangba' (*D. cayenensis-rotundata*) was the most stable. They also reported that 'Daminangba' (*D. alata*) had the clearest gel, was the most stable during storage at 4°C and 'Esculenta 7' (*D. esculenta*) had the weakest value of syneresis at –20°C. The gel of *D. dumetorum* was the strongest under acidic conditions, 'Bodo' (*D. alata*) gel had good resistance to shearing and 'Sopèrè', 'Lopka' and 'Kponan' (*D. cayenensis-rotundata*) had the strongest viscosities in all the treatments. In Thailand starch from *D. hispida* tubers is made into paste, which is then deep fried or boiled (Fig. 3.4).

Starch extraction from yam tubers is commonly achieved by grating, mixing with water and sieving to separate the starch slurry from the residual mass. The starch is recovered by decantation or centrifugation. However, starch extraction can be difficult due to the high viscosity of the slurry caused by non-starch polysaccharides. Yam starch suspensions have good stability to thermal stresses, with an absence of a peak of maximum viscosity during heating, good water retention capacity and high gelatinization temperature (Wang *et al.*, 2006a, 2006b). Karam *et al.* (2006) found that the amylose content in *D. alata* starches was between 21 and 29%, but they have limitations when they are subjected to extreme mechanical, thermal or chemical stresses. Also, they can show a tendency toward retrogradation and syneresis, which can restrict their use in some applications (Falade and Ayetigbo, 2015). Jairo *et al.* (2018) tested lintnerization of *D. alata* cultivar 'Diamante 22', which involved its chemical modification with hydrochloric acid to produce starch for the food, paper and textile industries. They found that lintnerization improved the stability of yam starches during heating and decreased their tendency to retrograde, thus producing starch that was suitable for the food processing industry.

Fig. 3.4. *D. hispida* paste. Photograph taken at Sarinya's yam farm in Chachoengsao province in the eastern region of Thailand where *D. hispida* were growing in 2018.

Extrusion

Extrusion technology is increasingly being employed for the development of novel food products such as snacks, baby foods, breakfast cereals and modified starch. Extrusion of yam tuber extracts has been investigated and there is still a need to optimize the process and establish the influence of extrusion parameters on the characteristics of extruded yam products. Extrusion requires first drying and milling the tubers in a flour after which the flour can be combined with any other material before being extruded. Hazarika *et al.* (2013) used a single screw extruder to prepare a ready-to-eat breakfast product from a combination. *D. alata* and sweet potato (*Ipomoea batatas*). To optimize the processing parameters screw speeds were varied from 132 to 468 rpm and the barrel temperatures from 103°C to 137°C. The optimized conditions obtained from response surface methodology were a screw speed of 400 rpm, barrel temperature of 130°C and 150 g of sweet potato flour per kilogram of yam flour. In Ghana, Adams *et al.* (2019) incorporated boabab pulp (*Adansonia digitate*) and tamarind kernel (*Tamarindus indica*) powders into an extruded puffed snack from *D. alata* tuber flour. The extruded snack containing 40% tamarind kernel flour in addition to yam flour was most preferred based on a consumer acceptance test.

Extrusion allows for the development of yam products with improved nutritional quality. Chiu *et al.* (2012) fortified corn extrudates with flour from three yam varieties – *D. alata* L. var. Tai-nung No.1 (TN1), *D. alata* L. var. Ta-shan (TS) and *D. doryophora* var. Hang-chun. They evaluated the influence of extrusion on the antioxidant activity of the finished corn-yam product and reported that extrusion processing increased the antioxidant activity – i.e. the thiobarbituric acid reactive substances (TBARS) inhibition ability for all yams without a negative impact on 1,1-diphenyl-2-picryl-hydrazyl (DPPH) free radical scavenging, oxygen radical absorbance capacity (ORAC), ferrous ion chelating (FIC) ability and 2,2-azino-bis-(3-ethylbenzothiazoline6-sulphonic acid) (ABTS) free radical scavenging activity for most yams in corn-yam extrudates.

Frozen and freeze dried

Abe (1973) described freeze-drying of whole fresh yam tubers. Frozen slices were also described that had been prepared and stored in sealed plastic bags. The portions were sliced into 1–2 cm thick pieces and dipped in a solution of 1% metabisulphite to prevent oxidation. The slices were then pre-cooked at 40°C for 15 minutes and frozen at –40°C for 30 minutes. The frozen slices could then be stored in a freezer at –3 to –5°C, giving an acceptable product that was ready to cook and eat. The use of an antioxidant improved the colour and avoided discoloration for up to 3 months of storage. As mentioned above Quansah *et al.* (2010) successfully produced frozen yam chips. Miyoshi *et al.* (2011) tested Sanyaku (a freeze-dried powder produced from yam tubers in China) extracted from various species of yam containing diosgenin and found that there were indications that the diosgenin could be ingested to prevent colon carcinogenesis in humans.

Functional Properties

The functional properties of yams comprise all the attributes related to their processing and cooking, which influence the quality or characteristics of the finished product. Amani *et al.* (2004) compared functional properties among 21 types of native starches representative of the major yam species cultivated in Côte d'Ivoire (Table 3.9). This included *D. alata*, *D. cayenensis-rotundata* complex, *D. dumetorum* and *D. esculenta*. They found a relationship between the functional properties of the yam starch and its physio-chemical properties. Of the four *Dioscorea* species they tested, Rašper and Coursey (1967b) found that *D. alata* had the strongest gel while *D. rotundata* gave the most viscous solutions, which they considered to be most likely because they had larger granule sizes (10–70 µm) compared to the other two, *D. esculenta* and *D. dumetorum*, which were smaller (1–5 µm). Other differences exist between different *Dioscorea* species, and there are also differences among cultivars of the same species. Alves *et al.* (2002) and Nadia *et al.* (2014) both reported significantly different functional characteristics among different cultivars of *D. alata* from Brazil and Indonesia. Processing conditions such as temperature and pre-treatment or substitution with other components can also affect the functional properties of yam starch. Olu *et al.* (2012) substituted yam flour

Table 3.9. Functional properties of four yam varieties. Modified from Amani *et al.* (2004).

	D. alata	*D. cayenensis-rotundata*	*D. dumetorum*	*D. esculenta*
Amylose content (%)	25.3–27.4	25.2–28.8	16.6	14.1–17.1
Particle size (µm)	18.6–29.3	20.4–30.9	2.2	5.8–6.2
Intrinsic viscosity ($cm^3\ g^{-1}$)	173–207	178–217	128	116–134
Gelatinization onset temperature (°C)	74.7–75.8	69.9–6.9	81.7	72.3–73.6
Enthalpy change ($J\ g^{-1}$)	16.1–18.1	13.7–15.9	16.7	16.7–20.3
Crystalline pattern (% B-polymorph)	65–100	65–100	–	65–100
Crystallinity (%)	31–41	27–45	37	26–35
Swelling power (gg^{-1})	13.8–16.0	10.8–16.4	13.7	13.9–14.3
Solubility ($g\ dm^{-1}$)	7.3–13.5	6.1–11.4	12.4	7.3–12.5
Pasting temperature (°C)	79.3–85.4	76.3–86.4	87.5	76.6–82.8
Viscosity (mPa s)	219–364	99–311	61	76–178

(*D. rotundata*) at varying proportions with soybean flour up to 30% to develop a more nutritionally balanced product. The data showed that the functional properties, swelling power (2.70–3.34%) and solubility (16.16–20.23%) increased with increasing substitution levels of soybean, with a corresponding decrease in the bulk density (0.84–0.71), dispersibility (60.50–52.50) and water absorption capacity (267.76–260.62%). The partial substitution of soy flour with yam flour also affected the rheological properties of the flour. The rate of water absorption of the yam flour decreased as soy flour substitution levels increased. These inherent functional properties of yam can enhance its use as food ingredient for different applications.

Food Uses

Tubers of many *Dioscorea* species have been used for food for millennia and are still major staple food crops in many tropical countries with a few used for food in temperate and subtropical countries. These *Dioscorea* species are discussed in more detail under the separate species headings in Chapter 8. Yams are commonly eaten after being freshly cooked in boiling water and their food quality factors include mealiness, colour and taste. There appears to be a general preference for boiled yam but the consistency, the colour and the stickiness determine the general preference of the pounded yam. Yams are sometimes dried and made into flour; this is often the case with damaged yams or with yams that are surplus to requirements, essentially as a method of storage. Other products have been developed including freeze-dried yams (Abe, 1973), canned yams and yam soup from Puerto Rico and yam flakes from Barbados. In general, these attempts have not been commercial successes, largely owing to the high cost of the raw material (Kay, 1987). Yams are prepared in many different ways in the Caribbean and similarly in Guyana, including mashed or cream yams that are peeled, cut into small chunks or cubes and boiled until soft, then mashed with milk or butter along with salt and other condiments for taste. Another method is where yams are peeled, cut into small pieces and boiled, then sautéed in olive oil with chopped onions, parsley, hot pepper and salt added for flavour. Mashed or cream yam can be eaten by itself or with various preparations of fish and meat. Yams are also included in soups instead of potatoes or in addition to potatoes and with other root crops like cassava, sweet potato and eddoe (Shamina Maccum, personal communication, 2019). Berisio (2010) described preparation of *D. cayenensis* tubers in Sierra Leone. She reported that they are poisonous if eaten raw and must be cooked for a long time before being used in traditional dishes. Preparation is by cutting them into large pieces and boiling them in salted water for 20–25 minutes. Cooked tubers become very soft and are ready to be used in many different ways including baked, fried or added to soups and stews.

Fufu

Fufu is dough made from boiled and ground starchy crops such as cassava, cocoyams (*Colocasia esculenta*), unripe plantains and yams and is used as a staple food in parts of West and Central Africa. In Ghana, boiled pieces are pounded together in a large wooden mortar using a wooden pestle. In between blows from the pestle, the mixture is turned by hand and water gradually added until it becomes a sticky slurry. The mixture is then formed into a ball or a rounded slab as shown in Fig. 3.5, which is eaten with the fingers usually after being dipped into an accompanying soup or sauce. In Ghana fufu made from yams is not common because cassava is cheaper. In Nigeria, Ekwu *et al.* (2005) tested three cultivars of *D. rotundata*, 'Ozibo', 'Okpebe' and 'Nwopoke', on their suitability for making fufu: 'The tubers were peeled, sliced, blanched and dried in an oven set at 60°C to constant weight. The chips were milled into flour, passed through American Standard sieve number 10.' They found that there were significant differences in the properties of the fufu from the three cultivars, but

'Nwopoke' was generally preferred by consumers followed by 'Okpebe'. They also found that although 'Okpebe' had good fufu-making qualities, its stretchability potential could present difficulty in pounding. Manual pounding is time and energy consuming. A prototype electric pounding machine was developed at Kwame Nkrumah University of Science and Technology in Ghana in 2004 to speed up the pounding process and it was reported that pounding could be achieved in just one minute (Joy Online, 2005). Recently, instant fufu powders have gained popularity and provide a convenient way of preparing fufu without pounding.

Amala

Amala is a popular starchy food that is largely eaten by ethnic Yorubas of south-western Nigeria. However, its popularity was reported to have been increasing among non-ethnic Yoruba consumers in Nigeria and people in some other West African countries (Orkwor and Asadu, 1998). Amala is a thick paste that is prepared by stirring flour from yams, cassava, cocoyams (*Colocasia esculenta*), sweet potatoes or unripe plantains in hot water to form a smooth paste. The most common type of amala is made from *D. rotundata* tubers. In West Africa, reconstituted yam flour is traditionally processed in the home to make amala but the process is very laborious and time-consuming. Making amala usually involves peeling, washing and slicing the tubers then drying the slices in the sun for several days. The sun-dried slices are crushed in a wooden mortar with a pestle and sieved. Coarse particles are returned to the mortar and again crushed and sieved. This recycling continues until the coarse particles are reduced to a minimum. The flour is then reconstituted in boiling water to give a brown, rather plastic dough, with more flour added if necessary, and stirred until all the water is absorbed (Fig. 3.6). More hot water is added, then the dough is left to simmer for approximately 5 minutes. This dough is kneaded until it has the desired texture. Kneading the dough into a smooth paste is the most difficult part of making amala. Amala is commonly consumed with vegetable stew, meat and condiments (Abe, 1973).

Fig. 3.5. Pounded fufu made from yam mixed with cassava and plantain.

Yam fries

Fried yam is a popular street food in West Africa that is typically sold by street vendors set up in kiosks or wooden stalls. The yam tubers are peeled and cut into elongated pieces that are deep fried in oil until a golden-brown colour (Fig. 3.7). Fried yam is often eaten with fresh pepper paste or concentrated tomato stew and fried fish or sausage. Traditionally, the yam tubers are freshly prepared, however, in order to increase their availability throughout the year and to help with food security, the yam tubers can

Fig. 3.6. Amala in Ghana.

Fig. 3.7. Fried yam at a local West African market.

be dried and stored as chips for later use (Akissoé *et al.*, 2001). In Ghana, Opata *et al.* (2009) studied the suitability of 15 *D. alata* varieties or cultivars including two local types for French fries. The tubers were minimally processed with the aim of increasing their shelf life, availability and enhancing their usage. Opata *et al.* reported that some of the varieties of *D. alata* could be used to produce very good French fries and small size fried chips were the preferred choice of consumers.

Quansah *et al.* (2010) processed two cultivars of *D. rotundata*, 'Puna' and 'White Yam' into frozen chips that could be used fried like frozen potato chips. They tested different preparation methods. One method was to blanch tubers for 0, 2.5 or 5 minutes at 70, 80 or 90°C. They found that blanching at 90°C for 2.5 minutes gave products that were preferred by sensory panellists when compared to commercial potato chips. Graham-Acquaah *et al.* (2013) investigated the effects of blanching and frying on the production of French fries from two cultivars of *D. rotundata*. They found that blanching temperature and frying time had a greater impact on the texture and appearance of fried chips than blanching time. They also subsequently used response surface methodology to characterize the effects of blanching temperature and time and frying time on texture and appearance of *D. rotundata* chips. They reported that the effects of these factors on the quality characteristics of fried yam chips was similar to French fries made from potatoes. Increases in moisture content during blanching has been reported for *D. schimperiana* (Quansah *et al.* 2010; Leng *et al.* 2011). When the tubers were blanched the starch may have gelatinized at the higher temperature. Also blanching can affect nutritional content. For example, Leng *et al.* (2011) reported that there were more than 50% ascorbic acid losses after blanching and drying of *D. schimperina*. In Thailand, *D. hispida* has traditionally been used to make a snack. For these snacks, tubers are pounded in a similar way to that described above (Fig. 3.8).

Yam crisps

Crisps, also called chips, are mostly made from potatoes, but they have been successfully made from many other crops including sweet potatoes, taro, cassava, plantains, bananas, carrots as well as yams. Tortoe *et al.* (2014a) made crisps from tubers of two *D. rotundata* cultivars, 'Pona' and 'Dente', by slicing them 2 mm thick and frying them for 5 minutes at 140°C. They used different flavourings and all the crisps appeared to be highly acceptable from the responses of a sensory evaluation panel. Martin and Ruberté (1972) cut yam tubers into thin slices (about 1.5 mm thick) with a simple kitchen slicer and fried them in corn oil at 191°C until they were crisp. The crisps were cooked to a pleasant colour, then rated by an informal sensory evaluation panel. All the crisps from *D. rotundata*, *D. esculenta* and *D. bulbifera* contained bitter or acrid substances making them unacceptable for crisps, but one variety of *D. trífida* yielded acceptable crisps. Crisps produced from *D. alata* were of varied quality with 'Forastero' and 'Farm Lisbon' and a few other cultivars producing excellent crisps. The tubers that produced the best crisps had white, compact flesh with minimum graininess and had a low tendency for flesh oxidation.

In Thailand, crisps made partly from yam flour (probably *D. alata*) are sold commercially. For example, one brand had the following ingredients: wheat flour 48%, vegetable oil 35%, yam 12%, starch 2%,

Fig. 3.8. Snack food in Thailand made from *D. hispida* tuber flakes with peanuts in rice flour and coconut milk batter, then deep fried in a mould.

potato powder, flavouring, sugar, iodized salt, INS170 (anticaking agent), INS500 (sodium carbonate, sodium bicarbonate), INS471 (iron oxide), plus either chicken or tomato flavouring (Fig. 3.9).

Boiled yam

Boiled yam is prepared by cooking medium-sized slices of yam in salted water (Fig. 3.10). This is the most common food product made from yams and it can be served with a variety of soups and sauces. In Ghana, it is often accompanied by tomatoes, gravy, kontomire stew, palava sauce or garden egg stew. Otegbayo *et al.* (2005) studied the textural quality of boiled yam. Histologically, the starch granules in varieties of *D. rotundata* were loosely packed whereas those from *D. alata* were more densely packed. The texture of the cooked yams from most varieties of *D. rotundata* was described as mealy and resulted from cell separation and rounding off of cells. The texture of cooked yams from the *D. alata* varieties was described as waxy and showed partial retention of textural cell integrity. Polycarp (2017) compared the textural characteristics of boiled yam produced from seven different *Dioscorea* species and reported that cooked samples of *D. esculenta* had the lowest hardness. *D. rotundata*, *D. alata* and *D. bulbifera* on the other hand had 'tolerable textural properties' appropriate for efficient industrial and food process applications.

Pounded yam

Pounded yam is also very popular in West and Central Africa. It is basically made from pounding and kneading boiled yam into a glutinous dough or shaping it into a ball (Fig. 3.11) and is very common in Nigeria. It is served with different types of soup, stew or sauces especially with peanut sauce or agushie stew. Agushie is made from dried, ground seeds of squash, melons or gourds. This meal is often served during festivities. The quality and texture of pounded yam depends on the choice of yam species and variety.

Fig. 3.9 Yam chips marketed in Thailand, June 2020. Photo: Pikunthong Nukthamna.

Roasted yam

Roasted yam is another convenience street food enjoyed in West Africa. The yam is cut into pieces and roasted over an open flame (Fig. 3.12). Roasting yams has been reported to result in increased levels of polycyclic aromatic hydrocarbons. This class of compounds may have carcinogenic or mutagenic effects and Ogbuagu and Ayoade (2012) reported that there were up to 0.0093 mg kg^{-1} of anthracene in roasted yam. They attributed an increase in polycyclic aromatic hydrocarbon levels to longer roasting times.

Yam pottage

Yam pottage is prepared by cutting the yam into small pieces and cooking in a tomato soup with other condiments. It can be consumed with or without an entrée (Fig. 3.13). There is a variation of this yam dish where the yam is mashed instead of being cut into pieces. This is usually referred to as 'mpotompoto' in Ghana.

Yam couscous

Yam couscous is a rice-like food prepared by milling dried yam that has been previously boiled and grated (Fig. 3.14). In West Africa, this food is called 'wasawasa' and is popularly consumed in Benin, typically being made from

Fig. 3.10. Boiled yam (*D. rotundata*) in salted water (a), and boiled *D. alata*, from a farm in Chachoengsao province, Thailand, 2020 (b).

Fig. 3.11. Yam fufu ready for serving at a local 'food joint' in Ghana.

Fig. 3.12. Roasted yam pieces served with fish and groundnuts as a street food in Ghana.

Fig. 3.13. Yam pottage in Ghana.

D. rotundata. However, Oduro *et al.* (2009) studied the suitability of producing couscous from *D. alata* with the aim of promoting underutilized yam species in Ghana. They reported acceptable consumer preferences and a storage period for prepared wasawasa of up to 24 weeks. In Côte d'Ivoire, Souleymane *et al.* (2018) pre-treated flour from *D. cayenensis-rotundata* with 1% lemon juice and pre-cooked it before producing the couscous. However, pre-cooking the flour resulted in reducing levels of protein, total sugars and reducing sugars during subsequent cooking, though it did not influence the swelling and solubility of the flour and the couscous was generally found to be acceptable by a sensory panel.

Fig. 3.14. Yam couscous served with ground pepper and fish in West Africa.

Bread

Yam flour has potential for inclusion in making of bread in order to improve its nutritional value and its good processing properties (Amandikwa *et al.*, 2015; Eke-Ejiofor *et al.*, 2015). Compared with wheat flour, yam flour is abundant in polysaccharides, phenolics, flavonoids and allantoin compounds, and has high antioxidant activity (Chiu *et al.*, 2013). Tsen and Tang (1971) made bread with pure yam flour, using *D. alata*, *D. rotundata* or *D. esculenta*, or mixtures with wheat flour and evaluated them in comparison to 'high-quality American-type bread'. Hard bread rolls made with 50% yam flour were not as good as American-type bread made with wheat flour. The dough of the yam flour tended to rise very slowly, which could have been because of the large starch granules of yams, or possibly because of damage to the dough structure from the dilution of gluten. Colours of the bread made with yam flour were sometimes objectionable, but this depended largely on personal opinion. None of the bread made with yam flour in these trials was as good as desired, but usually was equal in flavour to the American-type bread. Only one bread of very poor flavour was encountered, which was made from 100% *D. alata* 'Forastero' flour. When yam flour was substituted for part of the wheat flour in American-style bread, a dough conditioner (sodium stearoyl-2 lactylate) was found necessary to maintain loaf volume. The bread made from 20% yam flour was satisfactory but the bread was not acceptable when 40% yam flour was used. From 20% yam flour the bread was 'soft and good-tasting, but somewhat gummy'. No significant difference ($p = 0.05$) was noted between the bread prepared from 20% *D. alata* cultivars 'Farm Lisbon' or 'Forastero' flours, both of which gave the most satisfactory results of all the species and cultivars tested.

Previous studies have reported that yam and wheat notably affect the functional properties and quality of bread, which varies depending on the level of substitution. Hsu *et al.* (2004) reported that wheat substitution with yam flour resulted in bread with high antioxidant capacity and low digestibility and that bread made with no more than 20% yam flour was not significantly inferior in quality and acceptability. Li *et al.* (2020) tested wheat flour substituted with levels of *D. opposita* flour from 0 to 25% in bread and found that as the proportion of yam flour increased the addition of yam flour destroyed the gluten network structure in the dough. The addition of yam flour decreased the content of α-helix and β-sheet in gluten. With the increase in the proportion of yam flour, the specific volume and overall acceptability decreased whereas the total phenolics content, polysaccharides content, total flavonoids content, allantoin content, 1,1-diphenyl-2-picrylhydrazyl radical scavenging capability, fractal dimension and hardness all increased. They concluded that bread made from wheat flour replacement with up to 15% yam flour was of a high quality and had higher antioxidant properties. Seguchi *et al.* (2012) developed gluten-free bread, baked with *D. japonica* flour, wheat flour, sucrose, compressed yeast and water. The bread showed similar properties, such as bread height and specific volume, as bread made entirely with wheat flour containing no yam flour.

Flour

A traditional way of producing flour from yam tubers in West Africa is: washing, peeling, washing, slicing, parboiling, drying, milling and bagging. In Nigeria pounded yam instant flour is commonly known as Poundo yam. Poundo yam is prepared from *D. rotundata* tubers, and is marketed ready prepared (Fig. 3.15) so it can be easily used in cooking.

Yam peel

During processing, the outer layers of the yam tubers are peeled and often discarded as waste. Akinmutimi and Onen (2008) tested yam peel meal, made from *D. rotundata*, and found that it consisted of 89.25 dry matter, 12.17 crude protein, 9.3 ash, 6.3 crude fibre, 1.05 crude fat and 2.98 kcal g^{-1} gross energy. Yam peel is often used for animal feed but it is also prepared for human consumption, for example by drying and steam cooking where it appears as dark brown spherical particles of about 2–5 mm in diameter (Fig. 3.16). It is popular in northern Ghana where it is consumed by people of all ages and is served in almost every setting including as a daily meal, at funerals, in marketplaces, at parties and other social gatherings. It is mostly served by mixing it with oil, fried onions and pepper, accompanied by a stew and salad. Yam peel also has non-food applications and has been considered for production of the enzyme amylase. Uguru *et al.* (1997) isolated *Aspergillus niger* from yam peel to produce amylase and showed peak amylase activity after 4 days of fermentation. Oyeleke *et al.* (2010) isolated *Bacillus megaterium* from discarded yam peel, which was also used to produce amylase.

Other yam-based products

Besides the traditional uses of yam, the development of novel food products from yam tubers has been explored. These products include snacks, beverages, bakery products, baby food and other convenience foods. Interest in more convenient forms of food has increased as more developing countries become modernized and Wireko-Manu *et al.* (2014) indicated the potential for developing different instant foods from *D. alata* based on its physio-chemical properties and consumer acceptance. They reported that *D. alata* could be used as a composite with wheat to make such products as instant noodles and pasta (Fig. 3.17). It was also possible to produce high quality yam flour as a carbohydrate source for both domestic and industrial application akin to 'high quality cassava flour'. Yam flour can also be used to make a variety of baked products such as cupcakes, biscuits, yam fruit cakes and some desserts. Such products from yam flour have been reported in South East Asia, the Pacific islands and the Caribbean. Beverages have been developed from different yam species, mostly through fermentation. These beverages are often described as

Fig. 3.15. Instant yam flour on sale in a market in Huddersfield in November 2020.

Fig. 3.16. Dried and cooked yam peel served as a convenient street food in Ghana.

'functional foods' due to their potential health properties. Jin and Sun (2009) developed a functional beverage from the *D. polystacha* combined with red dates. Choi (2012) also developed a functional beverage using *D. opposita* where the processing technology was further optimized for commercial production.

In 2020, Praphan Pinsirodom developed some (unpublished) new recipes for yams in Thailand using a light purple variety of *D. alata* from Chachoengsao province (Fig. 3.18). Praphan Pinsirodom commented that 'yams are not a common ingredient in a modern Thai cuisine. We only can find simple homemade dishes in local areas. Here are some more familiar menus I created using the yam as a main ingredient. It can perfectly be used to replace potato or taro'.

Fig. 3.17. Yam pasta developed by the Kwame Nkrumah University of Science and Technology in Ghana.

Non-food Uses

Toxins and pharmaceuticals

Some 30 wild *Dioscorea* genotypes have been reported to be used for industrial purposes in the production of the sex hormones dioscorine and diosgenin (Kenyon *et al.* 2001). In Thailand the dioscorine content of dried tubers of *D. hispida* from 14 areas had an average content of 0.72 ± 0.07% w/w (Nonglapat Sasiwatpaisit *et al.*, 2014) and in Australia diosbulbin A was 0.07 mg g^{-1} in *D. bulbifera* var *rotunda* tubers (Webster *et al.*, 1984). Diosbulbin is a major constituent of the furano-norditerpenes in *D. bulbifera*, and exhibits potential antineoplasmic activity and hepatotoxicity. Diosgenin and dihydrodioscorine, which also occur in some species, are saponins and were first discovered in the 1930s in Japan in *D. tokoro* (Fujii and Matsukawa, 1936). Diosgenin has been subsequently detected in over 100 *Dioscorea* species (Table 3.10) (Martin, 1969). In Colombia, Raz *et al.* (2013) reported that *D. coriacea*, *D. lehmanni* and *D. polygonoides* all tested positive for sapogenins, with the highest yields occurring in *D. coriacea*, in which tissue culture studies had also been initiated. In Nepal the bitter components of *D. bulbifera*, *D. versicolor*, *D. deltoidea* and *D. triphylla* were identified as diosbulbin A (0.037g kg^{-1})

Fig. 3.18. Recipes made from *D. alata*: (a) yam salad using the same recipe as Japanese style potato salad; (b) yam in sweet coconut milk; (c) Yam croquettes. Courtesy of Dr Praphan Pinsirodom, dean, Faculty of Food Industry, King Mongkut's Institute of Technology Ladkrabang, Thailand.

Table 3.10. Diosgenin content in selected *Dioscorea* species modified from Martin (1969). It is not clear why rhizomes and tubers are listed separately, but they are quoted here from the original research.

Species	% diosgenin yield	Plant material	Habitat
D. belizenzis	Positive	Tubers	Honduras
D. bulbifera	4.5	Tubers	Asian tropics
D. bulbifera	0	Tubers	Asian tropics
D. bulbifera	0.08	Tubers	Asian tropics
D. burkilliana	0.61	Tubers	African tropics
D. caucasica	0.4		Caucasus
D. caucasica	1.5	Rhizomes	Caucasus
D. composita	0-13		Mexico
D. composita	Positive	Tubers	Mexico
D. deltoidea	0.9–1.72.3–3.7	Tubers	West Himalayas
D. deltoidea	Positive	Tubers	India
D. deltoidea	3.5–6	Whole plant	India
D. floribunda	0.2–4.0		Mexico
D. floribunda	Positive	Tubers	Mexico
D. hirtiflora	0.03	Tubers	African tropics
D. lobata	5.0		Mexico
D. mexicana	0.3–0.8		Mexico
D. minutiflora	0.07	Tubers	Cameroon
D. multiflora	0.4–1.4		Argentina
D. nipponica	0.3–2.0		Japan
D. praehensilis	0.18	Tubers	African tropics
D. prazeri	2.1		India, Myanmar
D. preussii	0.21	Tubers	Cameroon
D. zanzibarensis	0.05	Tubers	African tropics
D. spiculiflora	0.7–1.5	Tubers	Mexico
D. sylvatica	0.2–0.7	Tubers	South Africa
D. sylvatica	2.0–3.4	Tubers	South Africa
D. tepinapensis	0.4–0.7	Tubers	Mexico
D. testudinaria	5.0		Mexico
D. togoensis	0.12	Tubers	Africa
D. tokoro	1.0	Seeds	Japan
D. zingiberensis	5.93	Rhizomes	China
D. zingiberensis	6.13–16.15	Rhizomes	China
D. zingiberensis	4–8	Rhizomes	China

and diosbulbin B (0.314g kg^{-1}). Dioscorine was not detected in these tubers, but they contained cyanogens (as HCN equivalent) in the range of 3.2 to 6.0 ppm, which were reported as below safety limits. The inflammation and occasional toxicity observed from consumption of these tubers could possibly be due to the presence of high level of oxalate (Bhandari *et al.*, 2003; Bhandari and Kawabata, 2005).

Other examples include Shajeela *et al.* (2011) who gave the following analysis for antinutritional factors in *D. esculenta* tubers per 100 g^{-1} fresh weight as: total free phenolics 0.79 ± 0.07 g; tannins 0.20 ± 0.01 g; hydrogen cyanide 0.21 ± 0.03 mg; total oxalate 0.33 ± 0.02 g; amylase inhibitor 7.80 AIU mg^{-1}; soluble starch trypsin inhibitor 1.92 ± 0.07 TIU mg^{-1} protein. However, many of these substances can also have beneficial uses especially in medicine. A review of the pharmacological effects of diosgenin is included in Kanika Patel *et al.* (2012). As noted at the beginning of this chapter, Botany VM (2018) have stated that edible species of *Dioscorea* have opposite leaves while toxic species have alternate leaves. However is difficult to objectively understand the reasons for this, nor how many *Dioscorea* species were tested and exactly what the degree of toxicity was in all cases.

Berisio (2010) reported that *D. cayenensis* tubers are poisonous if eaten raw. One class of toxins found in many *Dioscorea* species are the steroidal saponins that are used in pharmacy, industry and cosmetics (Fenwick and Oakenfull, 2006). The bitterness and toxicity of many yam species may be caused by high level of saponins. Such is the case with *D. tokoro* in Japan, in which saponins were reported to be responsible for its bitterness (Webster *et al.*, 1984). Steroidal saponins are also used in the manufacture of progesterone and other steroid drugs and as oral contraceptives (Chen and Wu, 1994). These steroidal saponins and other chemicals from yams that are used or may be used as pharmaceuticals include the following.

Diosgenin ($C_{27}H_{42}O_3$) is closely related to human sex hormones, corticosteroids, vitamin D and cardiac glycosides. Diosgenin, a steroidal sapogenin has been isolated from species of various plant families including Agavaceae, Dioscoreaceae, Liliaceae, Solanaceae, Scrophulariaceae, Amaryllidaceae, Leguminosae and Rhamnaceae, and can be extracted from various plants including some *Dioscorea* tubers. Diosgenin is the product of hydrolysis, by acids, strong bases or enzymes of saponins. Dougall (1979) tested diosgenin production in tissue cultured *Dioscorea* plants and found that it can be at least as high as in intact plants. For example, the yield of diosgenin found in cell cultures of *D. deltoidea* (26 mg g^{-1} dry wt) was higher than that found in whole plants (20 mg g^{-1} dry wt). Diosgenin has been shown to be present in species originating in North America or in Asia including *D. bulbifera*, *D. villosa*, *D. opposita*, *D. hypoglauca*, *D. composata*, *D. deltoida*, *D. parazeri*, *D. mastrostachya*, *D. floribunda* and *D. barbasco*. Diosgenin levels in various yams tubers are given in Table 3.10. Baker *et al.* (1966) tested the distribution of diosgenin in *D. deltoidea* and *D. sylvatica* and found that it occurred in the endosperm of the dormant seeds as well as the tubers, roots, stems and leaves from the seedling stage onwards. Actively growing leader shoots appear to be the sites of formation from which the diosgenin is translocated to the tubers, where it is evenly distributed throughout the tubers. The level of diosgenin can vary considerably in different species, for example, in their analysis, Yi *et al.* (2014) found that *D. collettii* and *D. septemloba* had an average diosgenin content of 14.37 mg g^{-1}, *D. zingiberensis* up to 19.5 mg g^{-1}, *D. nipponica* up to 12.52 mg g^{-1}, *D. panthaica* up to 5.29 mg g^{-1} and *D. opposite* zero.

Dioscorine ($C_{13}H_{19}O_2N$), a water-soluble storage protein of yams, has a molecular weight of 221.29546 g mol^{-1} and Broadbent and Schnieden (1958) reported that it was probably first isolated from tubers of *D. hirsuta* by Boorsma (1894) and from the tubers of *D. hispida* by Levya and Guttierrez (1937). Dioscorine is an alkaloid toxin that has been used as a monkey poison in some African countries and as an arrow poison to aid in hunting in several parts of Asia. Dioscorine was reported to inhibit angiotensin converting enzyme activity which plays an important role in management of hypertension (Hsu *et al.*, 2002).

Dioscine ($C_{13}H_{21}O_2N$) is an alkaloid derived from a diosgenin that was isolated from the tubers of *D. dumetorum* by Bevan *et al.* (1956). It has a role as a metabolite, antifungal agent, antiviral agent, antineoplastic agent, anti-inflammatory agent, hepatoprotective agent, apoptosis inducer and tyrosinase inhibitor (Cho *et al.*, 2013; Wang *et al.*, 2013). In a comparison of dioscorine and dioscine Broadbent and Schnieden (1958) found that they had similar properties, but in most respects dioscorine was the more potent.

Diosbulbins are a furanoid norditerpene and include diosbulbin A = $C_{20}H_{24}O_7$, diosbulbin B = $C_{19}H_{20}O_6$, diosbulbin C = $C_{19}H_{22}O_7$, diosbulbin D = $C_{19}H_{20}O_6$, diosbulbin E = $C_{19}H_{22}O_6$. They are furanoid norditerpenes that occur in *Dioscorea* species. Diosbulbin B is the most abundant component in *D. bulbifera* tubers (Lin *et al.*, 2014) and diosbulbin A was reported in *D. bulbifera* var *rotunda* tubers (Webster *et al.*, 1984); 8-epidiosbulbin ($C_{21}H_{24}O_7$) is also a furanoid that can exhibit broad spectrum plasmid curing activity against multidrug resistant bacteria. A plasmid is a DNA molecule within a cell that is physically separated from chromosomal DNA and can replicate independently and is usually found in bacteria. Shi *et al.* (2018)

reported that 8-epidiosbulbin E acetate and diosbulbin B were hepatotoxicity related markers. Shriram *et al.* (2008) also reported that 8-epidiosbulbin E was a potential plasmid curing agent. Rakotobe *et al.* (2010) reported that 8-epidiosbulbin E was toxic to Japanese rice fish embryo larvae. Both *D. antaly* tubers (Rakotobe *et al.*, 2010) and *D. bulbifera* bulbs and tubers contain 8-epidiosbulbin E (Shi *et al.*, 2018; Shriram *et al.*, 2008).

Allantoin is an abundant and active constituent of some yam species, and has been shown to have medicinal benefits, including possible prevention of inflammation and ulcers in humans. Fu *et al.* (2006) analysed *D. batatas* 'Hualien no. 3' and *D. pseudojaponica* 'Keelung' for their allantoin and allantoic acid content. The total amount of allantoin and allantoic acid (mean ± RSD, mmoles g solid^{-1}) were: tubers of *D. pseudojaponica* 0.370 ± 6.8% (pulp), 1.130 ± 3.4% (peel), tubers of *D. batata* 0.278 ± 7.8% (pulp), 0.714 ± 9.1% (peel) and bulbils of *D. batata* 0.179 ± 7.9% (pulp), 0.297 ± 3.4% (peel). The allantoin and allantoic acid content in the peel of *D. pseudojaponica* was 305% higher than in the pulp and in *D. batatas* 257% higher in the peel than in the pulp. The tubers of some *Dioscorea* spp. have been reported to contain a higher allantoin content than any other plant, for example, the level in *D. batatas* tubers ranged from 4.1 to 7.1 mg g^{-1} dry weight (Go *et al.*, 2015).

Allantoin has the potential to be used as a therapeutic agent that can increase the smoothness of the skin, promote cell proliferation and has wound healing characteristics. Allantoin has also been reported to protect stomach tissues, inhibit tumour growth, reduce plasma glucose and has anti-diabetic effects, antioxidant activity and can modulate oxidative stress, as well as improve kidney and liver functions while maintaining insulin and glucose levels (Go *et al.*, 2015; Kim *et al.*, 2018; Lebot *et al.* 2019). Ma *et al.* (2018) investigated the effects of water extracts of allantoin from *D. batatas* tubers on skeletal muscle cells. They found the allantoin stimulated myoblast differentiation into myotubes and increased energy production through the upregulation of mitochondrial biogenesis regulators. They concluded that allantoin extracted from yams can help to prevent skeletal muscle dysfunction through the stimulation of energy metabolism. Wu *et al.* (2015) bred a new *D. alata* cultivar called 'Wenshanyao No.1', which had 48% higher allantoin than *D. alata* landraces.

Bevan *et al.* (1956) compared the effects of dioscorine and dioscine on isolated guinea pig ileum and found that their actions were qualitatively similar, but dioscorine was the more toxic and had greater local anaesthetic activity, antidiuretic activity and depressant actions. Both resemble cocaine in causing potentiation of the pressor action of adrenaline on blood pressure. Broadbent and Schnieden (1958) also compared the effects of dioscorine and dioscine and found that their toxic effects included convulsant activity (sudden and involuntary muscle contractions), central nervous system stimulation, local anaesthetic activity, adrenaline potentiating action, antidiuretic effect (helps to control fluid balance) and antiacetylcholine action and in most respects dioscorine was the more potent of the two. They also reported that aqueous solutions of dioscine were less stable than aqueous solutions of dioscorine.

Flavour and taste

The bitterness of many yam species is commonly caused by high levels of saponins. Such is the case with *D. tokoro* in Japan, in which saponins were responsible for its bitterness (Webster *et al.*, 1984). The yam species that are unpalatable (taste bitter) produce inflammation and toxicity but can still be used as foods if preparation methods take this into account. Domestic cooking methods were found to be very efficient in removing bitterness from *D. bulbifera*, *D. versicolor*, *D. deltoidea* and *D. triphylla* tubers, thus making them palatable (Bhandari and Kawabata, 2005). Bhandari and Kawabata tried boiling in water for 30 minutes and discarding the water after boiling, pressure cooking for 15 minutes and baking in aluminium foil at 180°C for 45 minutes. Boiling reduced the bitter principles (diosbulbins) in the range 75–100% and pressure cooking and baking were less, in the range 50–75%.

In Nepal the bitter components of *D. bulbifera*, *D. versicolor*, *D. deltoidea* and

D. triphylla were identified as the furanoid norditerpenes diosbulbins A and B (Bhandari and Kawabata, 2005). Histamine was reported as the principal allergen, causing mild inflammation and itching after consumption of some in some *Dioscorea* species (Schmidt and Moult, 1983). The bitter substances in some yam species have been reported as furanoid-norditerpene groups of compounds (Kawasaki *et al.*, 1968; Martin and Rubeste, 1976).

Traditional medicines

Yam tubers are used in various traditional medicines in many countries, but particularly in China, Korea and Japan where they have been used for traditional medical purposes including stimulation of the appetite, relief of bronchial irritation and coughs. PFAF *D. japonica* (nd) listed several species and rated their medical uses (Table 3.8) with only *D. batatas* having the highest rating. Mollica *et al.* (2013) showed that *D. trifida* extract could have an effect on food allergies and they tested its potential in the treatment of ovalbumin induced food allergy in Balb/c mice. They found that supplementing the mice's food with *D. trifida* extract reduced immunoglobulin E, a type of antibody that has only been found in mammals, which showed the potential for inclusion of *D. trifida* extract in food to prevent or treat this disease.

Mental disorder treatments

As well as the range of medical uses described above, numerous studies have reported that diosgenin is useful in the prevention and treatment of neurological diseases. Its therapeutic mechanisms are based on the mediation of different signalling pathways (Chen *et al.*, 2015). Diosgenin and its derivatives have been used as therapeutic agents for multiple neurological disorders along with their various mechanisms in the central nervous system. In particular, those related to therapeutic efficacy for Parkinson's disease, Alzheimer's disease, brain injury, neuroinflammation and ischemia (Li *et al.*, 2018). Diosgenin can facilitate the repair of axonal atrophy and synaptic degeneration and improve memory dysfunction in a transgenic mouse model of Alzheimer's disease. It has also been demonstrated that animal experiments have shown that diosgenin is active in the treatment of nervous system diseases including both Parkinson's disease and Alzheimer's. Chiu *et al.* (2009) attempted to access the neuroprotective effect of extracts of yam tubers (*D. pseudojaponica*) on senescent mice, whose senescence was induced by D-galactose administered orally at 20, 100 or 500 mg kg^{-1} for four weeks, starting at the sixth week. The yam diosgenin level was 5.49 mg g^{-1}. Mice treated with yam were found to have significantly (p = 0.05) improved learning and memory abilities compared to those treated with D-galactose (200 mg kg^{-1} for 10 weeks). In addition, treatment with the yam extract was also found to increase the activities of superoxide dismutase and glutathione peroxidase and decrease the malondialdehyde level on the brains of D-galactose treated mice. It was concluded that yams, as traditionally used in Chinese medicine, had the potential to be a useful treatment for cognitive impairment. It was found that the beneficial effect may be partly mediated via enhancing endogenous antioxidant enzymatic activities. Tohda *et al.* (2017) also tested the effects of extracts from yam tubers on mice. They found that among the 12 individual standard cognitive sub-tests, diosgenin-rich yam extracts significantly (p = 0.05) improved their semantic fluency, with no adverse effects. They concluded that diosgenin-rich yam extract treatment appeared to safely enhance cognitive function in healthy adults. Earlier, Ghayur *et al.* (2011) reported that the effects of diosgenin on the human central nervous system was potential generation of cortical neurons and big potassium channel activity improvement. They also showed that it increased intracellular calcium in the human cortical neuronal-1A cells and affected acetylcholinesterase inhibitory activity. They concluded that diosgenin may affect the activity of cortical neurons by acting on these channels.

4

Cultivation

Propagation

Degras *et al.* (1977) reported that the species of yam commonly used for food do not produce viable seeds with the exception of *D. trifida*, and *D. cayenensis-rotundata*. Almost all yams are invariably propagated vegetatively, usually from tubers. However, some species, such as *D. bulbifera* and some varieties of *D. alata*, produce bulbils in the axils of leaves that are also used for propagation. Whole tubers can be planted, especially for species that produce several small tubers on each plant, such as *D. trifida* and *D. esculenta*. For species that produce only a single large tuber, the tuber is commonly cut into pieces that are then planted. Pieces near the top of the tuber (proximal or 'head end') produce the earliest and most vigorous growth with vigour decreasing the nearer to the distal or 'tail end' the pieces are taken. In Jamaica, *D. cayenensis* is often harvested by digging up each tuber and then cutting off the head end, where the stems are still attached, and then replanting these heads for the next crop. In several countries recommendations for treating the planting material include dusting with a fungicide or anhydrous lime before planting, especially those tubers with an exposed cut surface, to prevent the entry of pathogenic organisms. The time from planting to emergence can be affected by physiological age, size, the part of tuber used (proximal, middle or distal), dry matter and nutrient content. Variability in the seed tubers may lead to a population with disparate plant ages and also tuber yield (Ferguson, 1973; Orkwor *et al.* 1998). Mignouna *et al.* (2016) reported that in West Africa less than 10% of yam growing households used certified seed and that yam planting material can take up to 50% of the total production costs.

In Nigeria, many farmers obtain their planting materials from the previous year's harvest either by cutting ware yams into setts for planting as described above or through 'milking' (also called the double harvest system), while the leaves of the plant are still green. Milking involves the harvesting of a yam tuber before it is fully mature and replanting the head to enable it to grow into a small tuber that is big enough for planting in the following year (Asumugha *et al.*, 2009) as described above for *D. cayenensis* in Jamaica. Milking is usually done between 4 and 5 months after shoot emergence and involves opening the mound in which the tuber is growing, from one side, then detaching the tuber from the base of the vine while keeping most of the roots intact and then re-moulding the mound.

DOI: 10.1079/9781789249279.0004

The plant initiates the production of a new tuber from the base of the vine, which often bulks into an amorphous shape with protuberances or digitations that can be detached and planted separately at harvest. The second harvest is at the end of the season when the plant senesces (1 to 3 months after the first harvest) for regenerated tubers are used solely as seed (Onwueme,1973; Aighewi *et al.*, 2015).

Mini-setts

The mini-sett technique is an alternative propagation technique to the production of seed yam through milking of ware yam and can be used for commercial production of seed yams. Eyitayo *et al.* (2010) compared the traditional method of production of *D. rotundata*, *D. alata* and *D. cayenensis* using 100 to 200 g setts and comparing them with mini-setts cut into different sizes (25, 20, 15 and 10 g) cut from mother tubers. They concluded that the mini-sett technique required only between 6 and 33% of the number of tubers needed for the traditional planting system and resulted in plants superior to those produced by the traditional method in all the production attributes they tested, especially in the case of *D. alata* and *D. rotundata*. However, previously, Campbell-Chin-Sue *et al.* (1994) tested various agronomic practices on *D. cayenensis* and found that mini-sett size and type did not affect subsequent total tuber yield. Igwilo and Okoli (1988) tested 25 g mini-setts of several cultivars of *D. rotundata*, *D. alata*, *D. cayenensis* and *D. dumetorum* and achieved sprouting percentages between 65 and 100%. Under good management about 1 tonne of mini-setts planting material gave yields of between 8 and 24 tonnes ha^{-1}, depending on the cultivar, with the best performance from cultivars of *D. alata*. They also reported that from mini-setts as small as 20–30 g it was possible to harvest seed tubers weighing up to 1 kg only 5 to 6 months after planting. Otoo *et al.* (1987) reported that in most breeding programmes the multiplication of yam plants in the field to obtain planting material was carried out routinely using mini-setts.

Olivier *et al.* (2012) tested two methods of producing yam mini-tubers from two *D. cayenensis-D. rotundata* complex cultivars 'Krengle' and 'Kponan'. The first method was to initiate micro-tubers *in vitro* and then transplant them into soil to generate plants that were either directly planted as micro-tubers in the soil, or germination was induced using various treatments before they were transferred to soil. In the second method plantlets were first produced *in vitro* and then transplanted into soil for further development and tuber production. Olivier *et al.* found that using the first method, the overall length of the process to produce mini-tubers from micro-tubers was 32 weeks, while in the second method the time was within 20 weeks when plantlets were directly transferred to soil. There were differences between the two cultivars with 'Kponan' responding better to micro-tuberization than 'Krengle' and in both methods the presence of jasmonic acid in the culture medium was found to be essential for yam tuberization, as well as for the 'germination' of yam micro-tubers.

Tissue culture

Balogun and Gueye (2013) and Aighewi *et al.* (2015) reviewed *in vitro* tissue culture techniques for yam propagation. They concluded that they were the most rapid methods of multiplying disease-free propagules, but their limitations included their high costs, need for skilled personnel and specialized equipment. A common method used involves taking nodes from plants that are surface-sterilized and transferring them to tissue culture media. Different compositions and incubation in a controlled light and temperature environment (14 h light at 29°C and 10 h darkness at 25°C) have been tested. The medium that gave the most consistent regeneration rates for meristem tips was: 30 g L^{-1} sucrose, 20 mg L^{-1} cysteine, 100 mg L^{-1} inositol, 80 mg L^{-1} adenine, 0.2 mg L^{-1} naphthaleneacetic acid, 0.15 mg L^{-1}

6-benzyl-aminopurine and 0.08 mg L^{-1} gibberellic acid (SPYN, 2003). Meristem culture techniques have been used successfully to produce high yielding pathogen-free yam plants (Balogun, 2009). Li *et al.* (2014) successfully developed a protocol for *in vitro* production of micro-tubers from *D. opposite*, which was a two-step procedure using 60 g L^{-1} sucrose. Lauzer *et al.* (1992) propagated *D. abyssinica* and *D. mangenotiana* using tissue culture from nodal cultures and found that segments shorter than 5 mm were less suitable for micropropagation. Nkere *et al.* (2011) described micropropagation of *D. rotundata* and found that adding cassava starch to the culture medium maintained the overall percentage survival, which ranged from 87 to 100% after 60 days in culture and was much better than the cultures without cassava starch. Multiplication of yams by *in vitro* growth of nodal segments is also used for the rapid propagation of virus-free clonal material (Mantell *et al.*, 1978; Mantell and Hugo, 1989).

Ovono *et al.* (2010) tested the control of field tuber dormancy in *D. cayenensis-D. rotundata* complex of micro-tubers obtained by *in vitro* culture. They observed no dormancy when micro-tubers were harvested after nine months of culture. However, if the micro-tubers were stored dry in jars in sterile conditions for 18 weeks, they sprouted more rapidly than those that had been stored for only 8 weeks. The size of the micro-tubers used for storage had great influence on further sprouting. Ovono *et al.* found that larger micro-tubers sprouted better and those stored at 25°C sprouted quicker than small tubers stored at 18°C, and that exposure to light during storage had no effect on time of sprouting.

Several workers have investigated growth stimulation to improve micropropagation of yams. Large sized micro-tubers were induced on nodal segments in 8-hour day length in media containing high levels of sucrose by Otusanya and Jeger (1996). They tested the effects of infection of *D. alata*, *D. rotundata* and *D. esculenta* seed yams with *Aspergillus niger* on shoot emergence and vine development. Shoot emergence was affected on all three species with *D. esculenta* most affected. *D. alata* seed yams, inoculated with *A. niger* after bruising, produced a higher number of shoots and it was concluded that bruising may have stimulated the release of β-sitosterol, an antifungal compound known to be present in the periderm of *D. alata*. Shah and Lele (2012) tested callus production during *in vitro* propagation of *D. alata* var. *purpurae* explants of tubers, leaves and nodes using Murashige and Skoog media (a plant growth medium used for cultivation of plant cell culture) at 25 ± 2°C under fluorescent tubes with a 16-hour photoperiod. The addition of 6-benzylaminopurine (0.2 ppm) and 2,4-dichlorophenoxyacetic acid (2 ppm) to the culture media gave the best results and all parts of the plants gave improved callus induction. Growth regulators have been shown to have stimulatory effects on yams.

Kim *et al.* (2003a) applied growth regulators to *D. opposita* plants and found that mepiquat chloride at 600 ppm induced early bulbil formation and increased bulbil yield compared to the control. Gibberellic acid at 600 ppm increased tuber size and tuber yield per hectare. Chen *et al.* (2007) tested various growth regulators on micro-tuber induction and showed plantlet regeneration in *D. nipponica* was strongly increased by a combination of n-benzyladenine and α-naphthalene acetic acid. Murashige and Skoog salt media supplemented with sucrose n-benzyladenine and α-naphthalene acetic acid gave the highest frequencies of shoot induction with sucrose being the single most significant medium constituent for micro-tuber growth. Lee *et al.* (2018) found that the highest efficiency of the micro-tuber induction in *D. cayenensis* plants was obtained from treatment with 7% sucrose, whereas glucose had no effect on micro-tuber formation.

In Ghana, various attempts have been made to improve the efficiency of the technique. The effect of sequential treatment (6, 8 or 10 weeks) with tuberization medium (T-medium) followed by Murashige and Skoog basal medium on the sizes and weights of micro-tubers in shoot cultures of *D. rotundata* was evaluated on three local cultivars ('Puna', 'Punjo' and 'Kablete').

Culturing the plants on T-medium for 8 and 10 weeks, followed by Murashige and Skoog basal medium, gave the highest induction of micro-tubers, the best frequency and yield and highest individual micro-tuber weights (Klu *et al.*, 2005). Bömer *et al.* (2019) evaluated next-generation sequencing for virus detection in yam plants that had been propagated by tissue culture and showed it could be used in diagnosing yam viruses, particularly viral species of the genera *Badnavirus* and *Potyvirus*.

Temporary immersion bioreactor (TIB)

There are different designs of TIB systems, but the most common are the twin flask types where one flask contains the medium and the other the cultures (Adelberg and Simpson, 2002; Watt, 2012). Several bioreactor systems have been successfully applied for cultivation of differentiated plants in *in vitro* systems (Steingroewer *et al.*, 2013). López *et al.* (2018) described a TIB system where the upper compartment, where the plants are, is placed in a raft support and the lower one containing 200 mL of the culture medium, is below. The compartments are connected through a small tube and the raft support placed at the bottom of the upper compartment. When the pressure in the small tube is increased, the culture medium goes up the tube and immerses the plants in the upper compartment. During immersion, the airflow continuously agitates and oxygenates the medium, allowing an inner atmosphere renewal. In this system, TIB immersion was for 3 minutes every 6 hours after which the pressure was reduced and the medium drained into the lower compartment by gravity. The air pump supplying the pressure was controlled by a timer that regulated the frequency and the length of the pressure increases. As the liquid passed it was filtered through 0.2 μm filters on entry and exit.

Watt (2012) reported that TIB systems have been successfully used for tissue culture in many crops where there was better growth, compared to continuous immersion in the medium, due to improved aeration. Escalona (2006) also reported that in TIB, growth was enhanced compared to other tissue culture methods because the immersion in the liquid medium is sporadic and not continuous. In *D. alata*, it was originally used by Balogun *et al.* (2014) for acclimatization of plantlets from tissue culture in order to automate micropropagation and scale-up and enhance micro-tuber production. Balogun *et al.* successfully tested single node cuttings using TIB. Cabrera *et al.* (2011) also successfully used TIB for propagating *D. alata*. In Cuba, Jova *et al.* (2008) reported that TIB was effective when used for propagation of *D. cayenensis-D. rotundata* complex. Jova *et al.* (2011), working with *D. alata*, compared TIB with two other systems and concluded that TIB gave better results than the ‘constant immersion system’, where aeration was through continuous bubbling in the culture medium and also better than the ‘static liquid system’.

Aeroponics

Aeroponics is the process of growing plants in an air or mist environment without the use of soil or an aggregate medium. Atomizing nozzles ensure the most effective delivery of nutrients, as they turn the water into a fine mist, and plants absorb the nutrients through their roots by absorption of compounds through cell walls.

One-node pre-rooted *Dioscorea* vine cuttings of approximately 2 months old, as well as two-node vine cuttings, freshly cut from yam seedlings, were transplanted into aeroponics boxes established in a screenhouse. The initial results of both pre-rooted vines and direct planted vines showed that plants continued growing normally and developed new roots and shoots. By the third week after planting, 85–100% of the direct planted vine cuttings produced roots in the aeroponics system. There were differences in response of the 13 *Dioscorea* species and varieties tested in terms of rooting and survival percentage (IITA 2013; Aighewi *et al.*, 2015).

Stem cuttings

Yams can be propagated by stem cuttings, but may take a long time to produce a crop, and Coursey (1967) concluded that this was not a viable method. The use of rooted stem cuttings for the production of planting setts was subsequently developed in IITA (Akoroda and Okonmah, 1982; Wilson, 1982), which not only accelerated propagation of selected clones but, if carried out with sterilized substrate, produced mini-tubers free of nematodes. The best results were obtained when the cuttings were taken from plants before tuber initiation. IITA recommended cuttings with four or five nodes with the leaves from the bottom three nodes removed and the laminas of the upper ones reduced by half. They recommended cuttings should be planted slanting in trays filled with steam-sterilized shredded coir or carbonized rice husk and maintained in a healthy condition by spraying fungicides at regular intervals. They found that vine cuttings can be used to produce mini-tubers within 100–120 days, which can, in turn, be replanted to produce tubers that are large enough for planting. They also found that some genotypes performed better than others (Acha *et al.*, 2004; Shiwachi *et al.*, 2005a, 2005b).

TNAU (2019) reported that *D. floribunda* can be propagated by single node leaf cuttings consisting of a single leaf with petiole and about 8 mm of the stem. These cuttings were dipped in 5,000 ppm indole-3-butyric acid and planted in a mist propagation chamber. After about 8–10 weeks, the rooted cuttings were potted up and planted in the field after 5–6 months. Carbonized rice husk was reported to be an effective medium for rooting cuttings and when combined with sandy, loamy soil produced the highest number of roots in *D. rotundata* cuttings (Agele *et al.*, 2010).

True seed

As indicated above, *Dioscorea* species commonly cultivated as food crops do not produce viable seeds, with the possible exception of *D. trifida* and *D. cayenensis-rotundata* (Degras *et al.*, 1977). Ferguson (1972) also commented that propagation of *Dioscorea* species by true seed is unusual as some species rarely produce viable seed and the progeny may vary undesirably from the parents. In addition, Coursey (1967) observed that plants from true seed can take up to 3 years to produce a viable crop. Andriamparany and Buerkert (2019) tested the regeneration ability of true seeds of three wild *Dioscorea* species from Madagascar that had been exposed to pre-germination treatments consisting of storage for 2 weeks at 5°C, submersion in an ultrasound water bath at 40 kHz for 10 or 20 minutes or soaking in running water for an hour. The optimum treatment was ultrasound water bath at 40 kHz for 10 minutes that gave 71% germination for *D. alatipes* and 91% germination for *D. fandra*, but soaking in running water resulted in the highest germination percentage for *D. nako*. Exposure to 5°C induced earlier germination in *D. fandra*. In the 1970s there was interest in improvement of *D. rotundata* through hybridization, but the low rates of flowering, fruit setting and seed germination were major obstacles (Okoli, 1975; Sadik and Okereke, 1975).

Field Practices

There are various methods of growing yams including the following.

1. On flat land, where plants are commonly planted about 45 cm apart within rows and about 1 m apart between rows, but this depends on species and soil conditions.
2. On ridges that are commonly 25–40 cm high and about 1 m between plants, either way.
3. On individual mounds. An individual tuber is planted in the top of each mound at a depth of 5 to 10 cm and covered with soil. If the conditions are hot and dry, mulching the surface with dried grass or other organic material will protect the young plant as well as the soil from erosion. Kang and Wilson (1981) made mounds about 30 cm high and

compared the yield of *D. rotundata* growing on flat land at different locations in southern Nigeria. They found that the average tuber yield, without fertilizers, was 7.83 tonnes ha^{-1} on the flat compared with 9.44 tonnes ha^{-1} on mounds. With fertilizer application, tuber yields were 7.43 tonnes ha^{-1} on the flat and 11.30 tonnes ha^{-1} on mounds. Working with the *D. alata* cultivars 'Dente' and 'Pona', Ennin *et al.* (2009) compared growing on mounds with growing on ridges. They found that 'Pona' on mounds yielded 22.2 tonnes ha^{-1} and on ridges they yielded 36.3 tonnes ha^{-1}. For 'Dente' on mounds they yielded 51.1 tonnes ha^{-1} and on ridges they yielded 60.9 tonnes ha^{-1}. These increases were due to both the increased mean tuber size and also to increases in tuber number per hectare. Bosrotsi *et al.* (2017a) developed a mechanical harvester for yams and reported that tuber damage levels during harvesting were lower for yams grown on ridges compared to those grown on mounds.

4. Minimum tillage allows vines to grow flat along the ground without a trellis. Lugo *et al.* (1993) tested three tillage levels: conventional, deep or minimum, with no-tillage on *D. alata* in Puerto Rica. The highest yields were from conventionally tilled plots but differences among tillage treatments were significant only on certain types of soil. At these sites, yields under no-tillage were significantly (p = 0.05) lower than those under conventional tillage. Enesi *et al.* (2018) working with culture effects on *D. rotundata* in Nigeria found that tuber yields varied between years. They found that in 2013 neither tillage nor fertilizer affected tuber yields. In 2014 tillage increased yields by about 25% and fertilizer reduced yields by about 10.5%. In 2015 tillage increased tuber yields by 8%. Agbede and Olasekan (2012), also working on *D. rotundata* in Nigeria, compared zero tillage, manual ridging, manual mounding, ploughing + harrowing and ploughing + harrowing + ridging. They found that ploughing + harrowing + ridging gave higher tuber yields by 12.3–12.9%, 12.8–13.5%, 25–34.9 and 45.5–50.7% compared to the other four methods, respectively.

TNAU (2019) reported that planting *D. alata* in India as 'mini-setts weighing 25 g [is] recommended for planting directly in the field or raising in a nursery and transplanted after 60 days. Planting is done in beds or in ridges or in mounds or in rows 75 cm apart either way' using a seed rate of 1,875–2,500 kg ha^{-1}. They recommended farmyard manure at 25 tonnes ha^{-1} at the time of last ploughing and a fertilizer schedule of 40:60:120 kg NPK ha^{-1} as basal and 50 kg N and 120 kg K ha^{-1} 90 days after planting followed by earthing up. They also recommended *Azospirillum*, a nitrogen-fixing bacterium, at 4 kg ha^{-1}, mixed with 40 kg of soil, 30 days after planting.

Environment

The most suitable environment for growing yams can vary considerably between species, particularly in terms of ambient temperature and precipitation, but they have many things in common especially the quality and fertility of the soil environment (Orkwor and Asadu, 1998; Diby *et al.*, 2009). Production conditions for *D. alata* in India were reported by TNAU (2019) as 'warm and humid conditions with a mean temperature of 30°C and a well distributed annual rainfall of 1,200–2,000 mm. Sandy loam soil with a pH of 6.0 to 6.5 is preferred with good drainage and cool weather'. The optimum temperature for growth of *D. bulbifera* was reported by Martin (1974a) to be around 25–35°C with high humidity and more than 1,000 mm annual rainfall at low to middle elevations in both partial or full shade. He also reported that the optimum soil conditions were loamy soil that has good drainage and is high in organic matter. Kay (1987) reported that *D. divaricata* prefers moist soil, but can be successfully grown on light (sandy), medium (loamy) and heavy (clay) soils. However, well-drained soils were preferable in semi-shade (light woodland) or no shade. Kay also asserts that *D. divaricata* plants are hardy to at least –18°C and can take 3–4 years to reach full maturity, although year-old tubers of well grown plants can weigh more than 500 g.

Staking

Since almost all yam species are climbing plants, they are commonly provided with some form of support on which they can grow and twine. This staking provides improved access of sunlight throughout their growth, development and yield. Methods used include: individual staking (e.g. bamboo stakes) with one stake per plant; pyramidal staking, where the top of the stakes is slanted to form a peak; and trellising where wire is strung between strong posts and stems are trailed along the wires. In Papua New Guinea Toktok (2006) described several staking methods including a traditional method where live stakes or small trees are used for *D. numularia*, because of the long duration of the crop, which is 1 to 3 years. More recently this method of using small trees or live stakes has been used for other species including *D. rotundata*.

Beckford *et al.* (2011) described a system where live sticks were used. Appropriate sticks should be cut from trees (*Spathodia companulata*, *Gliricidia sepium* and *Erythrina corallodendrum* were reported to be suitable), sharpened at the lower end and stuck in the ground alongside each yam plant. When the crop is harvested the sticks are lifted and the roots and leaves of the yam plant are cut off and the stick replanted immediately. Sticks may remain alive for many years. Also, for this method, farmers may plant two or three yam plants around the base of a sapling. *Gliricidia sepium* is often preferred since its root system does not grow vigorously and limits the competition with the yam plant. Makame (1994) also staked *D. alata* plants in Zanzibar with live *Gliricidia sepium* stakes and found that all staked yam plants yielded more than non-staked yam plants. The highest yield (2.75 kg plant^{-1}) was from staked plants on ridges and the lowest yield (0.63 kg plant^{-1}) was from non-staked plants on mounds, both with farmyard manure at 25 tonnes ha^{-1}.

The purchase of sticks and trellis can be expensive. Ndegwe *et al.* (1990) showed that staking costs for *D. rotundata* varied with the staking method, the costliest being the use of synthetic rope in a net arrangement, staking six plants to one stake. However, this method gave the highest net cash return which was 136% more than non-staked plants. In Jamaica, Campbell-Chin-Sue *et al.* (1994) tested various agronomic practices on *D. cayenensis* and found that staking substantially increased yields. Enyi (1970) showed that total tuber production of *D. esculenta* was increased by about 50% when they were staked compared to non-staked plants. In Nigeria Ndegwe *et al.* (1990) showed that staking *D. rotundata* increased tuber yield between 34 and 105% compared to non-staked plants. In Sierra Leone Norman *et al.* (2015) reported that for *D. alata* staking resulted in about 30% mean yield increase compared to non-staked plants but this varied between cultivars. For example, in 'Makeni' the yield increase due to staking was 39%, with 'Njala' about 30% and 'Kenema' about 26%.

Breeding and variety evaluation have been undertaken to reduce the requirement for staking. For example, Danquah *et al.* (2015) compared vertical staking, trellis staking (50 or 30% number of stakes used in vertical staking) and no staking on yams in Ghana. They found that *D. alata* had similar tuber yields under all staking options, but *D. rotundata* under no staking had reductions in yield ranging from 37 to 65% compared to staking options. They also demonstrated that

> with the appropriate selection and breeding, there is hope of at least sustaining yam production on 30–50% number of stakes compared with the current optimum staking practice, if not under no staking. This would reduce the drudgery associated with the search for stakes and pressure on the forest to address the problem of deforestation.

Otto *et al.* (2008) described 'TDR95/19177' as a promising *D. rotundata* cultivar with a potential for high yields with no staking. Campbell-Chin-Sue (1995) described the mini-sett system used in Jamaica, which involves growing the yams on mounds that are covered with either plastic or organic mulch so that the vines trail along the mulch. This method therefore eliminates the need for staking, and the yams produced in this way were reported to be very straight with no toes, which has the advantage of facilitating grading and packing, especially for export. In Puerto Rico, Costas *et al.* (1968) showed that staking sharply increased yields of the 'Guinea Blanco' cultivar of *D. alata* mainly by increasing the size of tubers. Planting in raised beds also increased yields of staked yams. In Puerto Rico, Costas *et al.* (1968) showed that staking sharply increased yields of the 'Guinea Blanco' cultivar of *D. alata* mainly by increasing size of tubers. Planting in raised beds and planting approximately on the square, further increased yields. The highest yields (exceeding 26 tons of tubers per acre) were obtained with planting at 2 feet x 2 feet giving a plant population of 10,890 plants per acre. They found that combining closer planting, staking, and planting in beds could more than triple yields of yams. Planting approximately on the square further increased yields above those obtained with 4-foot rows. Highest yields, exceeding 26 tons of tubers per acre, were obtained with 10,890 plants per acre and 2 feet x 2 feet plantings. Combining closer planting, staking, and planting in beds more than tripled yields of yams.

Fertilizers

Many field trials on yams have shown positive impacts of N, P and K inputs on tuber yields (Carsky *et al.*, 2010; Susan John *et al.*, 2016), but as with other crops, responses depend on many factors including level of nutrients in the soil, weather conditions, cultivar, planting material quality, planting material weight, planting density, planting date, weeds, diseases and pests. Srivastava *et al.* (2010) recommended using high levels of organic manure and high levels of potash for optimum yields. In trials in Benin with *D. rotundata* they tested animal manure at the rate of 10 tonnes ha^{-1}, mineral fertilizer (N30:P30:K60, but the rate was not specified), combination of manure, mineral fertilizer or left with no application of fertilizer (control). In 2002 and 2003 the same combinations were made taking crop residues (at the rate of 10 tonnes ha^{-1}) from external sources as a source of organic matter in the place of manure. They found that application of manure did not have a significant effect on yield when applied on fields after fallow, whereas crop residue application had a positive significant effect on tuber production when applied after cotton and maize and with adequate rainfall amount and distribution. Hgaza *et al.* (2011, 2012) observed a positive effect of NPK input on *D. alata* yield in a low fertility savanna soil, but the fertilizer application did not change root growth, so it was concluded that the fertilizer had increased the rate of soil organic matter mineralization. They also observed a maximum N fertilizer recovery of below 30% in the tubers, which they explained by the coarse and superficial root system of *D. alata* and the low planting density (one plant m^{-2}), which is typical for West Africa. In field experiments in central Côte d'Ivoire on *D. alata* and *D. rotundata* by Diby *et al.* (2009) the clear importance of soil organic matter in productivity of yams was shown, which was not compensated in soils with low organic matter content by the use of inorganic fertilizers. They, however, found that in all situations, the yields of *D. alata* were significantly ($p = 0.05$) higher, but the beneficial impact of fertilizer on yields was significantly ($p = 0.05$) lower in the fertile forest soils. In Nigeria Kayode (1985) found that the optimum level of nitrogen for successful production of *D. rotundata* was 35 kg ha^{-1} in a forest Alfisol that had been under cultivation for at least two years. NPK application had no significant effect on starch content, but large applications of P and K significantly ($p = 0.05$) decreased dry matter accumulation. Koli (1973) working with *D. rotundata* in Ghana found that application of 33.6 kg ha^{-1} N gave an increase of 13.5% and 67.2 kg ha^{-1} gave a 22.1% increase in yield. Increasing N did not further increase yield, while the application of 33.6 kg ha^{-1} gave only 5.6% increase in yield. Manure application (at the rate of 10 tonnes ha^{-1}) did not have a significant effect on yam total biomass production nor on tuber yield when applied on fields after fallow.

However, crop residue application had a positive significant effect on yam tuber and total biomass production when applied after cotton and maize and with adequate rainfall amount and distribution. Regarding partitioning patterns in yam crops, no effect of different practices of fertilization has been observed. Akom *et al.* (2015) treated the soil with Biochar (a fine-grained charcoal made by burning wood and agricultural by-products slowly at low temperatures and reduced oxygen) on which they grew *D. rotundata*. They tested four Biochar application rates (0, 5, 10 or 15 tonnes ha^{-1}) and three inorganic fertilizer rates (0-0-0 N-P_2O_5-K_2O kg ha^{-1}, 30-30-30 N-P_2O_5-K_2O kg ha^{-1} or 60-60-60 N-P_2O_5-K_2O kg ha^{-1}) and found that dry matter production was significantly (p = 0.05) increased by fertilizer application with no effects of Biochar.

Harvesting

Harvesting yam tubers is labour intensive and physically demanding. In West Africa yams are typically harvested by hand using wooden tools, which are preferred to metal tools as they are less likely to damage the tubers. Otoo *et al.* (2013) reported that out of the total manually harvested tubers they tested, between 10–40% are damaged from cuts during harvesting and poor handling of tubers. Most yams grow naturally in the wild in forests so harvesting is made more difficult because of tree roots (Fig. 4.1); tubers may also be pierced by the roots of trees.

The period from planting to maturity varies from about 6–7 months to 7–10 months depending on the species. In some species, e.g. *D. rotundata*, *D. cayenensis* and *D. alata*, an early crop may be taken as well as the main harvest; in this case the tuber is carefully cut below the head and removed, leaving the top to grow again and produce another tuber or tubers. This harvest method is commonly practised even for the main crop – for example, in Jamaica for *D. cayenensis*, and leaves a large cut surface on the tuber often resulting in rapid postharvest deterioration because it facilitates attack by microorganisms as well as drying out of the tubers. For double harvesting, Martin (1984) recommended 4–5 month intervals and Bencini (1991) recommended the first harvest at 5–6 months after planting and then the second 3–4 months later. It has been shown that double harvesting does not result in a greater season's yield than single

Fig. 4.1. Harvesting *D. alata* var. *purpurea* at a small farm in Thailand using a standard digging hoe.

harvesting (Onwueme, 1977a, 1977b). Indeed, if the first harvest of double harvesting occurred at certain times, the total season's yield from double harvesting could be lower than that from single harvesting.

Mechanical harvesting

Current crop production practices pose hurdles to successful mechanization of yam harvesting, particularly for small-scale farmers. Itodo and Daudu (2007, 2013) reported that there was 19% tuber damage at harvest using a tractor-mounted mechanical yam harvester fitted with a collection unit, which has perhaps contributed to the failure of previous attempts of mechanizing yam harvesting (Opara, 2003; Itodo and Daudu, 2007).

The shape of yam tubers and also the depth at which they grow present challenges in designing suitable mechanical harvesting equipment. Tubers that grow shallowly in the soil are oval or round with a tough skin and several smaller tubers per plant rather than one large tuber. These are simpler to harvest and to design harvesting machines for. Martin *et al.* (1975) selected a cultivar of *D. alata* in Puerto Rico, 'Florido', that has a compact tuber size and shape, for mechanical harvesting experiments. It was introduced into Côte d'Ivoire for a harvest mechanization project but the project was never implemented. Bosrotsi *et al.* (2017a, 2017b) developed a harvester for the *D. rotundata* cultivars 'Lili' and 'Pona', which was operated at an average soil depth of 0.30–0.33 m and tuber spread of 0.18–0.22 m. The tuber length varied from 0.33 to 0.38 m and tuber depth of penetration also varied from 0.29 to 0.37 m irrespective of yam variety. The tractor harvesting speed and field capacity values were 1.04 to 1.31 m second^{-1} and 0.30 to 0.38 ha hour^{1} respectively. Tuber damage levels were 1.96% and 3.45% for ridge and mound harvesting respectively. Bosrotsi *et al.* found that mechanical harvesting reduced harvesting cost by 50%. The cost of US$ 1,500 for the yam harvester gave a break-even point of about 180 hours of use (about 64 hectares of harvested land area) when used for custom services at a cost of US$ 5.00 per hour. Mounding offered lower resistance to tractor movement coupled with reduced engine fuel consumption and lower tractor wheel slip, though ridging was better since it considerably reduced tuber damage during harvesting with the mechanical harvester, irrespective of yam variety harvested. Ridging methods of land preparation were generally better than mounding at reducing the total cost of production and harvesting. Bosrotsi *et al.* recommended the harvester for use by farmers to reduce drudgery and to expedite yam harvesting operations.

Martin and Gaskins (1968) suggested a large mouldboard plough and Martin *et al.* (1966) mention a 'lister bottom' plough as having possibilities for 'turning up' the tubers for subsequent collections by hand. Coursey (1967) described the construction and operation of a tuber-lifting plough after the use of which the tubers were 'raked' by hand from the loosened soil. It was designed to harvest the tubers from ridges, which is the common method of growing yams in West Africa. Efforts to select cultivars with regularly shaped roots for easier, mechanical harvesting were reported to be being made at that time. Onwueme (1977a) tested a potato spinner on harvesting *D. rotundata* and suggested these devices could be used when a number of small tubers were produced, but not for the typical large tubers.

Nystrom *et al.* (1973) tested mechanical harvesting of *D. composite* and found that it was possible to harvest them when they were distributed through large volumes of soil, by lifting the entire soil volume with the types of machines usually used for digging and grading road-building materials, and then employing screens that separated out the yam tubers from the main body of the soil. Their results showed that *D. composita* could be machine harvested for US$ 1.76 per tonne less than by hand (US$ 20.38 compared with US$ 22.14), when yields were about 40 tonnes per hectare. This cost advantage was due to 'more efficient root recovery'. Their analysis of costs showed that there would be a relatively greater cost advantage for machine harvesting at higher gross yields and that harvester costs per tonne of tubers harvested decreased proportionately as yield increased whereas hand harvest costs were relatively inelastic. They reported other advantages of mechanized harvesting including better control of the rate of harvest and less dependence on a seasonal supply of temporary labour.

Bosrotsi *et al.* (2017a, 2017b) reported that the 'proper design of harvester blade for yams is very important in order to increase its working life time and reduce the harvesting costs' since the blade is subjected to impact and high frictional forces that creates unbalanced forces that can result in blade wear. In their research, different geometric model concepts of blades were modelled from which they fabricated and assembled a harvester.

Dr Colin Hudson, of Carib-Agro Industries in Barbados, developed a root and tuber harvesting machine in the 1980s that was claimed to reduce crop damage by up to 75%. He reported that in Barbados yams are grown on a field scale, without stakes and about 25% of the crop was damaged during traditional harvesting when tubers were lifted using agricultural forks. He also reported that his harvester left only about 6% of the yams unmarketable and so was economical to use, and that the harvester 'floated' the crop out, by shaking the soil under the roots with an array of mechanical fingers after which the tubers were easily lifted from the soil by hand (CTA, 1989). Yam harvesting machines are available commercially, such as from Zhengzhou Yingwang Machinery Co. Ltd. in Henan, China.

Transport

In a study of sweet potatoes (*Ipomoea batatas*) in Tanzania by Tomlins *et al.* (2000), the handling and transport system used resulted in up to 20% and 86% of roots with severe breaks and skinning injury, respectively and a reduction in market value of up to 13%. The most severe impacts occurred during unloading and loading from road vehicles and ships. However, skinning injury and broken roots were correlated with a large number of minor impacts. In West Africa, yams (*Dioscorea* species) are often transported to market by the farmers themselves. They may be kept in a shed and then taken to market, while at the same time some may be offered for sale at the roadside (Fig. 4.2). Adejo (2017) evaluated postharvest practices in parts of Nigeria and found that 79% of the farmers were engaged in transporting their yams from farm to home and home to markets or farm to markets, as well as growing them. On a larger scale they may be transported loosely packed in the back of lorries (Fig. 4.3).

Fig. 4.2. *D. rotundata* tubers packed in a shed, awaiting transport (a), and being offered for sale at the roadside (b), in Tamale, Ghana.

Fig. 4.3. *D. rotundata* tubers packed in a vehicle ready for transport and distribution to various markets.

Yield

Mean yield of yams in 2017, on a worldwide basis irrespective of country and species, varied between 6.7 and 9.9 tonnes ha^{-1} (Chapter 7) and yield of yams in selected Caribbean and Latin American countries in 2017 showed even more variation (Table 4.1). This variation is clearly due to different yam species, varieties and cultivars as well as soil conditions, water availability and climate and weather conditions, but could also have been contributed to by the expertise of individual growers, pointing to the fact that there continues to be considerable scope for yield improvement.

Table 4.1. Mean yield of yams of various *Dioscorea* species in Caribbean and Latin American countries in 2017 (FAO Stats, 2019).

Country	Mean yield (tonnes ha^{-1})
Barbados	11.3
Costa Rica	15.0
Dominica	11.5
Dominican Republic	6.6
Haiti	7.9
Jamaica	16.5
Nicaragua	13.3
Puerto Rico	8.9
Saint Kitts and Nevis	9.1
Saint Vincent and the Grenadines	12.9

Coyne *et al.* (2005b) and Tchabi *et al.* (2010) reported that tuber yields of yams in West Africa were steadily decreasing due to declining soil fertility and an increase in pest and disease levels. However, Martin and Degras (1978) previously reported that on an experimental basis yields in Guadeloupe had reached 55 tonnes per hectare. In Vanuatu, with no fertilizers or special inputs other than normal hand weeding throughout the growth cycle, yields of more than 51 tonnes ha^{-1} of fresh tubers for *D. rotundata*, 82 tonnes ha^{-1} for *D. alata* and up to 128 tonnes ha^{-1} for *D. esculenta* were recorded (Van Wijmeersch and Bule, 1988). However, these very high mean yields were probably multiplied up from yields in small plots, which notoriously exaggerate yields. TNAU (2019) gave the yield of *D. alata* as 20–25 tonnes ha^{-1} of tubers in 240 days and for *D. floribunda* 50–60 tonnes ha^{-1} of tubers after 2 years with a diosgenin content of 3.0–3.5%. Enesi *et al.* (2018), working with cultivation effects on *D. rotundata* in Nigeria, found that tuber yields varied between years and in 2013 neither tillage nor fertilizer affected yields. In 2014 tillage increased yields by about 25% and fertilizer reduced yields by about 10.5%. In 2015 tillage increased tuber yields by 8%.

Genotype

Lebot (2019) reported that a *D. alata* germplasm collection in Vanuatu, composed of more than 300 accessions, showed that the average yield was related to ploidy. Plants with 40 chromosomes (diploids) yielded 2 kg fresh tubers plant^{-1} on average, plants with 60 chromosomes (triploids) yielded 2.5 kg fresh tubers plant^{-1} and plants with 80 chromosomes (tetraploids) more than 3 kg fresh tubers plant^{-1}. This was the case for two *D. alata* cultivars in New Caledonia, 'Nouméa Blanc' and 'Nouméa Rouge', both of which have 80 chromosomes and were reported to be very high yielding.

In Japan, Wu *et al.* (2015) bred a new *D. alata* cultivar, 'Wenshanyao No.1', which had 23.8% higher fresh weight yield (2,217 kg per 667 m^2), 36.9% higher polysaccharides, 48.3% higher allantoin and 20.9% higher dioscin than *D. alata* landraces. Interactions between genotype and environment were described for *D. alata* in Nigeria, where yield stability of six genotypes, across five major yam growing areas and two cropping seasons, was affected by both biotic and abiotic stresses (Egesi *et al.*, 2006). In Ghana, similar effects were shown on the tuber yield of eighteen *D. cayenensis* varieties (Otoo *et al.*, 2006a), and it was concluded that yam genetic improvement should focus on multiple disease and pest resistance, which it was claimed could guarantee stability of the crop (Otoo *et al.*, 2006b).

Soil fertility and fertilizer

Clearly the soils in which the yams are grown affect their yield. The degree of this effect will vary with many factors but can be extreme. For example, Diby *et al.* (2009) reported that in subsistence conditions yields of yams may be low, varying between 9 and 10 tonnes ha^{-1} in low fertility soils, compared with a potential yield of 51 tonnes ha^{-1} for *D. alata* and 27 for tonnes ha^{-1} for *D. rotundata*. Koli (1973), working with *D. rotundata* in Ghana found that the application of 33.6 kg ha^{-1} N gave an increase of 13.5% and 67.2 kg ha^{-1} gave a 22.1% increase in yield. Increasing N further did not further increase yield, while the application of 33.6 kg ha^{-1} gave only a 5.6% increase in yield. Application of animal manure (at the rate of 10 tonnes ha^{-1}) did not have a significant effect on yam tuber yield when applied on fields after fallow. However, it was found that crop residue application had a positive effect on yam tuber and total biomass production when applied after cotton and maize and with adequate amounts and distribution of rainfall. Regarding partitioning pattern in yam crops, no effect of different practices of fertilization was observed.

Srivastava *et al.* (2010) recommended using high levels of organic manure and high levels of potash. In trials in Benin with *D. rotundata* they tested animal manure at the rate of 10 tonnes ha^{-1}, mineral fertilizer (N30:P30:K60, but rate not specified), a combination of manure and mineral fertilizer or left with no application of fertilizer (control) on production of yams. In 2002 and 2003 the same combinations were made taking crop residues (at the rate of 10 tonnes ha^{-1}) from external sources as a source of organic matter in the place of manure. Manure application did not have a significant (p = 0.05) effect on yield when applied on fields after fallow, whereas crop residue application had a positive significant effect (p = 0.05) on tuber production when applied after cotton and maize and with adequate rainfall amount and distribution. Hgaza *et al.* (2011, 2012) observed a positive effect of NPK input on *D. alata* yield in low fertility savanna soils, but the fertilizer application did not change root growth so it was concluded that the fertilizer had increased the rate of soil organic matter mineralization.

Propagation

Enyi (1972a, 1972b) and Ferguson (1977) found that variability in growth and yield among *Dioscorea* species of the same clone could vary because of variability in the planting material. They found that variability could be due to the direct influence of the sett's dormancy status, the part of the tuber from where the sett was taken (whole or top, middle or tail), its nutrient content as well as cultural practices including staking, mounding and mulching. From an evaluation of propagation methods for *D. rotundata* it was recommended that minitubers weighing at least 100–300 g, planted slanting or vertically, gave higher yield than traditional methods (IFAD, 2016). Onwueme and Haverkort (1991) also tested propagation of yams with setts and found that sett size on early root growth appeared to have a larger and more reproducible impact on final yield than did fertilizer applications. In contrast, Campbell-Chin-Sue *et al.* (1994) tested various agronomic practices on *D. cayenensis* and found that minisett size and type did not affect subsequent total tuber yield. Variability in the seed tubers may lead to a population with disparate plant ages and also tuber yield (Ferguson, 1973; Orkwor *et al.* 1998). Ferguson found that in *D. alata* it appears that factors which promote the initiation of an excessive number of tubers could therefore result in low yield.

Cultivation

Yams are cultivated in various ways including on flat land and on mounds and ridges. Wilson (1982) tested mound size (about 30 cm high) at different locations in southern Nigeria on yield of *D. rotundata* and found that the average tuber yield without fertilizers was 7.83 tonnes ha^{-1} on the flat compared with 9.44 tonnes ha^{-1} on mounds. With fertilizer application, tuber yields were 7.43 tonnes ha^{-1} on the flat and 11.30 tonnes ha^{-1} on mounds.

Most species should be staked to improve yield, reduce weed competition and the incidence of diseases. Danquah *et al.* (2015) compared vertical staking, trellis staking (50% or 30% number of stakes used in vertical staking) and no staking on yams in Ghana and found that *D. alata* had similar tuber yields under all staking options, but *D. rotundata* under no staking had reductions in yield ranging from 37 to 65% compared to the other staking options. In Sierra Leone, Norman *et al.* (2015) reported that for *D. alata* staking resulted in about 30% mean yield increase compared to no staking but this varied between cultivars. With 'Makeni' the yield increase due to staking was 39%, with 'Njala' about 30% and 'Kenema' about 26%. In Nigeria, Ndegwe *et al.* (1990) showed that staking *D. rotundata* increased tuber yield between 34 and 105% over non-staked plants. Makame (1994) staked *D. alata* plants in Zanzibar with live *Gliricidia* stakes. All staked plants yielded more than non-staked ones. The highest yield (2.75 kg plant^{-1}) was from staked plants on ridges and the lowest yield (0.63 kg plant^{-1}) was from non-staked plants on mounds, both with farmyard manure at 25 tonnes ha^{-1}.

Diseases and pests

It has been estimated that an average of over 25% of the yield of yams is lost annually to diseases and pests (Ezeh, 1998). Nwankiti and Arene (1978) reported that anthracnose commonly occurs in southwestern Nigeria and is one of the most severe yam diseases. In *D. alata* infection can stop growth and depress yield by 67%. Agbaje *et al.* (2003) found that there were positive correlations between tuber yield and virus infection in *D. rotundata*. It was also suggested that resistance to leaf virus infection, as expressed by foliage vigour, is the most important trait for selecting high-yielding genotypes (Akoroda, 1983). Mudiope *et al.* (2012) showed that in general *Meloidogyne* species caused little galling to yam roots and tubers and limited reduction in their growth and yield.

Growth substances

Several growth substances have been tested on yams, but these are mainly used postharvest to control dormancy and sprouting (Passam, 1982). However, Kim *et al.* (2003a, 2003b) applied growth regulators to *D. opposita* plants and found that mepiquat chloride at 600 ppm induced early bulbil formation and increased bulbil yield compared to the control. They also found that gibberellic acid applied to yam plants at a level of 50 to 500 ppm increased tuber size and tuber yield per hectare compared to those not treated.

5

Storage

Due to the seasonal cultivation and harvest periods of yams, there is often a need to store them over a period of time, especially during off-seasons. However, postharvest losses of cultivated yams are a major problem during storage. These losses are caused by endogenous factors, which include transpiration, respiration and sprouting, or exogenous factors including insects, nematodes, rodents and rotting as a result of fungal and bacterial infections. While good management practices may help control some exogenous factors, adequate storage conditions are essential in reducing postharvest losses. There are three main conditions that need to be considered for adequate yam storage. These are aeration, temperature and regular inspection of the stored produce. Suitable, correct aeration prevents moisture from condensing onto the tuber and also assists with dispersion of heat from the tuber surface as a result of respiration. Losses from respiration, sprouting and rotting can also be reduced by lowering the temperature of the tubers through refrigeration. Also, the choice of storage facility conditions must inhibit the onset of sprouting, which can cause tubers to shrivel and rot from the loss of dry matter. Yam tubers that are in good condition after storage should clearly be edible and marketable. Good storage management should therefore prevent unacceptable moisture loss, spoilage by pathogens, attack by insects and other pests, as well as controlling sprouting.

Traditionally yam tubers are stored in structures that simply protect them from exposure to direct sunlight, allow air circulation and often enhance humidity. However, refrigerated storage has been tested in many countries with varying degrees of success. The tubers of most *Dioscorea* species are very sensitive to chilling injury at temperatures below about 12°C (Thompson, 1972a; Noon and Colhoun, 1981). General recommendations for refrigerated storage, which do not define a particular species, include: 16°C and 65% RH for 4 months (Mercantilia, 1989), 13.3°C and 85–90% RH for 50 to 115 days (SeaLand, 1991) and 16°C and 70–80 RH for 6–7 months (FAO, 2019). However, from the published data there appears to be little difference in the optimum refrigerated temperature between different species (see storage conditions under different species headings in Chapter 8). For example, Olorunda *et al.* (1974) found that *D. alata* could be stored without sprouting at 15°C for 18 weeks and Noon (1978) considered 15–16°C and about 70% RH to be near optimum conditions for storage of cured *D. rotundata* tubers for 4–5 months without sprouting. Little information is available in the literature on controlled

 Yams: Botany, Production and Uses (A.K. Thompson and I. Oduro)
DOI: 10.1079/9781789249279.0005

atmosphere storage of yams – that is, reduced O_2 and increased CO_2 in the storage atmosphere (Thompson *et al.*, 2018; Tongchai Puttongsiri and Thompson, 2019). However, one report indicated that controlled atmosphere storage had only a slight to no effect on the postharvest life of yams when used to enhance the effects of refrigeration (SeaLand, 1991). However, in Japan, Imakawa (1967) reported that fully mature *D. opposita* tubers could be stored for long periods of time under high O_2 tensions at 5°C, which was also shown to reduce browning of tubers.

Preharvest Factors

As would be expected, the conditions in which yams are grown can affect their postharvest life. For example, Asadu *et al.* (2008) showed that during storage there was a significant difference ($p < 0.001$) in loss of weight in *D. rotundata* tubers from fertilized and non-fertilized plots with the mean monthly weight loss being 44% higher in tubers that had been grown with the application of fertilizer (15:15:15 NPK) than those grown without fertilizer. The application of mineral fertilizer during production of yams was also reported by Vernier *et al.* (2000) and Baimey *et al.* (2006) to increase tuber weight loss and rotting during storage as well as to affect negatively the organoleptic properties of tubers.

Storage Structures

FAO (1998) reported that the widespread use of refrigerated storage of yams was not feasible mainly because of the high capital cost. However, considerable research has been done on defining the optimum temperatures for successful refrigerated storage of yams, for example Thompson *et al.* (1973a), Olorunda *et al.* (1974) and Noon (1978). However, yams are stored to a very limited degree in cold stores, and when they are, this is essentially to facilitate control of the marketing process over short periods or for export by sea in reefer containers.

In West Africa, yams are traditionally stored in ventilated barns in ambient conditions of about 30°C and 60% RH for several months (Tindall 1983). Wireko-Manu *et al.* (2013) found that dry matter, sugar content, swelling power and pasting viscosity of *D. alata* tubers, as well as flour made from the tubers, increased during development and also during storage for 5 months in a conventional yam barn. They concluded that these changes would improve the organoleptic and textural properties of the fresh tubers and the flour made from the tubers. In Nigeria the design of yam barns was shown to vary, but they usually consisted of live stakes driven into the ground with the tubers suspended from these by twine (Fig. 5.1).

The live stakes produce foliage that shades the tubers from the sun and provides a cooler, moist environment. Yams are also stored in pits dug into the ground. Knoth (1993) described one method where pits were dug and lined with straw and the top layer of yams was also covered with straw or similar materials and then protected with soil or a conical roof arrangement made of maize or millet stalks. Knoth also described one type of yam barn with a roof made from locally available organic materials such as straw or palm leaves, and the tubers stored on multi-level shelves. In a comparison between barn storage and pit storage Ezeike (1985) recorded weight losses over a 5-month period of about 60% for yams stored in barns but only 15–25% for yams stored in pits. In Ghana, Boakye and Essuman (2016) compared three storage methods: in a cold room (conditions not specified); in a pit; and in a barn for the *D. rotundata* cultivar 'Asobayere'. They found that the initial moisture content of the tubers was about 63%, but after 5 months storage this has fallen to 58% in the cold room, 55% in the barn but only to 61% in the pit. The initial carbohydrate content was about 30% but this increased during storage to 38% in the cold room, 34% in the barn and 32% in the pit, clearly because of loss of moisture. No other properties of the tubers appeared to be

Fig. 5.1. Yam barn in Nigeria showing the construction using live branches that produce leaves to provide a shaded humid environment (a); detail of how yam tubers are tied to the upright poles (b).

differentially affected by the different methods over the time period.

Other methods of storing yams include slatted wooden trays, but Lyonga (1985) reported losses of 29 to 47% in only 2 months in the coastal region of Cameroon using this method. Ravindran and Wanasundera (1993) cured six *D. alata* cultivars and three *D. esculenta* cultivars and then stored them for 150 days as thin layers on the floor of a well-ventilated room in ambient conditions of 24 to 28°C and 70 to 90% RH in West Africa. They found that crude protein, starch and ascorbic acid content decreased significantly during the first 60 days of storage, but the decrease was only gradual thereafter. There were also losses of some minerals and a significant reduction in levels of oxalates during storage.

Taylor and Bancroft (2002) described traditional barns that were commonly used by yam farmers of West Africa. The barns consisted of timber racks supported on a frame of vertically erected posts about 3 m high and set about 50 cm apart. Horizontal bars were attached to the posts to stabilize them. Sometimes, use was made of growing trees as uprights as they strengthened the structure and the foliage provided natural shade. Where trees could not be incorporated in the structure, the barn was provided with a roof. Siting of the barn to ensure maximum ventilation was considered important. It was reported that yams stored in traditional barns may be vulnerable to attack by rodents or even domestic animals. Attempts were therefore made to exclude animals by surrounding the barn with a fence. Taylor and Bancroft also reported that, alternatively, yams may be stored on simple shelving systems with rodent guards fitted to the legs.

A modified mud-walled bin with a thatched roof may be used for storage of yams. These structures have a door or opening and are found in the savannah areas of the yam belt in West Africa and provide good protection against direct sunlight and rain. However, as yams are piled on top of each other, ventilation is restricted and development of rots in the tubers could occur.

Knoth (1993) reported that underground storage may also be used for relatively short periods. For example, if labour at harvest time is limited, yams may be held temporarily in trenches or pits dug in or close to the field. The pits are usually lined with straw and the tubers stored on a layer of straw either horizontally on top of each other or with the tip vertically downwards beside each other. In areas where there is a pronounced dry season the pits may be constructed and filled in such a way that part of the stored crop is above ground level. In this example the exposed tubers are covered with straw and, in some cases, a layer of soil is also added. Although this system provides protection against respirational and transpirational weight losses, it has the disadvantage of reduced ventilation and direct contact between tubers. Consequently, the tubers may become warm and conditions may be conducive to the formation of rots.

Yam storage work in Côte d'Ivoire includes the work of Girardin *et al.* (1997) who stored *D. cayenensis-rotundata* complex and *D. alata* for 6 months in pits and in sheds, using modified traditional techniques. The temperatures and relative humidity varied more in the sheds than in the pits. The weight losses of *D. cayenensis-rotundata* were the same in pits and in sheds. *D. alata* lost more weight in sheds than in pits; the difference was 8% in 1993 and 2% in 1994. *D. cayenensis-rotundata* lost more weight than *D. alata* in both systems. The relative weight of the *D. cayenensis-rotundata* complex sprouts was significantly greater (p = 0.05) in the pits than in the sheds. For *D. alata*, sprout weight was the same in the pits and sheds. In each phase of storage, the fresh weight losses for *D. cayenensis-rotundata* complex were much greater than for *D. alata*.

In the early 2000s, over a series of consecutive storage seasons, rural farm trials were conducted in the yam growing regions of Ghana to determine the effects of a range of different low-cost pre-conditioning and storage regimes on the postharvest quality of local *D. rotundata* cultivars by Bancroft *et al.* (2005). The study suggested that beneficial curing protocols could be deployed in rural communities. For periods of 2 to 3 months, traditional pit storage resulted in less rots and less moisture loss of yams than storage above ground in barns. Eventually, however, pit-stored yams became more prone to nematode damage and sprouting. Curing in polyethylene bags was found not to be reliable. A more successful method was the use of temporary 'curing clamps', rooms or enclosures. Under field conditions, curing was enhanced by exposure to temperatures of about 32°C for 5 to 6 days. Such conditioning helped reduce the incidence of rots during subsequent storage but did not retard water loss. In contrast, superoptimal curing temperatures favoured the development of rots. Strict grading of the yam tubers prior to storage had a marked impact on the extent of postharvest losses, although chronic internal infections and disorders could compromise the subsequent marketability of the yams. Stored tubers were particularly susceptible to rots that originated at the site where the tuber was cut away from the vine.

Another storage method is used by the Igbo people in southeastern Nigeria, where yams are the most important food crop. They store the yams by tying them to a structure, or in a barn, to allow air flow. A popular Igbo proverb was quoted by Obidiegwu and Akpabio (2017) 'diji kọrọm anaghi eke ọbaji', meaning, in a practical sense, 'it is only the farm owner of the yam plot that devotes all his energy to ensure that all cultural practices in yam cultivation are carefully and timely implemented'. Yam is predominantly male, and 'women did not have the privilege of moving into the storage yam barn to collect yams for food preparation. Such privilege was arrogated to their husbands or fathers who were regarded as the chief custodians of the yam barn'.

In Ghana, Amponsah *et al.* (2015) designed and constructed 'affordable' yam storage barns using locally sourced material and expertise from smallholder farmers. Wood from the African fan palm (*Borassus aethiopum*) was used for the structure frame, including the pillars. *B. aethiopum* is a local palm mostly found in the forest-transition and savannah agro-ecological zones of Ghana and is commonly used by locals for many construction purposes. It was chosen because of its high strength and natural resistance to decay and insect attack. Amponsah *et al.* found that *B. aethiopum* wood could withstand loads of up to 3,800 kg. The walls of each structure were made from wooden slabs from sawmill cut-offs, and were used because of their durability, light weight and relatively low cost. Bamboo sticks were also considered as alternative material for the structure walls in areas where they were more readily available.

Inside the structure, shelves were constructed on which to place the yam tubers. These shelves were made from *Ceiba pentandra* (Kapok) wood because of its durability and resistance to harsh weather conditions. The cladding, or major covering material for the storage structure, was made from elephant grass (*Pennisetum purpureum*) woven into mats, commonly referred to as 'Zanna mats'. The structure's roofing consisted of thatch made from partially dried spear grass (*Heteropogon contortus*), because of its low cost and ready availability. The individual grass strands were bundled and tied together with nylon or fibre ropes. The entire structure was designed to hold up to 5,000 yam tuber pieces (Fig. 5.2).

Curing

In the context of agriculture, curing is treating crops after harvest in a way that can repair damage caused during harvesting and handling and also change their outer cells in such a way that their postharvest life is extended and losses are reduced. It has been successfully used on carrots (Garrod *et al.*, 1982), cassava (Booth, 1976; Rickard and Coursey, 1981), European chestnuts (*Castanea sativa*) (Botondi *et al.*, 2009), garlic (University of Massachusetts, 2013), ginger (Technical Bulletin No. 23, 2004), onions (Thompson *et al.*, 1972a; Downes *et al.*, 2009), potatoes (Priestley and Woffenden, 1923; Burton, 1989), shallots (Woldetsadik and Workneh, 2010), squash (Schales and Isenberg, 1963), sweet potatoes (Artschwager and Starrett, 1931; Walter and Schadel, 1983), tannia, new cocoyam (*Xanthosoma* spp) (Been *et al.*, 1975; Agbo-Egbe and Rickard, 1991), taro, cocoyam (*Colocasia esculentum*) (Passam, 1982; Agbo-Egbe and Rickard, 1991), Asian pears (*Pyrus* spp.) (Park *et al.* 2000), kiwi fruit (Lallu 1997; Retamales *et al.* 1997; Tonini

Fig. 5.2. Yam storage barns constructed with locally sourced materials: (a) Zanaa mat covering; (b) wooden slabs from sawmill cut-offs for structure walls; (c) bamboo sticks for structure walls; and (d) storage shelves made from wood from the Kapok tree (*Ceiba pentandra*).

et al. 1999), nectarines (Casals et al., 2012), peaches (Casals et al. 2010), grapefruits (Ben Yehoshua et al. 1989), lemons (Ben Yehoshua et al. 1989; Tariq, 1999), mandarins (Tariq, 1999), oranges (Tariq, 1999) and pomelos (Ben Yehoshua et al. 1989).

Curing is a generic term and can mean different processes for different crops. For example, in onions it simply means drying the outer scale leaves of the bulbs, making the inner scales less susceptible to drying out and attacks by microorganisms. In citrus fruits it appears to be associated with inhibition of fungal infection through lignification and an increase in antifungal chemicals in the peel of the fruit (Ben Yehoshua et al. 1989; Davies and Albrigo, 1994; Tariq, 1999). Root crops such as potatoes, yams, edible aroids and sweet potatoes have a cork layer over the surface of their tubers that develops as the tubers or corms develop in the field. These cork layers serve as a protection against microorganism infections and can reduce water loss. However, the cork layers can be damaged during harvesting and handling operations, exposing the tubers or corms to deterioration. Exposing these crops to high temperature and humidity for a few days directly after they have been loaded into the store can facilitate repair of the damaged tissue and reduce postharvest losses.

Curing was first described in detail for potatoes by Priestley and Woffenden (1923) and subsequently by Artschwager (1927), and for sweet potatoes by Artschwager and Starrett (1931). Both potatoes and sweet potatoes are dicotyledons and yams are monocotyledons, therefore it was not clear that the same curing process would apply to yams. Cambium tissue is common in the stems and roots of dicotyledons but rare in monocotyledons. Certain palms, which are monocotyledons, produce cambia in the stems but this is unusual. Therefore, experiments carried out in Jamaica on *D. rotundata* tubers in 1970 were somewhat speculative and the spectacular effective results were surprising (Thompson, 1972a). Curing has subsequently been shown to be effective on *D. alata*, *D. bulbifera*, *D. cayenensis*, *D. esculenta* as well as *D. rotundata* (Table 5.1). However, in a comparison between two of the species, Thompson (1972a) showed that curing was generally more effective on *D. rotundata* than on *D. cayenensis* particularly in terms of its effect on tuber weight loss (Table 5.2).

The curing process

When yam tubers are exposed to the optimum conditions for curing the first thing that happens is that the starch granules migrate to the surface from the broken cells of any area that has been damaged. These granules form a layer of starch at the cut surface within a few hours. Been et al. (1974) and Passam et al. (1977) also commented that the success of curing does not rely solely on the formation of suberin and

Table 5.1. Optimum curing conditions for yams. Modified from Rees and Hammond (2002) and Thompson et al. (2018).

Species	Temperature (°C)	Humidity (% RH)	Duration (days)	Reference
D. alata	29–32	90–95	4	Gonzalez and Rivera (1972)
D. alata	35	85–90	not given	Passam (1999)
D. bulbifera	26–28	High	5–7	Martin (1974a)
D. cayenensis	25–40	95–100	1–7	Been et al. (1976)
D. cayenensis	36–40	91–98	1	Thompson et al. (1977)
D. esculenta	29	84	5	Thompson (1972a)
D. esculenta	26–28	High	5–7	Martin (1974a)
D. rotundata	35–40	95–100	1	Thompson (1972a)
D. rotundata	25–30	55–82	5	Adesuyi (1973)
D. rotundata	25–40	95–100	1–7	Been et al. (1976)
D. rotundata	26	92	11–15	Nnodu and Nwankiti (1986)

periderm, but relates to the integrity and moisture content of the surface starch layer. Then a suberized layer and finally a cork cambium (periderm) are formed (Fig. 5.3), in a similar way to that reported for potatoes (Been *et al.*, 1976). There were considerable differences in the thickness of the periderm especially in those that had been cured in full sunlight, which produced a very thick periderm (Fig. 5.4).

Conditions for curing yams

Curing yam tubers after harvest, in order to extend their storage life and reduce postharvest losses, has attracted the attention of researchers for decades who have given various recommended conditions depending on species and other factors (Table 5.1). Effective curing requires high temperature and high humidity and simple curing methods have been tested that can be used, without special facilities, to trap self-generated heat and moisture from the tubers. Curing in the field can be carried out by piling tubers in a heap in the open and covering the heap with cut grass or straw as insulating materials, then covering the heap with canvas, burlap or woven grass mats. However, Ravindran and Wanasundera (1993) successfully cured *D. alata* and *D. esculenta*

Table 5.2. Effects of curing of *Dioscorea* species and storage conditions on the percentage weight loss, fungal score (0 to 5, where 0 = none and 5= maximum) and % internal necrotic tissue after 127 days' storage.

		Weight loss %		Fungal score		Necrosis %	
Species	Storage	Not cured	Cured	Not cured	Cured	Not cured	Cured
D. cayenensis	Ambient	45	37	0.2	0.0	62	58
D. cayenensis	13°C	32	26	2.9	2.3	96	46
LSD (p = 0.05)		7		0.5		40	
D. rotundata	Ambient	23	18	0.0	0.0	4	3
D. rotundata	13°C	10	8	1.4	1.2	9	10
LSD (p = 0.05)		4		0.2		ns	

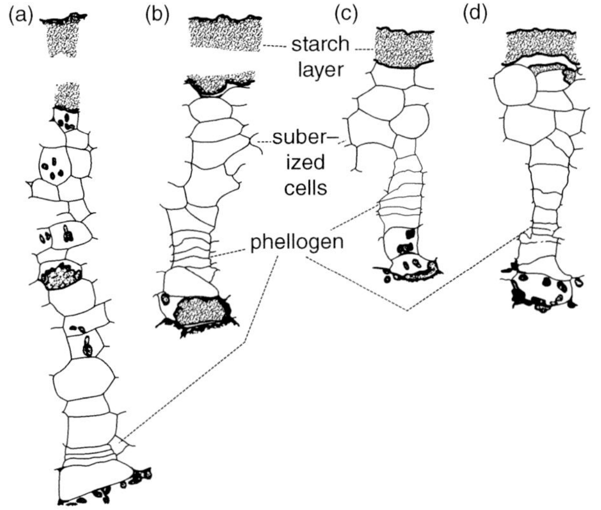

Fig. 5.3. Sections through the cut surfaces of *D. rotundata* tubers after 7 days curing and 8 days subsequent storage. Curing conditions were: (a) direct sunlight; (b) 26°C and 66% RH; (c) 30°C and 91% RH; (d) 40°C and 98% RH. Modified from Been *et al.* (1976).

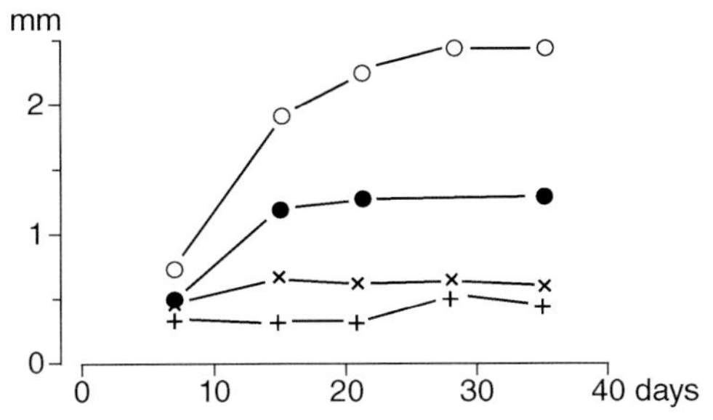

Fig. 5.4. Effects of curing *D. rotundata* tubers for 7 days on the thickness of wound healing tissue during subsequent storage at 26°C and 66% RH. Curing conditions were: O = direct sunlight; ● = 26°C and 66% RH; x = 30°C and 91% RH; and + = 40°C and 98% RH. Modified from Thompson (1972a) and Been *et al.* (1976).

tubers by exposing them to direct sunlight for 3 days after harvest in mean ambient conditions of 29°C and 84% RH. Tortoe *et al.* (2014b) also tested field curing using straw, polypropylene sheets or jute sacks on the *D. rotundata* cultivars 'Pona', 'Lariboko', 'Dente', 'Mutwumudoo' and 'Serwah' and the *D. alata* cultivars 'Matches' and 'Akaba' for 7, 14 or 21 days. Curing temperatures ranged between 27–40°C and humidity between 87–100% RH inside the heap, with the control treatment (with no covering) having a lower humidity of 60–80% RH. The material used to cover the heap, duration of curing, climatic conditions and yam species and cultivar influenced curing, but overall jute sacks were most suitable and 7 days was the optimum time.

Wilson (1980) previously described a similar method of curing *D. alata*, *D. nummularia* and *D. esculenta* tubers. The tubers were first stacked on the ground in a lightly shaded area and covered with grass or mats. Then a canvas tarpaulin was placed over the whole stack to cover the grass or alternatively another mat was used. It was emphasized that when a tarpaulin was used it was important to ensure it did not touch the yams. As an alternative, a simple wooden frame could be built and the canvas draped, like a tent, over the piled yams. Wilson indicated that plastic sheets should not be used for curing since they can make the yams too hot. If no canvas tarpaulin was available, Wilson found that several layers of sacks or mats could be used.

Been *et al.* (1974, 1976) cured *D. rotundata* tubers at 30°C or 40°C in a humidified oven, in uncontrolled ambient conditions in direct sunlight or in a storehouse in Jamaican ambient conditions. All curing conditions had similar effects on wound healing of tubers but the rate and thickening of periderm formation were different (Figs 5.4, 5.5). Only tubers cured at 40°C in a humidified oven were completely free from mould growth during subsequent storage and these tubers had the overall lowest storage losses, however they sprouted earlier. Gonzalez and de Rivera (1972) recommended curing *D. alata* at 29 to 32°C and 90 to 95% RH for about 4 days directly after harvest.

Effects of curing yams on weight loss

As has previously been shown for other root crops, cured yam tubers have a lower weight loss than those that have not been cured (Fig. 5.5) and specifically for *D. rotundata* tubers, the effect was considerable (Fig. 5.6). Also, this reduction continued throughout the storage period (Fig. 5.6). The reduction in weight loss due to curing was effective whether the tubers were stored in Jamaican ambient conditions or in cold storage (Fig. 5.7). However, cured tubers sprouted earlier by about 2 weeks when stored at tropical ambient temperatures, probably due to the increased cambial activity associated with curing (Thompson, 1972a).

Effects of curing yams on fungal diseases

Snowdon (1992) reported that *Cortidum rolfsii* can cause rotting of yam seed pieces in Nigeria and curing in a shaded barn resulted in fewer rots than if the yam pieces had been cured on the soil surface shaded with palm fronds. When *D. bulbifera* bulbils or tubers were left in the shade in tropical ambient conditions, harvest wounds healed quickly, which made them more resistant to fungal infections (Kay, 1987). Curing yam tubers at 35 to 40°C and 95 to 100% RH for 24 hours

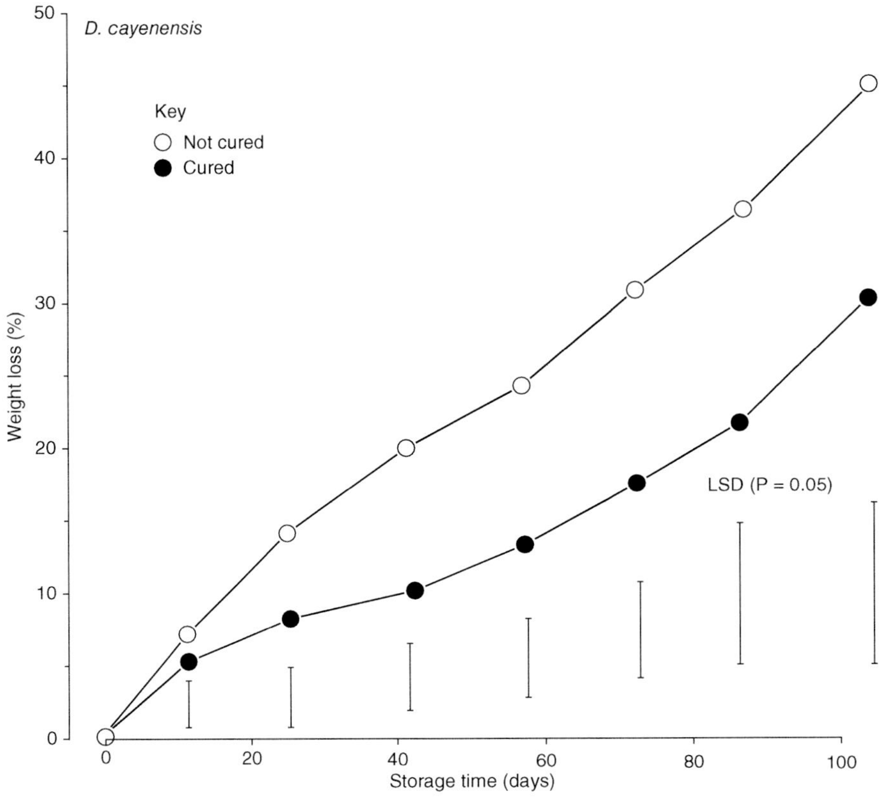

Fig. 5.5. Curing *D. cayenensis* tubers: subsequent weight loss during storage in Jamaican ambient conditions, where ○ = not cured, ● = cured.

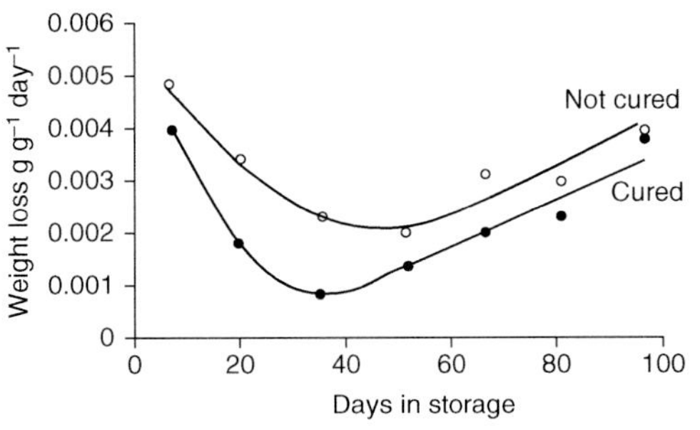

Fig. 5.6. Curing: weight loss of *D. rotundata* tubers.

was found to initiate the curing process and controlled storage rots. Curing at 40°C with 98% RH for 7 days was also shown to be effective, although sprouting was higher when the tubers were subsequently stored (Table 5.3). Although curing for 7 days in direct sunlight resulted in a high weight loss during the curing process, subsequent weight loss was lower, but again sprouting was increased at 40°C with 98% RH for 7 days or 35 to 40°C and 95 to 100% for a single day could well replace chemical fungicides for postharvest disease control (Thompson *et al.* 1977). In a comparison of the effects of curing on *D. cayenensis* and *D. rotundata* tubers at either ambient or 13°C, it was found that after 127 days storage there was little or no fungal growth on the tubers stored in ambient conditions. However, at 13°C fungal levels were 2.9 for non-cured and 2.3 for cured in *D. cayenensis* and 1.4 for non-cured and 1.2 for cured in *D. rotundata* with 14% of the surface of *D. rotundata* tubers after 75 days storage, compared with zero for those that had been cured (Figs 5.8, 5.9).

Other effects of curing

Curing was also shown to reduce the necrotic tissue that developed during storage on both

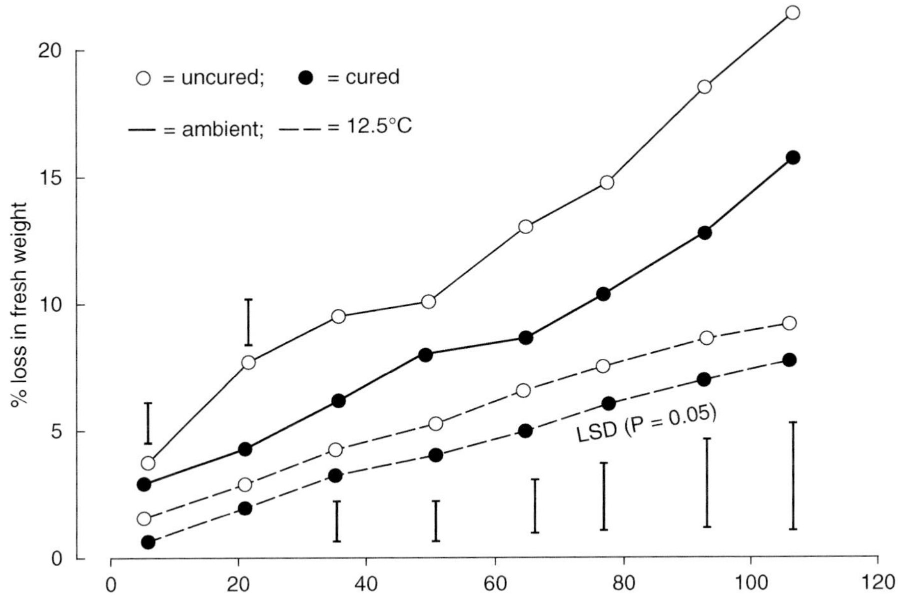

Fig. 5.7. Effects of curing and storage conditions on weight loss during storage of *D. rotundata* in Jamaica. Thompson *et al.* (1973c).

Table 5.3. The effects of curing conditions of *D. cayenensis* tubers for 7 days on their subsequent deterioration in storage in Jamaican ambient conditions. Figures followed by the same letter in each column were not significantly different (p = 0.05).

Curing conditions	Weight loss during curing (%)	Surface mould (mean of 11 weekly scores)	After 70 days storage at 24-30°C and 62-76% RH		
			Weight loss (%)	Sprouting (%)	Necrotic tissue (%)
Direct sunlight	11.0a	0.9b	22.5b	77a	7b
26°C + 66 RH	9.1b	1.0b	35.5a	33c	27a
30°C + 91% RH	2.1d	1.3a	36.1a	50b	21a
40°C + 98% RH	4.3c	0.0	20.9b	73a	6b
CV%*	13.7	19.5	3.7	30.0	23.3

*CV = Coefficient of Variation

D. cayenensis and *D. rotundata* (Table 5.2), presumably as a result of having lower fungal infections (Thompson, 1972b). Ikeemobi *et al.* (1986) showed that lipolytic acyl hydrolase activity decreased in tubers after wounding and disappeared by the fourth day. However, there was between two and nine times increased activity of lipoxygenase peroxidase and Polyphenylene oxide (PPO) with maximum values on the fourth and fifth days respectively after wounding. Lipoxygenase and PPO activities measured in the proximal half of tubers were much higher than corresponding activities in the distal half. Wilson (1980) reported that yams may sweeten during curing, which may not be to everyone's taste.

Commercial application of curing

When curing was applied commercially in Jamaica in the 1970s there were problems of bacterial soft rots developing on cured tubers during subsequent storage and marketing. Loading the tubers into the curing room, heating them until the surface of the tubers reached 40°C and then injecting

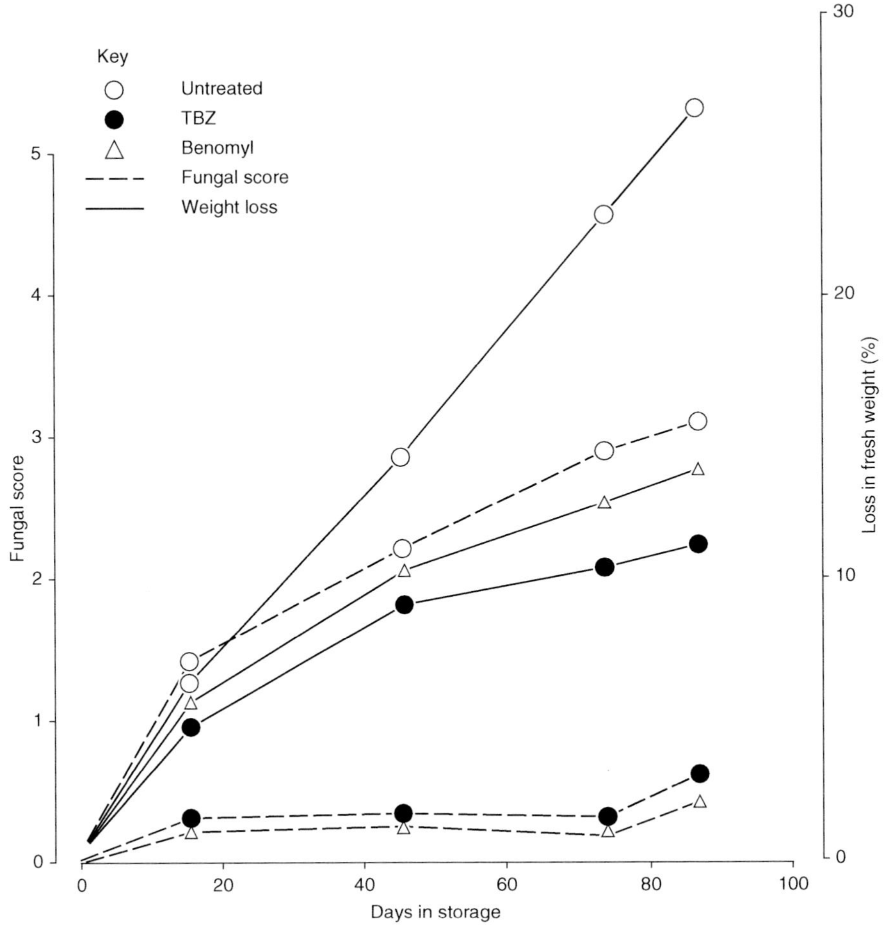

Fig. 5.8. Effects of fungicides of the surface fungi (mean score where 0 = none and 5 = surface entirely covered) and weight loss during storage of D. *rotundata* at 12.5°C and 95% RH. Thompson *et al.* (1977).

steam to increase the humidity overcame these problems. Tubers cured in this way had reduced weight loss, no fungal infection, no necrotic tissue and no bacterial rots after they had been transported by sea in reefer containers to the UK at 13°C and marketed through commercial retail outlets (J.E. Cecil *et al.* unpublished).

Sprouting

As tubers grow on the plant they develop and mature and when they reach a certain maturity they have the capacity to sprout. Lebot (2019) defined tuber maturity as when the distal meristematic area changed from light to dark and produced a cork layer and suberized cells. Mature, freshly harvested tubers cannot sprout and they enter a dormancy phase which can last for less than a month or up to 5 months according to temperature. Optimum sprouting conditions occur between 25 and 30°C and are delayed below 15°C and above 35°C; they are a major limitation to protracted storage (Thompson, 2015). As indicated above, dormancy and sprouting are affected by the maturity of the tubers at harvest as well as the postharvest conditions. In *D. rotundata* in ambient conditions in Jamaica, tuber sprouting occurred after only 4 weeks

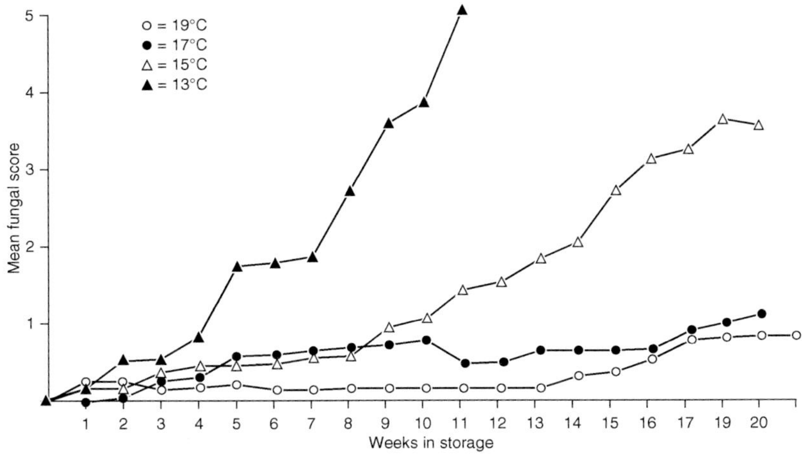

Fig. 5.9. The effects of storage temperature on the development of fungi on the surface of yam tubers: ○ = 19°C, ● = 17°C, Δ = 15°C and ▲ =13°C.

(Thompson, 1972a) while in Nigerian ambient conditions sprouting of *D. rotundata* occurred within 8 weeks (Coursey, 1961). Passam (1982) and Opara (1999) reported the dormancy period of *D. rotundata* from Nigeria as 12 to 14 weeks. In India, tubers of the *D. alata* cultivars 'Sree Keerthi' and 'Sree Roopa', the *D. rotundata* cultivars 'Sree Priya' and 'Sree Subra' and the *D. esculenta* cultivar 'Sree Latha', stored at room temperature of 30 to 32°C and 80 to 85% RH in the dark, all sprouted 70 to 80 days after harvest.

Panneerselvam *et al.* (2007), working with *D. esculenta* stored in yam barns in West Africa, found that starch was degraded and the enzymes (α-amylases and starch phosphorylase) showed a lower level of activity during early periods of dormancy, while sugar content and enzymes of carbohydrate metabolism increased rapidly during sprouting. The key enzymes of glycolysis, tricarboxylic acid cycle and the pentose phosphate pathway (aldolase, succinic dehydrogenase, malic dehydrogenase, glucose-6-phosphate dehydrogenase and 6-phosphogluconate dehydrogenase) increased even before the visible appearance of sprouting and their activity was at its maximum during sprouting. Ile *et al.* (2006) described three phases that occurred during the storage of *D. rotundata*. The first phase was when the tuber was dormant and did not sprout whatever the storage conditions. The second phase was when the meristem started to form in the parenchyma tissue just below the tuber's periderm. The third phase was when the tuber sprouted. They postulated that sprout suppressants can only act on *D. rotundata* after the start of phase two. As sprouting begins, the starch content of yam tubers in the apical and basal regions decreased while the sugar levels and α-amylase activity increased (Muthukumarasamy and Panneerselvam, 2000). Dormant yam tubers, in contrast to potatoes (*Solanum tuberosum*), do not have any external or internal shoot buds but have a layer of meristematic cells below the surface of the tuber (Fig. 5.3). Onwueme (1973) and Wickham *et al.* (1981) have shown, in both *D. rotundata* and *D. alata,* that when dormancy is broken, bud formation begins in this meristematic layer of cells before any external shoot sprout is visible on the tuber surface.

Ile *et al.* (2006) describe in detail the formation of sprouting in the meristematic tissue just below the tuber surface of *D. rotundata.* They also conducted field trials where the first *D. rotundata* tubers started sprouting 326 days after planting. The average duration from harvest at vine senescence (179 days after planting) to 50% sprouting was

160 days and tubers sprouted at 339 days after planting. Tubers harvested 30 days prior to vine senescence (149 days after planting) sprouted 187 days later (i.e. 336 days after planting). Therefore, date of harvest had no effect on the date to 50% sprouting of tubers. Passam (1982) indicated that dormancy duration varied with species as follows: *D. alata* (14–16 weeks); *D rotundata* (12–14 weeks); *D. cayenensis* (4–8 weeks); and *D. esculenta* (12–18 weeks). The most popular variety/cultivar of *D. cayenensis* in Jamaica is called 'Roundleaf' and this began to sprout some 90 days after harvest compared with the variety/cultivar 'Common' that sprouted after some 50 days. This study was at ambient temperatures of 25 to 34°C and 64 to 92% RH (Thompson, 1972b). These dormancy periods were influenced by temperature; at decreasing temperature there was increasing dormancy and vice versa. Similarly, high humidity promoted sprouting and low humidity prolonged dormancy. The storage duration of yam is affected by the breaking of dormancy period and is associated with formation of buds and subsequent sprouting.

Hamadina and Asiedu (2015) investigated the effects of provenance and storage on sprouting of *D. rotundata* in Nigeria. They found that tubers stored in the regions around Onne and Ibadan sprouted about 10 days earlier than those at Abuja, which was associated with slightly higher temperature and humidity at Onne and Ibadan but provenance had no effects. They also showed that there were differences in the date of sprouting between landraces within the species of *D. rotundata*. At the time of sprouting of *D. rotundata* tubers, Jaleel *et al.* (2007) showed that levels of protein, amino acid and sugars increased as did the activity of protease and α-amylase while starch content decreased.

Amponsah *et al.* (2007) conducted sprout control experiments during storage with *D. rotundata* at different stages of harvest maturity. The team compared two types of yam barn, the first of which was constructed by researchers at the Crops Research Institute in Kumasi in Ghana using locally available materials, which could hold 10,000 average sized yam tubers and was fitted with rodent guards made from aluminium sheets (Amponsah *et al.*, 2015). The second was a typical structure made by local farmers as was conventionally done in the study communities for storing yams. The materials used for constructing these yam barns consisted mainly of wooden pieces from the African fan palm (*Borassus aethiopum*), thatch and straw. After 120 days in storage, the traditional yam barn maintained a higher humidity compared to the improved yam barn and therefore the tubers had a lower weight loss. A total of 160 yam tubers, selected at random, were stored in each structure. When the tubers began to sprout, they were subjected to different sprout control treatments. These treatments were: snapping off the sprouts by hand (hand pick); cutting the sprouts off with a knife (full cut); and cutting the sprouts off halfway up to the first node (half cut). The hand snap sprout control treatment was found to be the best at limiting weight loss, tuber rot and rate of sprouting.

Irradiation

Irradiation has been shown to delay or prevent sprouting in a variety of root crops. The optimum absorbed dose for sprout inhibition of potatoes varied with cultivars, time of irradiation after harvest and post-irradiation storage temperature, but doses were usually between 0.07 to 0.1 kilograys (kGy). Irradiation at doses of between 0.075 and 1.5 kGy inhibited sprouting of *D. rotundata* during 6 months storage without inducing adverse changes in acceptability or physiological properties (Burton, 1989; Vasudevan and Jos, 1992). Differences in cultivar responses to γ-irradiation were also reported. Imeh *et al.* (2012) irradiated *D. alata* tubers with γ radiation at doses of 0, 60, 80, 100, 120 or 140 grays at an average dose rate of 3 grays min^{-1} and then stored them for 7 months. The non-irradiated tubers sprouted by the end of the third month of storage but 90% of tubers irradiated at 60 grays,

30% of tubers irradiated at 80 grays and only 10% of tubers irradiated at 100 grays sprouted within the 7-month storage period. None of tubers irradiated at 120 and 140 grays sprouted within the 7-month storage. Osunde (2008) reported that an average dose of 0.12 kGy and a dose rate of 0.114 kGy hour^{-1} were applied to the *D. rotundata* cultivar 'Asana', which were then stored for 6 months in a barn or on the ground in the open. Their results showed that irradiation reduced sprouting in both storage types. However, there was less rotting in the non-irradiated yams stored on the ground than the irradiated ones. Also, food products made from irradiated yams were judged to be better in quality than those made from the non-irradiated ones (Vasudevan and Jos, 1992).

Temperature

Low temperature storage can control sprouting. No sprouting occurred on tubers of *D. rotundata* and *D. cayenensis* at 13°C during 5-month storage, but there was chilling injury at temperatures below 15°C when they were stored for over a month (Thompson *et al.*, 1973a). Olorunda *et al.* (1974) found that storage of *D. alata* tubers at 10 or 15°C suppressed sprouting during 18-week storage, but at 10°C the yams deteriorated as a result of chilling injury. Noon (1978) also reported that yams suffered from chilling injury at temperatures below 12°C. Cool storage above the temperature that causes chilling injury is therefore generally inappropriate for controlling sprouting, except for short periods.

Chemicals

Sprout suppressing chemicals are used in storage for many root crops and for potatoes a range of chemicals have been shown to suppress sprouting (Burton, 1989). The most commonly used is CIPC, also called chloropropham (chloroisopropyl phenylcarbamate), which is vaporized and applied as a thermal fog to the stored crop. An emulsifiable concentrate and a granular formulation are also available for application to tubers as they are loaded into store, however this method of application can impede the curing process and result in increased levels of deep skin spot infections. IPC, also called propham (isopropyl phenylcarbamate), is also used either by itself or in a formulation with CIPC.

Regulations for the use of chemical sprout suppressants are mainly for potatoes. The CIPC maximum residue level (MRL) from the Environmental Protection Agency in the USA was revised in 1996 to 30 mg kg^{-1} (Kleinkopf *et al.*, 1997). The regulatory requirements for 2017–2018 in the UK were 36 g CIPC tonne^{-1} of potatoes for processing and 24 g tonne^{-1} for the fresh market. Regulatory requirements for 2019–2020 for potatoes in the EU remained at 24g CIPC tonne^{-1} of potatoes as the total dose for the fresh market and 36 g CIPC tonne^{-1} of potatoes total dose for processing, including fish & chip shops and peeling (European Food Safety Authority, 2019). Lentza-Rizos and Balokas (2001) found that, during storage, the mean concentration of CIPC in individual potato tubers was 3.8 mg kg^{-1} after 10 days, 2.9 mg kg^{-1} after 28 days and 2.2 mg kg^{-1} after 65 days. Peeling removed 91 to 98% of the total residue and washing reduced residues by 33 to 47%. Detectable residues were found in boiled potatoes and the boiling water, and in French fries and their frying oil (Lentza-Rizos and Balokas, 2001). Alternative chemicals for sprout suppression in stored potatoes in the UK include ethylene, maleic hydrazide and spearmint oil. Other chemicals were reported to have been tested including 3-decen-2-one (approved as a sprout suppressant in USA in 2013 where it is exempted from an MRL), diisopropylnaphthalene, dimethylnaphthalene, 1,4-dimethylnapthalene, caraway oil, clove oil and orange oil. Working on seed potatoes, Njogu *et al.* (2015), showed that combining benzylaminopurine with gibberellic acid could have positive effects on tuber sprouting.

In yams, Wickham *et al.* (1984a, 1984b) tested various chemicals on sprouting of *D. rotundata*, *D. alata* and *D. esculenta* tubers during storage. They found that abscisic acid, kinetin, indole acetic acid and 2-chloroethyl-trimethyl ammonium chloride (CCC) had little or no effect on either shoot or root production from tubers, while ethrel greatly promoted root production from tubers with little or no effect on shoot development. Treatment with 2,4-dichlorophenoxy acetic acid (2,4-D) caused massive root and callus development but delayed shoot development, while gibberellic acid delayed sprouting. However, Tschannen (2003) reported that gibberellic acid applied to *D. rotundata* directly after harvest prolonged dormancy by 9 to 11 weeks and reduced respiration rate at the time of sprouting. Girardin *et al.* (1998) prolonged the dormancy period of *D. cayenensis-rotundata* and *D. alata* tubers with immersion of the tubers in gibberellic acid at 75 mg L^{-1} for 2 hours or 0.5 mg L^{-1} for 5 hours.

Passam (1982) found that gibberellic acid could prolong dormancy of *D. alata* tubers but its effect was critically dependent on its time of application. When applied just after harvest, it delayed sprouting, but when applied later in storage it had virtually no effect. A proprietary potato sprout suppressant containing CIPC and IPC in a dust formulation had no effect on sprouting of *D. rotundata* tubers (Thompson *et al.*, 1973a). CIPC was applied to *D. alata* at 0, 0.45 or 0.90 g kg^{-1} during storage for 18 weeks at 10, 15, 20 or 25°C. At 20°C sprouting was not suppressed, but at 25°C it prevented further sprouting of de-sprouted tubers (Olorunda *et al.*, 1974). They concluded that these findings suggest that CIPC does not significantly extend the storage life of yams.

Orhevba and Osunde (2006) also reported that CIPC did not have any effect on sprouting of *D. rotundata* tubers. The effect of maleic hydrazide in controlling sprouting in yam showed that soaking the tubers in 1,000 μL L^{-1} solutions for 10 hours before

Table 5.4. Effects of some sprout suppressant chemicals on the length of yam tuber storage. TCNB = tetrachloronitrobenzene; PCNB = pentachloronitrobenzene; IPPC = isopropylphenylcarbamate; CIPC = isopropyl-N-(3-chlorophenyl) carbamate. Modified from Passam (1982).

Species	Chemical	Effects on storage
D. alata	TCNB	None
D. alata	PCNB	None
D. alata	IPPC	None
D. alata	maleic hydrazide	Reduced sprouting
D. alata	CIPC	Delayed sprouting less than 1 month
D. esculenta	maleic hydrazide	None
D. rotundata	TCNB	None
D. rotundata	maleic hydrazide	Slightly delayed sprouting

Table 5.5. Growth-regulating chemical effects on yam tuber storage life. Methyl-α-NAA = methyl ester of naphthaleneacetic acid; β-NAA = β-naphthaleneacetic acid; IAA = indole-3-acetic acid; kinetin = 6-furfuryl-aminopurine. Modified from Passam (1982).

Species	Compound	Storage life
D. alata	methyl-α-NAA	Up to 1.5–2 months, phytotoxic
D. alata	2-chlorethanol + thiourea	Up to 3 months
D. alata	2-chlorethanol	Promoted sprouting
D. alata	gibberellic acid	Up to 4–7 weeks
D. esculenta	gibberellic acid	Up to 6 weeks
D. rotundata	methyl-α-NAA	None
D. rotundata	β-NAA	None
D. rotundata	gibberellic acid	None
D. rotundata	IAA	None
D. rotundata	kinetin	None

storage reduced the rate of sprouting by 8% (Ramanujam and Nair, 1982). However, Osunde (2008) reported that maleic hydrazide was not effective in inhibiting sprouting in *D. rotundata* tubers. Nonanol placed inside polyethylene bags with tubers reduced and delayed sprouting but increased the level of flesh necrosis. Osunde and Orhevba (2011) reported that yam tubers treated with either slurry or bark extract made from neem trees (*Azadirachta indica*) had less sprouting than those not treated. *D. rotundata* tubers were dipped in triadimefon (TDM) fungicide at 20 mg L^{-1} and stored at 30±2°C. In the control tubers sprouting occurred after 70–80 days, but treatment with TDM led to earlier sprouting (Jaleel *et al.*, 2007). Passam (1982) tested a range of potential sprout suppressants on *D. alata*, *D. esculenta* and *D. rotundata* (Tables 5.4, 5.5) and found some positive effects.

However, before any of these postharvest chemical treatments are used in commercial practice, permission must be obtained from the appropriate regulative authorities for their use on yams.

6

Diseases and Pests

Field Fungal Diseases

Fungi reported to have been isolated from the foliage of *Dioscorea* species (modified from Zohouri, 1998 and SPYN, 2003) include: *Alternaria* sp., *Ascohyta* sp., *Botryodiplodia theobromae*, *Cercospora* sp., *Choanephora cucurbitarum*, *Cladosporium* sp., *Colletotrichum capsici*, *Colletotrichum gloeosporioides*, *Corticium rolfsii*, *Corynespora cassiicola*, *Curvularia* sp., *Diplodia* sp., *Fusarium* sp., *Fusarium oxysporum*, *Glomerella cingulata*, *Goplana dioscoreae*, *Heminthosporium* sp., *Heterosporium luci*, *Illosporium* sp., *Lasiodiplodia theobromae*, *Macrophomina phaseoli*, *Monochaetia* sp., *Mychosphaerella dioscoreicola*, *Oidium* sp., *Periconia* sp, *Pestalotia* sp., *Pestalotiopsis cruenta*, *Phoma* sp., *Phomopsis* sp., *Phyllosticta* sp., *Pythium* sp., *Rhizoctonia* sp., *Sphaeropsis* sp., *Sphaerostilbe repens*, *Stachybotrys* sp., *Stemphyllium* sp. and *Vermicularia* sp.

Mignucci *et al.* (1985), working on yams in Puerto Rico, identified the damaging diseases occurring from planting to harvest to include seed-tuber rot (*Penicillium* spp, *Fusarium oxysporum* and *F. solani*), root rot (*F. oxysporum* and *Pythium* sp.), stem basal canker (*F. oxysporum*), vascular wilt (*F. oxysporum*), anthracnose (*Colletotrichum gloeosporioides*), leaf spots (*Curvularia eragrostidis*, *C. geniculata* and *Cercospora* sp.), rectangular leaf spot (*Aphelenchoides ritzema-bosi*) and stem blight (*Botryodiplodia theobromae*). They described possible control methods as including chemical control, use of varietal resistance and cultural practices.

Postharvest Fungal Diseases

Fungi that have been reported to have been isolated from tubers of *Dioscorea* species include: *Aspergillus flavus*, *Aspergillus niger*, *Aspergillus ochraceus*, *Aspergillus flavipes* *Aspergillus tamari*, *Aspergillus tamari*, *Aspergillus tamarii*, *Botryodiplodia theobromae*, *Cladosporium* sp., *Colletotrichum gloeosporioides*, *Corynebacterium* sp., *Fusarium* sp., *Fusarium moniliforme* var. *subgluti-nans*, *Fusarium monoliforme*, *Fusarium oxysporum*, *Fusarium solani*, *Hendersonula toruloidea*, *Lasiodiplodia theobromae*, *Marcophomina phaseoli*, *Mucor circinelloides*, *Penicillium chrysogenum*, *Penicillium cyclopium*, *Penicillium gladioli*, *Penicillium oxalicum*, *Penicillium sclerotigenum*, *Pennicillium digitatum*, *Rhizopus nodosus*, *Rhizopus stolonifer*, *Sclerotium rolfsii* and *Trichoderma* sp. (Okafor, 1966; Ogundana *et al.* 1970; Asare-Bediako *et al.*, 2007; Ogbo Frank and Agu Kingsley, 2014a; Ogunleye

DOI: 10.1079/9781789249279.0006

and Ayansola, 2014). In addition to the fungi described above, Ogunleye and Ayansola (2014) reported that infections of the *D. alata* and *D. rotundata* tubers they tested during storage were caused by several fungi including *Lasiodiplodia theobromae* (about 12%), *Penicillium* sp. (about 6%), *Alloymyces arbuscula* (about 4%) and *Rhodotorula* sp. (about 1%).

Snowdon (1992) described 15 microorganisms causing postharvest diseases in yams as mainly fungi. Four species were isolated: *Aspergillus niger*, *Fusarium oxysporum* and *Penicillium citrinum*, all of which were found to be mildly pathogenic, and *Fusarium moniliforme* [*Gibberella fujikuroi*] which did not produce any rot symptoms. *Aspergillus niger* and *Penicillium oxalicum* were the most common fungi on stored tubers with the optimum temperature for growth of 26 to 30°C (Ogundana *et al.*, 1970), but there was no growth observed on undamaged tubers (Ogali *et al.*, 1991). Amusa (1997) reported that *Colletotrichum gloeosporioides*, *Curvularia pallescens*, *C. eragrostides*, *Pestalotia* sp. and *Rhizoctonia solani* were parasitic on yams in Nigeria. Tubers may be infected with *Botryodiplodia theobromae*, *Lasiodiplodia* sp. and *Fusarium* species, causing postharvest diseases (Kay, 1987). Amusa and Baiyewu (1999) identified the following microorganisms associated with yam tubers from the tropical forest region of southwestern Nigeria during marketing and storage: *Botryodiplodia theobromae*, *Aspergillus tamari*, *Penicillium oxalicum*, *P. cyclopium*, *P. italicum*, *Fusarium oxysporium*, *F. solani*, *Rhizopus nigricans*, *Sclerotium rolfsii*, *Mucor circinelloides*, *Trichoderma viridae* and *Erwinia carotovora*. Ogaraku and Usman (2008) identified the following fungi from samples of *D. rotundata*, *D. cayenesis* and *D. alata* from the southern Guinea savannah region of Nigeria: *Aspergillus niger*, *Aspergillus flavus*, *Rhizopus stolonifer*, *Sclerotium rolfsii*, *Fusarium oxysporium* and *Rhizoctonia* species. Nwankiti and Arene (1978) identified the following fungi associated with postharvest rots in Nigeria: *B. theobromae*, *Fusarium moniliforme*, *Penicillium sclerotigenum* and *Aspergillus niger*.

Aboagye-Nuamah *et al.* (2005) collected *D. rotundata* tubers from a yam barn and from selected markets in Accra, Ghana, to identify the most predominant pathogens associated with rots. Nine fungal spoilage microorganisms, *Aspergillus flavus*, *Aspergillus niger*, *Botryodiplodia theobromae*, *Fusarium culmorum*, *Fusarium oxysporium*, *Fusarium* sp., *Penicillium brevi-compactum*, *Penicillium* sp. and *Rhizopus stolonifer*, and a bacterium species *Erwinia carotovora* were identified. The mean incidence of occurrence of the organisms on rotten tissues ranged from 1.2% to 28.5%. Of the ten microorganisms isolated, *B. theobromae*, *F. oxysporium* and *R. stolonifer* were the most frequently encountered spoilage microorganisms in the markets. *E. carotovora*, *Fusarium solani* and *Penicillium* sp. were relatively sparse (incidence not exceeding 3%) compared to the other microorganisms. Although there was generally no significant difference in the severity of rots induced by the different microorganisms in the tubers, *R. stolonifer* commonly induced more rots in the tubers compared to *B. theobromae* and *F. oxysporium*.

Yusuf and Okusanya (2008) found *R. stolonifer*, *Penicillium oxalicum* and *A. niger* on *D. rotundata* in the Yola markets of Nigeria. The mean percentage incidence was 45% for *R. stolonifer*, 38% for *A. niger* and 22% for *P. oxalicum*. The optimum temperature for growth of the three isolates was the same at 35°C, while rot development was inhibited at 15°C and below. The growth of the fungi increased with increase in humidity. The most rapid fungal rot was at 100% RH, while at 32.5% RH the fungal rot was much reduced.

Diseases are a major cause of postharvest losses in yams. Different species and varieties of *Dioscorea* show different levels of susceptibility to various diseases as do storage methods (see below). For example, Nyadanu *et al.* (2014) stored the *D. rotundata* cultivars 'Labalkor', 'Kplondzo', 'Olordor' and 'Fushiebila' in barns, pits, on a platform or in heaps. 'Olordor' and 'Kplondzo' had highly significantly ($p < 0.001$) lower internal microbial rot than 'Labalkor' and 'Fushiebila' and those stored in barns or on a platform had significantly ($p < 0.05$) lower internal rot than those stored in pits or heaps. *Aspergillus niger* (52.3%), *A. flavus* (44.6%), *A. ochraceus* (16.9%), *A. tamari*, *Cladosporium* sp. (16.6%), *Corynebacterium* sp., *Fusarium* sp. (20.0%), *Penicillium* sp., *Rhizopus stolonifer* (38.5%), *Sclerotium rolfsii* (30.8%) and *Trichoderma* sp. were identified infecting mini-setts of the *D. rotundata* cultivars 'Pona' and 'Dente', with 'Pona' suffering more severe rot than 'Dente' (Asare-Bediako *et al.*, 2007). *S. rolfsii* caused the most severe rotting followed by *A. niger* and *Fusarium* sp. while lower levels of rotting were caused by *R. stolonifer*, *Trichoderma* sp. and the bacterium *Corynebacterium* sp.

Ogundana *et al.* (1970) and Okigbo (2003) reported that the major microorganisms associated with storage rots of yams in Nigeria included *Botryodiploidia theobromae*, *Penicillium oxalicum*, *P. sclerotigenum*, *Fusarium moniliforme*, *Aspergillus niger*, *A. tamarii*, *Rhizoctonia* and *Serratia* spp. Noon (1978) listed over 50 fungi associated with rotting in yams although their degree of pathogenicity was not always identified. The organisms identified as causing soft rot in stored yams were *Botryodiplodia theobromae*, *Fusarium moniliforme* var. *subgluti-nans*, *Penicillium sclerotigenum* and *Aspergillus* spp. Ogbo Frank and Agu Kingsley (2014a) stored *D. dumentorum*, *D. alata* and *D. rotundata* in an experimental barn at approximately 30°C and 95% RH. The tubers had been obtained immediately after harvest and were examined at intervals during storage for rots. All three species of yam studied suffered fungal rots during storage, predominantly dry rots. Seven distinct fungal isolates were recorded as causing these rots (*Aspergillus tamari*, *Fusarium solani*, *Lasiodiplodia theobromae*, *Aspergillus niger*, *Mucor circinelloides*, *Aspergillus flavus* and *Aspergillus* sp.). In all yams, storage rots reduced shelf life and resulted in increased weight loss.

Diseases in yam tubers are mainly caused by fungi entering through wounds and natural openings on the surface. Ogundana *et al.* (1970) reported that optimum rots caused by pathogens occurred at 90% RH and at 26–30°C, but the extent of rotting varied with both the fungal species and the yam species. USNAS (1981) listed six and seven fungi in Nigeria and Côte d'Ivoire respectively causing soft rot in yam tubers. These were *Aspergillus niger* (52.3%), *A. flavus* (44.6%), *A. ochraceus* (16.9%), *A. tamari* and *Cladosporium* sp. (16.6%), *Corynebacterium* sp. and *Fusarium* sp, (20.0%) and *Penicillium* sp. and *Rhizopus stolonifer* (38.5%). Postharvest rots of yams are common, especially where they have been damaged during harvesting and handling. Thompson (1972a) reported that the mean fungal scores during storage was more than double that for tubers that had been cut compared to those that had not been cut (1.9 compared to 0.9) and fungal levels progressively increased with increasing height, when tubers were dropped onto a hard surface (Table 6.1).

Specific Fungal Diseases

Anthracnose

Anthracnose was reported to be the most serious disease affecting yams (Winch *et al.*, 1984; Onyeka *et al.*, 2006a, 2006b). Yam anthracnose is caused by field infections by the fungus *Colletotrichum gloeosporioides*, which is the sexual stage (teleomorph). The asexual stage (anamorph) is *Glomerela cingulate*. Initial infection results in small brown spots on young leaves that become larger, coalesce and may develop pale yellow margins as the disease increases and the leaves grow. Nwankiti and Arene (1978)

Table 6.1. Effects of dropping *D. trifida* tubers from various heights onto a concrete floor on storage losses in Jamaican ambient conditions of about 28°C for 86 days. Fungal score is 0 = no fungi to 5 = surface covered. Thompson (1972b).

Height from which tubers were dropped (cm)	0	30	60	90	120
Fresh weight loss %	29	43	42	55	64
Fungal score	0	0.1	0.2	0.3	0.4
Necrotic tissue in the flesh %	24	55	53	65	84

reported that anthracnose commonly occurred in southwestern Nigeria and was one of the most severe yam diseases: in *D. alata* infection can stop growth and depress yield by 67%. Infected leaves are usually shed. Stems may also be infected or the fungus can spread from leaves to stems resulting in similar symptoms. Infection can also spread to buds, which terminates their growth. Infections can also spread to the flesh of tubers giving small blisters under the corky layer that can develop into deeper rots. In the Pacific region, *C. gloeosporioides* has been detected in yam tubers from Vanuatu and Papua New Guinea, confirming that the fungus is able to infect, and survive in, tuber tissues from season to season. Infected tubers, used as planting material, might then act as a primary source of inoculum, playing an important role in the epidemiology of the pathogen in the field. The disease has also been shown to be tuber-borne in both West Africa and the West Indies (Green and Simons, 1994). Infection starts on tubers in the field and develops in storage. Tuber-borne infections could have severe implications for production, both as a source of infection and as a means of dispersal. The mechanism of spread of *C. gloeosporioides* from tuber to canopy was reported by Lebot (2019), but it is yet to be understood fully. The pathogen has not been re-isolated or detected from shoot parts of plants raised from heavily infected tubers, ruling out any systemic spread of *C. gloeosporioides*.

D. alata is thought to be more susceptible to anthracnose than other *Dioscorea* species, but susceptibility varied between varieties. Malapa *et al.* (2005, 2006) studied 331 accessions of *D. alata* collected throughout the Vanuatu islands and identified 35 groups of morphotypes including some resistant or tolerant to anthracnose. Two of the groups consisted of diploid varieties that were late maturing (7–9 months) and tolerant to anthracnose, and two other groups that included late maturing (9–12 months) tetraploid varieties that were also tolerant to anthracnose. They also reported that *D. nummularia* and *D. transversa* are closely related to *D. alata* and both resistant to anthracnose. Unfortunately, the most popular cultivars, or those which are particularly adapted to commercial production, were also the most susceptible to anthracnose. In Barbados, anthracnose severely reduced the production of *D. alata* (McDonald *et al.*, 1998). In Puerto Rico, an evaluation of the marketable yield and the natural reaction to anthracnose of the *D. alata* cultivars 'Florido', 'Diamantes', 'Forastero' and 'Kabusah' showed no significant yield differences among the last three cultivars. Their average marketable yields of 26.2, 25.6 and 24.8 tonnes ha^{-1}, respectively, were significantly higher than that from the widely used 'Florido' cultivar (2.9 tonnes ha^{-1}), but 'Florido' had the highest disease severity (Gonzalez, 2006).

Aspergillus rot

Several species of *Aspergillus* have been shown to cause rotting in yam tubers, but *A. niger* is the most widely reported and important (Adenui, 1970; Ogundana *et al.*, 1970; Noon and Colhoun, 1979). Other species reported causing rotting in yam tubers include *A. flavus* (Noon and Colhoun, 1979), *A. tamarii*, *A. flavipes* (Adenui, 1970) and *A. ochraceus* and *A. lilacinum* (Ogunleye and Ayansola, 2014). Ogunleye and Ayansola also reported that infections caused by *A. niger*, *A. flavus*, *A. ochraceus* and *A. lilacinum* during storage occurred in about 19, 14, 2 and 0% cases, respectively of the *D. alata* and *D. rotundata* tubers they tested. Infection of tubers can affect flesh that becomes fawn or brown, sometimes tinged with purple. Infection takes place via injuries sustained during harvest, handling or storage. Lesions generally remain fairly firm, unless there is secondary infection by soft rot bacteria. Decay is especially rapid at temperatures between 25 and 35°C (Ikotun, 1983).

Awuah and Akrasi (2007) reported that wood ash gave good control of *Aspergillus niger*, but control was improved when a synthetic chemical fungicide was added to the wood ash. They also found that *Rhizobacterium* (isolate ESI) isolated from a yam rhizosphere soil, collected in the Ashanti region of Ghana, gave good control of *Aspergillus niger*. The following plant extracts were tested for their control of *A. niger* in yam tubers by Obilo *et al.* (2005): *Vernonia amygdalina*, *Azadirachta indica* (neem), *Ocimum gratissimum*, *Pergularia species*, *Citrus aurantifolia* (limes), *Allium sativum* and *Capsicum annum* (red pepper). Ethanol extracts of *Azadirachta indica* and *Ocimum gratissimum* were observed to have the highest percentage fungi growth inhibition, having 68.8% and 65.2% respectively, while *Allium sativum* and *Capsicum annum* had the lowest with 36.3% and 40.6% respectively. *Azadirachta indica* was also observed to have the highest severity and *Allium sativum* had the lowest. However, yam tubers treated 2 days before inoculation were observed to have a more inhibitory effect on the fungal pathogen than those treated 2 days after inoculation. Okpogba *et al.* (2019) reported that the extracts from the leaves of *Xylopia aethiopica* and *Syzygium aromaticum* had high antimicrobial activity against several plant pathogens. They also showed that these extracts at 50% concentrations inhibited the growth of *Aspergillus niger* on *D. alata* and *D. rotundata* tubers.

Blue and green mould

Penicillium species infection is often associated with damaged surfaces of yam tubers, but serious rotting of the flesh can sometimes occur with no external symptoms. Rotted tissue is pale to dark brown in colour, and may be firm or soft. *Penicillium* species can survive in the soil on crop debris and their spores are dispersed both by wind and water. Tubers can be infected at harvest through damage, and Adenui (1970) showed that postharvest decay was especially rapid at between 15 and 20°C. Kim *et al.* (2008) found that during a survey in May and June 2007 of postharvest diseases of yams in Korea, severe tuber loss was caused by infections with *Penicillium* species. They isolated *Penicillium sclerotigenum* and *P. polonicum* from the infected tubers.

Several methods of control of blue and green mould have been reported, including chemical fungicides and wood ash (Snowdon, 1992), and Okpogba *et al.* (2019) showed that extracts from the leaves of *Xylopia aethiopica* and *Syzygium aromaticum* at 75% concentrations inhibited the growth of *Penicillium chrysogenum* on *D. alata* and *D. rotundata* tubers. Plumbley *et al.* (1985) reported that a strain of *Penicillium sclerotigenum* infecting yam tubers in Jamaica was resistant to benzimidazole fungicides, but imazalil gave good control.

Botryodiplodia rot

Botryodiplodia theobromae was, perhaps, first reported on rotten stored yams by Dade and Wright (1931) and was reported to be common in Ghana (Piening, 1962). Infections of tuber tissue by *B. theobromae* tend to be dark brown or black in colour and the margin between diseased and healthy tissue is usually a distinct brown line (Ogundana, 1983). At storage temperatures below 20°C decay by *B. theobromae* on yam tubers was substantially reduced (Adenui, 1970). *B. theobromae* can survive for many months in the soil and on crop debris, and infection of tubers can occur in the field through injuries. Field infections with *B. theobromae* can also cause wilt diseases of yam plants, which may result in premature death of 45–70% of the crop (Nwankiti and Arene, 1978). Markson *et al.* (2012) tested several fungicides (Forcelet, Ridomil, Caocobre, Nordox, Hydrox and Borax) at different concentrations from 20 to 100 g L^{-1} on the control of *B. theobromae* on *D. rotundata* tubers and found that they all gave good control, with Coacobre giving the best results. Amadioha and Markson (2007) previously showed that some plant extracts (seed and leaves of *Piper nigrum* and *Aframomum melegueta*) could be used to control postharvest infections of cassava roots infected with *Botryodiplodia acerina*.

Fusarium rot

Fusarium species infection of yam tubers results in dry, pale tissue bordered by a brown margin. In tissue damaged by chilling, infections caused by *Fusarium* species can be differentiated by their texture being more likely to be soft and moist. Also, in a humid atmosphere the surface of infected tubers can become covered with tufts of dense white mould (Nwankiti and Arene, 1978; Noon and Colhoun, 1979). Ogundana and Dennis (1981) recommended control of Fusarium rot by curing directly after harvest to promote wound healing and subsequently storing the tubers at a 'moderate temperature'. As also reported for *B. theobromae*, field infections with *Fusarium oxysporum* can cause wilt diseases of yams, which may result in premature death of 45–70% of the crop (Nwankiti and Arene, 1978).

Pseudophloeosporella leaf spot

Hong *et al.* (2010) report that *Pseudophloeosporella dioscoreae*, synonymous with *Cercospora dioscoreae* and *Cylindrosporium dioscoreae* (Braun, 1993), causes leaf spot in *D. batatas* in Korea. The first symptoms were white or pale grey circular mycelial masses on tubers, usually limited to less than 3 mm in diameter, and as the disease progressed, the white masses disappeared, resulting in circular to irregular lesions with pale yellow edges and blackish brown acevuli in the centre. Four of the fungicides tested by Emua and Fajola (1983) (Captan, Captafol, Mancozeb, and Phaltan) were consistently effective in inhibiting germination of the conidia in *in vitro* tests and inhibiting the disease in greenhouse and field experiments, but the authors recommended Mancozeb because of its effectiveness, low phytotoxicity and low cost.

Sclerotium rot

Sclerotium rolfsii [*Athelia* (*Cortidum*) *rolfsii*] was reported to cause soft rots of yam seed pieces as well as tubers in Nigeria (Osai and Ikotun, 1994). Curing in a shaded yam barn resulted in fewer rots than if pieces are cured on the soil surface shaded with palm fronds (Snowdon, 1992). *S. rolfsii* has been found on both *D. alata* and *D. rotundata* tubers, but none of the 20 cultivars tested by Osai *et al.* (1996) were resistant to *S. rolfsii*. Ogunleye and Ayansola (2014) reported that infections caused by *S. rolfsii* during storage occurred in about 14% of the *D. alata* and *D. rotundata* tubers they tested. Field infections with *S. rolfsii* can also cause wilt diseases of yams, which may result in premature death of 45–70% of the crop (Nwankiti and Arene, 1978).

Rhizopus rot

Diseases of yam tubers caused by *Rhizopus oryzae* and *R. stolonifer* are sometimes also known as 'watery rot'. Infected tissue is mottled brown in colour and soft. In a humid atmosphere the cut surface is soon covered by a copious growth of coarse white mycelium. Ogunleye and Ayansola (2014) reported that infections caused by *R. stolonifer* during storage occurred in about 14% of the *D. alata* and *D. rotundata* tubers they tested. *R. stolonifer* has been reported by Adeniji (1970) and Noon and Colhoun (1979) to be highly pathogenic but Noon (1978) considered it also as a secondary invader. Control measures include careful handling in order to minimize injury to the tubers. Application of fungicide in hot water or wax has been shown to give partial protection against invasion (Snowdon, 1992).

Pythium rot

Zhang *et al.* (2018) reported that the stem base of *D. polystachya* seedlings showed water-soaked lesions, and their fibrous roots also showed hygrophanous lesions, which they identified as being caused by *Pythium ultimum* var. *ultimum*. Ogunleye and Ayansola (2014) reported that infections caused by *Pythium* spp. during storage occurred in about 4% of the *D. alata* and *D. rotundata* tubers they tested. *P. spinosum*was was shown to cause a brownish decay in which there was a clear distinction between healthy and rotted tissue (Raí *et al.*, 1986). Development of the fungus is rapid at temperatures between 25 and 30°C and, especially in humid conditions, a white mould forms on the surface. *P. sylvaticum* can also cause shallow, sunken, elliptical lesions, occasionally with lengthwise wrinkles. Rotted tissue was soft and brown (Harada *et al.*, 1982). Yams that have been wounded or whose cells have been weakened by chilling injury are predisposed to infection by *P. spinosum* and *P. sylvaticum* (Harada *et al.*, 1982; Raí *et al.*, 1986).

Control of Fungal Diseases

Amusa *et al.* (2003) described strategies for controlling the diseases that occur on growing yams in the field. These strategies include: the use of resistant cultivars; crop rotation; fallowing; planting of healthy material; and the destruction of infected crop residues. For storage diseases Amusa *et al.* recommended pre-treatment with chemical fungicides (e.g. thiabendazole, locally made dry gin or wood ash). Various methods that have been tested for disease control in yams are discussed below.

Traditional methods

Yam tubers are traditionally dusted with wood ash or sometimes lime (calcium hydroxide) directly after harvesting. These methods are used by small-scale farmers in Nigeria and other West African countries to reduce subsequent storage rots (Coursey, 1961). Similar treatments were also reported to be used by farmers in Jamaica (Thompson, 1972b). In Nigeria tubers were treated with either calcium oxide or local gin (fermented palm wine or sap from *Elaeis guineensis*) by Ogali *et al.* (1991) and stored at 25 to 32°C with 62 to 90% RH for 32 weeks. After 24 weeks the tubers that had been treated with gin showed no rotting, while 22.5% of the yams that had been treated with lime were rotten. After 32 weeks, the level of rotting was comparable in all treatments. Tubers of the *D. rotundata* cultivar 'Iyawo' inoculated by spraying with a conidiospore suspension of *Trichoderma viride* in potato dextrose broth showed a large reduction in the range and number of mycoflora, including the pathogens *Rhizopus* sp., *Rhizoctonia* sp., *Aspergillus niger*, *A. flavus*, *Penicillium oxalicum*, *Botryodiplodia theobromae*, *Fusarium oxysporum*, *F. solani*, *Neurospora* sp. and *Choanephora* sp. on the tuber surface during 5 months of storage in a traditional yam barn in Nigeria (Okigbo and Ikediugwu, 2001).

Chemical control

Use of chemical fungicides to control the postharvest rots of various *Dioscorea* species has previously been reported in numerous publications including Plumbley *et al.* (1985), Nnodu and Nwankiti (1986) and Snowdon (1992). Postharvest dipping in a benzimidazole fungicide effectively controlled storage rots (Thompson, 1972b), but its use is now restricted in many countries. In Jamaica benzimidazole fungicides were used for many years and were effective (Fig. 6.1), but during commercial application a rot caused by infection with *Penicillium sclerotigenum* was frequently observed on the treated tubers (Plumbley *et al.*, 1984, 1985). In *in vitro* tests this organism was found to be tolerant to benomyl, and highly susceptible to the fungicide Imazalil, which gave good control of the disease (Table 6.2). This tolerance or resistance commonly occurs with the benzimidazole group of fungicides on many crops and against many organisms, especially when frequently used. For example, commercial postharvest treatment of bananas to control crown rot alternates between thiabendazole and Imazalil to prevent the build-up of resistance. Also, Dichloran treatment was found to actually increase rotting of yams (Thompson *et al.*, 1977), probably due to synergy between the various organisms that were infecting the tubers (Fig. 6.2).

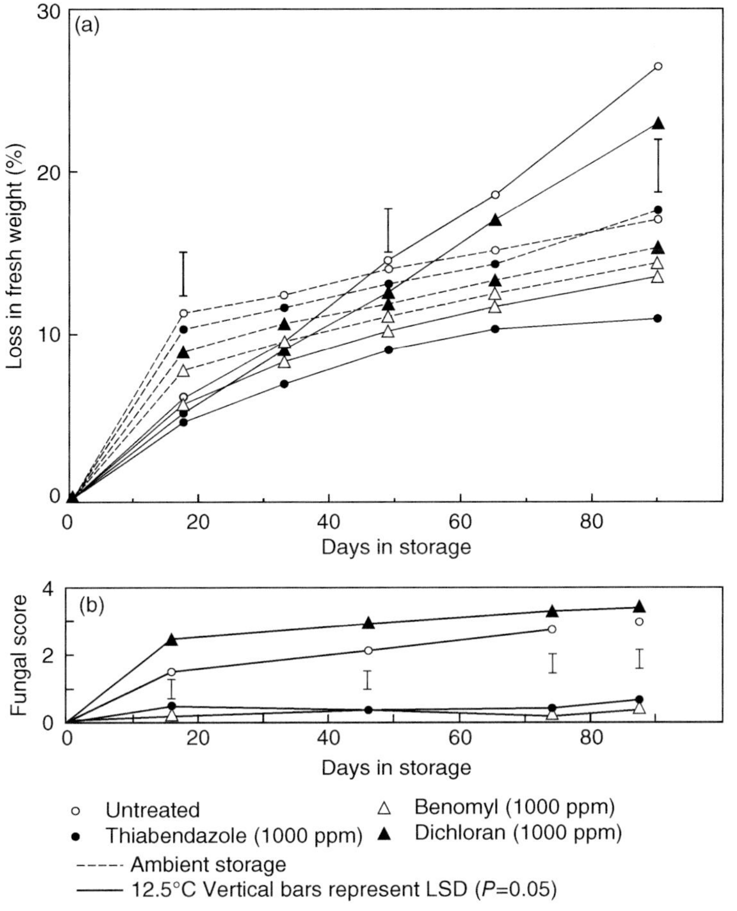

Fig. 6.1. Effects of postharvest treatment with fungicides on yams during storage at 12.5°C where: o = not treated, ● = 1,000 ppm thiabendazole, Δ = 1,000 ppm benomyl, ▲ = 1,000 ppm Dichloran. Vertical bars represent LSD ($p = 0.05$). Thompson *et al.* (1977).

Biological control

Biological control of diseases, using various organisms, has been tested on yams. Okigbo (2005) showed good or total control of disease when yam tubers were inoculated with the Gram-positive bacterium *Bacillus subtilis* (obtained from soil close to where yams were being grown) the day before they were inoculated with *Aspergillus niger*, *Botryodiploidia theobromae* or *Penicillium oxalicum*, giving good control of all three fungi. These were compared to those not inoculated or inoculated at the same time with *B. subtilis*, which all showed considerable rotting. Swain *et al.* (2008)

Table 6.2. Effects of benomyl and Imazalil on the *in vitro* growth in cm of a benomyl tolerant strain of *Penicillium sclerotigenum* on yam tubers (*D. cayenensis*) during storage at 20°C. Modified from Plumbley *et al.* (1984).

	Storage time in days				
Treatment	0	7	14	21	28
Control	0	1.10	2.20	3.19	3.66
Benomyl 500 µL L^{-1}	0	0.83	1.87	2.05	3.49
Benomyl 1000 µL L^{-1}	0	0.60	1.34	2.35	2.99
Imazalil 500 µL L^{-1}	0	0	0	0	0

isolated *B. subtilis* from cow dung microflora and tested it both *in vitro* and *in vivo* against *Fusarium oxysporum* and *Botryodiplodia theobromae*. They showed that in

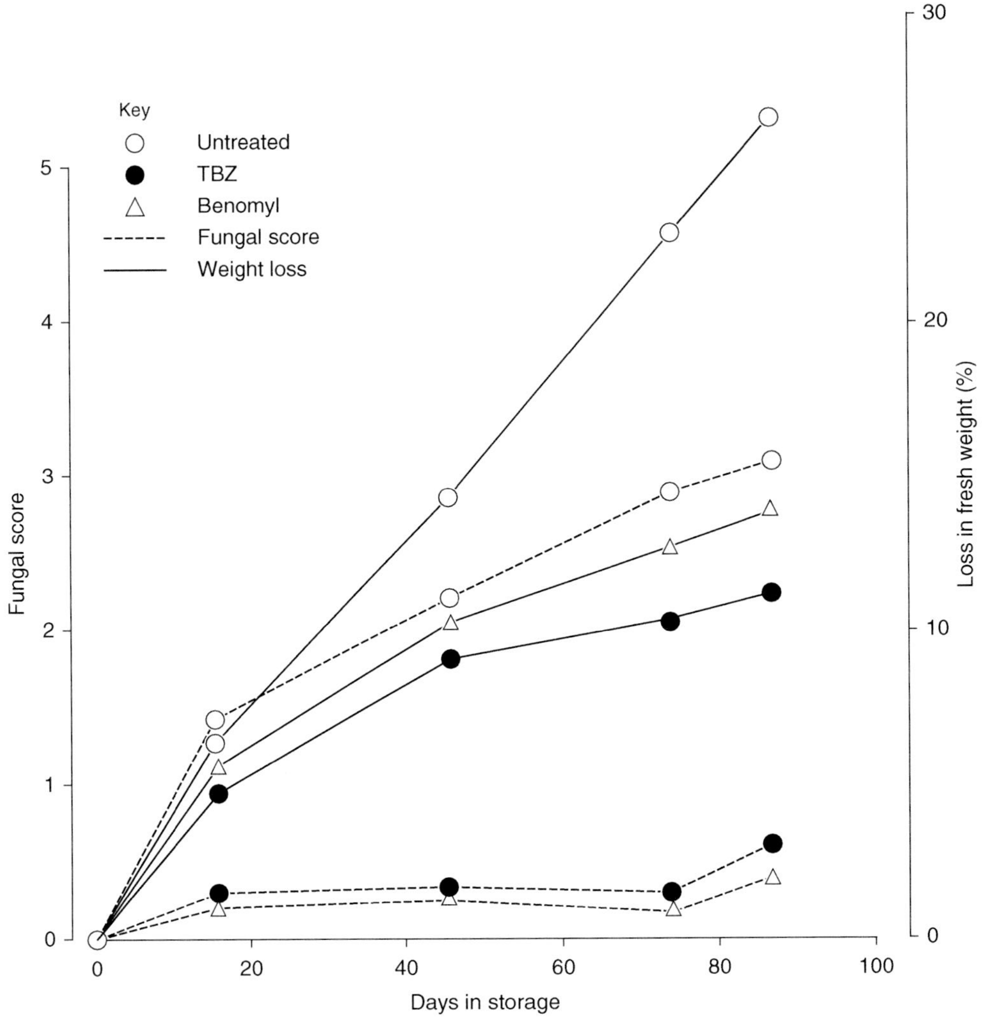

Fig. 6.2. Effects of fungicides on weight loss ——— and surface fungi ---------- (mean score where 0 = none and 5 = surface entirely covered) of yellow yam tubers (*D. cayenensis*) during storage in Jamaican ambient conditions where: ○ = not treated, ● = 1,000 ppm thiabendazole, Δ = 1,000 ppm benomyl. Thompson (1972a).

in vivo tests, *B. subtilis* inhibited their growth by up to 83% in wound cavities of yam tubers. From *in vitro* and *in vivo* screening tests for antagonism by isolates of *Trichoderma* against postharvest pathogens of yam tubers, Okigbo and Ikediugwu (2000) selected an isolate of *T. viride* as the most promising candidate for the biocontrol of postharvest rot of yams. Inoculation of *D. rotundata* tubers with conidia spores of *T. viride* and subsequent storage in a traditional yam barn resulted in a large reduction in the frequency of occurrence of the normal tuber surface microflora over a 4-month (December–April) storage period, while levels of *T. viride* on the tubers were maintained. Up to 52% rotting was found among groups of tubers that were artificially inoculated with *Aspergillus niger*, *Botryodiplodia theobromae* or *Penicillium oxalicum*. Also, in the group that was not inoculated with any pathogenic organism, but was inoculated with *T. viride*, rotting was either totally suppressed or only at a very low level.

Gwa and Abdulkadir (2017) tested *T. harzianum in vitro* against *Penicillium purpurogenum* isolated from rotted *D. rotundata* tubers. They reported that the mean percentage growth inhibition of the fungus increased up to 69% when *T. harzianum* was introduced 2 days before inoculation of *P. purpurogenum*, followed by introducing them both at the same time that gave about 39% inhibition of *P. purpurogenum*. Okigbo and Emeka (2010) tested *T. harzianum*, *Pseudomonas syringae* and *P. chlororaphis* for the control of postharvest diseases caused by *Botryodiplodia theobromae* and *Fusarium solani* on *D. alata* tubers. *T. harzianum* was the most effective in reducing the fungi by up to 85% for *B. theobromae* and 87% for *F. solani* infections.

Bacterial Diseases

Pre-harvest and postharvest rots of yam tubers can be due to infection by bacteria as well as by fungi. Bacteria have often been observed when soft rot of yam tubers develops into wet rot. Many bacterial rots are secondary infections after damage by primary fungal infections and Coursey (1967) advanced the argument that internal bacteria with latent infectibility could contribute to tuber rot. *Erwinia* sp. has been described as a destructive secondary invader in yam tubers. However, *Erwinia* species have also been widely reported to be primarily responsible for watery soft rot of yams (Burton, 1970; Noon and Colhoun, 1979). Noon (1978) also found a dry rot caused by *Corynebacterium* sp. Other bacteria associated with soft rots of yams include *Serratia* spp. (Okafor, 1966; Burton, 1970; Noon and Colhoun, 1979), *Pseudomonas* and *Proteus* spp., *Clostridium* sp., *Vibrio* sp. and *Bacillus tenus* (Obi, 1981). Lebot (2009) also reported that *Corynebacterium* sp. can cause a dry rot of yam tubers. *Serratia* species have been identified infecting tubers from Nigeria and Puerto Rico (Zohouri, 1998). In Nigeria a pigmented bacterium, tentatively identified as '*Serratici* sp.', was found to cause an extensive primary rot (Okafor, 1966). Bacteria can also cause damage to parts of yams other than the tubers. For example, Hseu *et al.* (2010) reported a bacterial leaf spot disease, caused by *Bacillus pumilus* on *Dioscorea* species, in central and northern Taiwan. The symptoms on the leaves were brownish, water-soaked lesions that later turned black with yellow halos.

Besides causing diseases certain bacteria have been applied to yams for their beneficial effects. For example, as reported above, *Trichoderma viride* has been identified for its antagonistic properties against postharvest pathogens of *Dioscorea* species during storage, especially *D. rotundata* on *Aspergillus niger*, *Botryodiplodia theobromae* and *Penicillium oxalicum* (Okigbo and Ikediugwu, 2008). Michereff *et al.* (1994) reported that *Bacillus subtilis* showed promise in controlling leaf spot on *D. cayenensis*, caused by *Curvularia eragrostidis* in Brazil and Okigbo (2002) reported that *Bacillus subtilis* greatly reduced rotting of *D. rotundata* tubers throughout a storage period of 5 months in Nigeria.

Virus Diseases

About 17 viruses have been identified in yams that are cultivated in different parts of the world for their edible and medicinal purposes. Two potyviruses, yam mosaic virus (YMV) and yam mild mosaic virus (YMMV) and several badnaviruses (YBVs) have been detected in yams in Cameroon, Côte d'Ivoire, Nigeria, Benin, Togo and Ghana. Potyviruses have been reported to be a major limiting factor for the successful cultivation of *D. rotundata*, *D. cayenensis* and *D. trifida* in certain parts of the world. Virus infections have been identified on *D. cayenensis-rotundata* complex, *D. dumetorum*, *D. bulbifera* and *D. esculenta*, but are most prevalent in *D. rotundata* and *D. alata* (Kenyon *et al.*, 2001). Bousalem and Loubet (2008) reported that virus prevalence data supported the presence of endogenous pararetroviral sequences in *D. rotundata*, but not in *D. abyssinica*, *D. alata*, *D. praehensilis* or *D. trifida*. They found five yam badnavirus species characterized by a wide host range that seemed to be of African origin. Seven other yam badnavirus species, with a limited host range, were probably of Asian-Pacific origin. It was concluded that the risks linked to yam plant introduction and exchanges were high. In West Africa, YMV is the most important in distribution, incidence and its effect on plant growth.

Yam plants infected with viruses show mosaics, distortions and, in severe cases, strap-like leaves, poor growth and smaller tubers than those from non-infected plants. Viruses infecting yams are spread by planting material (seed tubers) as well as aphids. The most important management techniques are those using new varieties from breeding programmes, controlling weeds and on-farm selection of healthy plants as seed tubers (Eni *et al.*, 2008). *In vitro* biotechnology has been tested and in Korea – Shin *et al.* (2013), working with *D. opposite*, reported up to 90% of the plants regenerated from cryopreserved shoot tips were YMV-free, whereas only 40% of those regenerated using meristem culture were YMV-free. Dallot *et al.* (2001) reported that YMMV is widespread on *D. alata* in Africa, the South Pacific and the Caribbean islands and was found on *D. alata* and *D. cayenensis-rotundata* in Colombia.

Phillips *et al.* (1999) identified a bacilliform virus from *D. alata*, designated *Dioscorea alata* bacilliform virus (DaBV), from Barbados and West Africa and from other *Dioscorea* species in West Africa, the Caribbean islands, Asia and South America that is transmitted by mealybugs as well as by mechanical transmission. Bömer *et al.* (2019) tested next-generation sequencing as a method of detecting viruses in tissue-cultured yams and detected both *Badnavirus* and *Potyvirus*, confirming its utility as a method for diagnosing yam viruses and contributing towards the safe distribution of germplasm.

Physiological Disorders

Internal brown spotting (IBS) is a physiological defect in which small brown spots, due to cell death, appear in the flesh of the tuber. Degras (1993) commented that IBS can have serious effects on the yam industry, particularly on the export trade, since the symptoms cannot be detected on the surface of the tuber and spotted tubers are therefore indistinguishable from spot-free ones, unless they are cut and examined. In the 1960s losses due to IBS reached up to 50% of yam exports from Barbados to the UK. IBS has been found only in *D. alata*. In Barbados, other species of *Dioscorea* were examined including *D. trifida*, *D. rotundata*, *D. cayenensis*, *D. esculenta* and *D. bulbifera* and they, and all of the 11 cultivars of *D. alata* ('Ashmore', 'Barbados', 'Bottleneck', 'Coconut Lisbon', 'Harper', 'Moonshine', 'Oriental', 'St Vincent Red', 'Seal Top', 'Smooth Statia' and 'White Lisbon') examined were found to be susceptible to IBS (de Courcey and Headley, 1967; Chandler *et al.*, 1983). In the Lesser Antilles IBS was associated with virus infection. The affected tubers developed hard brown nodules in the flesh, often surrounded by necrotic areas, and the foliage had (not always easily discernible) mosaic symptoms. It was also found that the yield of affected plants may be reduced by about half. No vector was identified but a meristem culture technique and the production of virus-tested yams was developed and implemented on a commercial scale in Barbados (Chandler *et al.*, 1983).

Chilling injury is another physiological disorder that can occur at storage temperatures of about 12°C or lower, depending on exposure time (Chapter 5) (Thompson, 1972a; Passam, 1982).

Insect Pests

Among insect, mite and vertebrate pests of yams, Reddy (2015) identified the following pests: scale insects *Aspidiella hartii*; aphids *Aphis gossypii*; coffee bean weevil *Araecerus fasciculatus*; yam beetle *Heteroligus meles*; mealybugs *Ferrisia virgata*, *Planococcus citri*, *P. dioscoreae*, *Pseudococcus citriculus*, *Rhizoecus* sp., *Geococcus coffeae*, *Phenacoccus gossypii*; defoliating caterpillars, *Loxura atymnus*, *Theretra nessus*, *Tagiades gana*; crickets *Gybnogryllus lucens*; yam weevil *Palaeopus costicollis*; yam moth, *Euzopherodes vapidella*, *Chrysomelids*, *Crioceris livida*, *Lema armata*, *Diaprepes abbreviates*, *D. famelicus*; mealworm *Tenebrio guineensis*, sawfly *Senoclidia purpurata*; termites *Coptotermes* sp., *Amitermes evuncifer*, *Protermes minutus*; and red spider mites *Tetranychus* spp.

Korada *et al.* (2010) reviewed a total of 73 insect species associated with *Dioscorea* species in different parts of the world. They reported that yams were infested by 48 species when the crop is growing in the field and by 27 in storage. The majority of these insects belong to the order Coleoptera (35 species) followed by the orders Hemiptera (15), Lepidoptera (13), Isoptera (5), Hymenoptera (2), Diptera (1) and Thysanoptera (1). Yam scales, mealybugs and a few beetles caused significant losses to tubers both in the field and in storage. The greater yam beetle (*Heteroligus meles*) attacks the tubers and can kill seed tubers after planting and also makes holes in tubers that affect their postharvest life and marketability. Yam scale (*Aspidiella hartii*) infests tubers in the field, which also affects them postharvest, and sometimes they attack foliage resulting in poor growth. Mealybugs (*Rastrococcus* spp.) may infest the tubers, especially during storage. Moth pests of yams include *Dasyses rugosella*, *Euzopherodes vapidella* and *Decadarchis minuscula*. *D. alata* was found to be the most susceptible species to moth infestation among those tested by Ashamo (2010). Among the pests of stored yams, the insects *Araecerus fasciculatus* and *Decadarchis minuscula* have been identified in Nigeria as infesting tubers of *D. dumetorum* (Plumbley and Rees, 1983). *Dinoderus porcellus* was found to be the most important yam storage pest in Benin (Vernier *et al.*, 1999, 2005).

As a method of counteracting pest losses in several countries, particularly in West Africa, yam tubers are dried and stored as chips in order to increase their availability throughout the year and reinforce food security (Akissoé *et al.*, 2001). However, yam chips can also be attacked by insects, which sometimes reduce them to powder in a few months. Loko *et al.* (2013), in a survey in 25 villages of northern Benin, found that damage of yam chips during storage was mainly due to insects (64% of responses) and they identified 12 insect pest species including *Dinoderus porcellus*, which was by far the most important species and the most widely distributed (97% of the samples).

Among traditional insect control methods for dried yam tubers, Loko *et al.* (2013) recommended regular inspection and exposure of chips to sunlight to repel insects, using insect-resistant varieties and insecticides and/or insect repulsive plants including the leaves of *Piliostigma thionnigii* and *Khaya ivorensis*. Heavy infestations of tubers with mealybug (*Rastrococcus* spp.) have been traditionally treated with horticultural oils or soapy solutions. Synthetic chemical insecticides have been used to control insect infestation in yams. For example, Tobih (2014) evaluated the application of the insecticide Chlorpyriphos E.C. at various stages of crop development on the control yam beetle (*Heteroligus* spp.). They found that yield increased from 13 to 61% and the best time for application was at tuberization.

Nematode Pests

Probably the most serious pests in yams are nematodes. Nematode infestation can be a major constraint on yam production (Bridge *et al.*, 2005; Kwoseh *et al.*, 2007). Several genera of nematodes are known to be pathogenic on yams including *Scutellonema*, *Pratylenchus*, *Meloidogyne*, *Helicotylenchus*, *Aphelenchus*, *Tylenchus*, *Hoplolaimus* and *Rotylenchulus*. From a survey of parasitic nematodes associated with yams in the Western State of Nigeria, Bridge (1973) found that 47% of yam tubers were infected with *Scutellonema bradys*, 9% with *Pratylenchus brachyurus*, 1% with *Rotylenchus reniformis* and 1% with *Meloidogyne incognita*. Nzeako *et al.* (2015) found five nematode genera in yams in the Rivers State in Nigeria: *Aphelenchus* (8.5%), *Practylenchus* (14.7%), *Meloidogyne* (22.7%), *Scutellonema* (53.9%) and

Tylenchus (0.24%). Of these, 83% of the nematodes were found in *D. rotundata*, 16% in *D. cayenensis*, 0.48% in *D. alata* and 0.24% in *D. dometurum*. Tubers collected from six states in Nigeria between 1973 and 1974 showed infestation with *Meloidogyne incognita* and *Meloidogyne javanica* being predominant. It was found that for *D. alata* infestation varied between10 and 90%, *D. rotundata* infestation varied between 3 and 70% and infestation was found to be zero for *D. cayenensis* (Adesiyan and Odihirin. 1978). It was also found that *Scutellonema bradys* was associated with infestations of *Meloidogyne* on *D. rotundata* in southeast Nigeria. Infestation with *Hoplolaimus*, *Pratylenchus* and *Scutellonema* species gave rise to dark necrotic lesions on tubers, which could also be seen as skin cracking with a corky appearance (Adesiyan *et al.*, 1975b). Quenerhervé (1998) reported that *Scutellonema bradys*, *Practylenchus coffeae* and *Meloidogyne incognita* were the most economically important. *Scutellonema bradys* has been recorded in West Africa, the Pacific islands and the West Indies. De Almeida *et al.* (2019) reported that *Scutellonema bradys*, *Pratylenchus coffeae* and *Pratylenchus brachyurus* were the prominent nematode species in Brazil.

Nematodes occur in the soil around yam plants, and Adesiyan and Badra (1982) found sizeable populations of nematodes in the soil at the beginning of the yam growing season. They also reported that in spite of them occurring in the soil, they were primarily distributed through yam planting material. De Almeida *et al.* (2019) evaluated weed species in yam fields and found 43 species of which 23 were infected with at least one of the three nematode species (*Scutellonema bradys*, *Pratylenchus coffeae* and *Pratylenchus brachyurus*) with *Pratylenchus coffeae* being the most prevalent in the fields. Nematodes are mainly found as a migratory endoparasite of roots and tubers. Bridge (1982) reported that nematodes primarily attack roots growing from the head of a tuber and can enter developing tubers through their growing points. Invasion of tubers can also occur through cracks or damaged areas in the outer cells (Bridge, 1973) causing necrosis (dry rot) on tubers. Bridge (1982) reported that the maximum number of nematodes recorded in tubers infected with *Scutellonema bradys* was 62,000 nematodes 10 g^{-1} of tuber in Nigeria. Coyne *et al.* (2016) extracted *Scutellonema bradys* from necrotic outer layers of *D. cayenensis* tubers in Tanzania while they were still growing in the field and showed variable results with some tubers having low densities and some with densities up to 240 g^{-1} of tissue. Atu *et al.* (1983) reported infestations in the field of *D. rotundata* with *Meloidogyne incognita* consisting of 1,250 eggs per plant that were contained largely in galls. Plants that were heavily galled had a reduced market value by about 40%. In Uganda, Mudiope *et al.* (2012) showed that in general, *Meloidogyne* species caused little galling to yam roots and tubers and limited reduction in their growth and yield. Of three yam species tested, *D. rotundata* was the most susceptible and most heavily affected by *Meloidogyne* species, *D. cayenensis* was intermediate and *D. alata* was the least affected.

Nath and Mukherjee (2000) evaluated extracts from *D. floribunda* tubers against *Meloidogyne incognita* both *in vitro* and *in vivo* and found that the extract had strong nematocidal properties. Labuschagne *et al.* (2006) investigated the influence of chemical fertilizer application on *S. bradys* densities and damage in the field and in storage to two *D. rotundata* cultivars, 'TDr 131' and 'Ala', and one *D. cayenensis* cultivar, 'Kokoro'. Exposure to nematode populations of 1,000 per plant to fertilizer levels of NPK at 300 kg ha^{-1}, diammonium phosphate at 155 kg ha^{-1} or potassium chloride at 60 kg ha^{-1} did not significantly affect yield and the number of tubers per plant, but the fertilizers, especially diammonium phosphate, suppressed nematode multiplication in tubers. Acosta and Ayala (1976) compared the effects of infestation with *Pratylenchus coffeae* and *Scutellonema bradys* alone or in combination on *D. rotundata* tubers. They reported that when the two species were inoculated together, there was a 53% suppression of top growth, but when plants were inoculated only with *Pratylenchus coffeae* there was a 29% suppression and when plants were inoculated only with *Scutellonema bradys* there was a 29% suppression. The reproduction of *Scutellonema bradys* was greatly inhibited when both nematode species were together on the same plant in comparison with that on plants inoculated with *Scutellonema bradys* alone. *Scutellonema bradys* apparently did not affect the reproduction of *Pratylenchus coffeae*.

Nematodes that have infested yam tubers in the field will, of course, be in the tubers at harvest. Field infestation of *D. rotundata* tubers with *Pratylenchus coffeae* was shown to increase when the tubers were stored in tropical ambient conditions in Jamaica (Fig. 6.3) resulting in areas of necrotic tissue. However, when the tubers were stored at 13°C there was no increase in nematode population and no increase in necrosis (Thompson *et al.* 1973b). In one experiment, Thompson *et al.*, (1973b) found levels of viable *Pratylenchus coffeae* remaining in *D. rotundata* tubers after 111 days storage at 22–27°C. These levels were on average 208 nematodes per gram of tissue, while in those stored at 12–13°C the level was 0.1 nematodes per gram of tissue. Labuschagne *et al.* (2006) found that nematode multiplication rates were greater in tubers during the first 3 months of storage under ambient conditions, but declined in the fourth and fifth month. Cadet and Quénéhervé (1994) found that multiplication in the population of nematodes in tubers coincides with the onset of plant dormancy and was greatest during storage.

Besides the dry rot and other physical damage to tubers caused by parasitic nematodes, it has been shown that infestation can increase tuber weight loss during storage as well as delaying tuber sprouting and facilitating invasion by fungi and other micro-organisms (Osagie, 1992). Labuschagne *et al.* (2006) also showed that the damage to yam tubers caused by nematodes included increased weight loss during storage (the temperature and humidity conditions were not specified, but were presumed to be

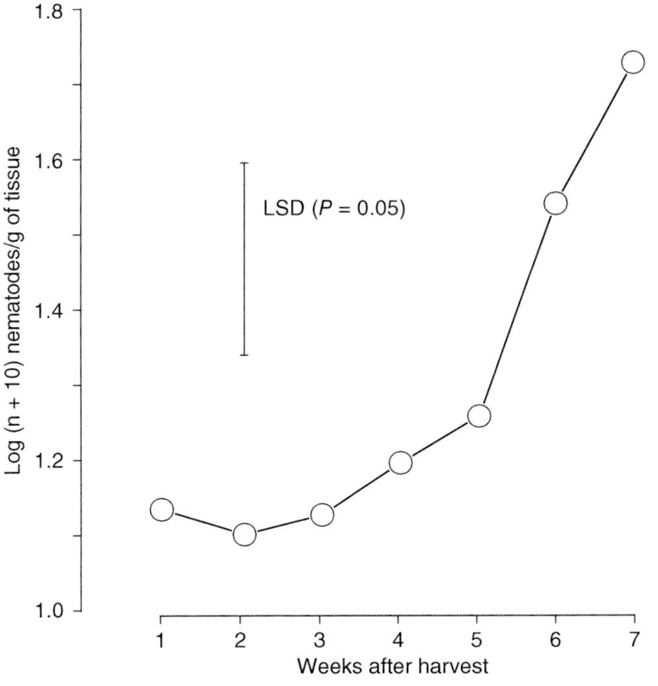

Fig. 6.3. Number of parasitic nematodes (*Pratylenchus coffeae*) extracted from *D. rotundata* tubers during storage in ambient conditions (about 25°C) in Jamaica. Thompson *et al.* (1973b).

Benin ambient), which was higher in tubers infested with *Scutellonema bradys* than tubers not infested. The carbohydrate constituents of *D. rotundata* tubers infected by *Scutellonema bradys* were compared with those of non-infected tubers by Adesiyan *et al.* (1975a). They found that the increase in monosaccharides recorded in the nematode infected yam tubers explained one of the ways by which the nematodes may predispose yams to infection by secondary invaders. This was supported by the findings of Ogundana *et al.* (1970) who showed that simple sugars have been shown to support the growth of fungi which invade the yam tissue. Acosta and Ayala (1976) found that tuber quality was reduced by 72% by *Pratylenchus coffeae* infestation, but only by 20% by *Scutellonema bradys*, and the two species together resulted in a reduction of 84% in 'storage-root quality'. *Scutellonema bradys* feeds intracellularly on the tuber tissues resulting in rupturing of cell walls, loss of cell content and formation of cavities. This damage is mainly confined to the sub-dermal, peridermal and underlying parenchymatous tissues in the outer 1–2 cm of tubers (Bridge, 1982). An increase of pectic enzymes has been detected in nematode infested tissue, which could be produced by the nematode and contribute to tissue damage (Adesiyan *et al.*, 1975b).

The life cycle of a plant parasitic nematode, for example *Scutellonema bradys*, was described by Bridge (1982). They lay eggs in roots and tubers or, less commonly, in the soil, where they hatch into juveniles. There are four juvenile stages and when each stage is completed they moult and finally reach the adult stage. The life cycle from egg to adult is completed in about 4 weeks depending mainly on temperature.

7

Production

Introduction

World production of yams for 2017 was estimated by the Food and Agriculture Organization (FAO) of the United Nations as over 73 million tonnes, produced from over 8.6 million hectares of land (FAO Stats, 2019). According to the FAO, in the year 2000, 53% of world yam production was in Africa, 29% in Asia, 17% in South America and 1% in the rest of the world with Nigeria accounting for some 70% of yam production in Africa (Table 7.1).

However, the crop is also of major importance in parts of East Africa, Asia, the Pacific area and the Caribbean islands. *D. rotundata* and *D. cayenensis* are the most commonly cultivated species and represent 95% of yam cultivation worldwide (Coursey, 1967; Kay, 1987; Nzeako *et al.*, 2015; FAO Stats, 2019).

According to Kumar *et al.* (2013) the following are countries where Dioscorea species are cultivated: Benin, Brazil, Cameroon, Central African Republic, Colombia, Costa Rica, Côte d'Ivoire, Cuba, Democratic Republic of Congo, Fiji, Guadeloupe, Haiti, Hawaii, India, Indonesia, Jamaica, Japan, Martinique, Mexico, Nigeria, Papua New Guinea, Philippines, Thailand, Trinidad and Tobago, Togo, USA, Venezuela and Vietnam. However, yams are grown in other countries but perhaps on a much smaller scale and the top twenty yam producing countries in the world were reported by Ike and Odjuvwederdie (2006) to be Nigeria, Ghana, Ivory Coast, Benin, Togo, Burkina Faso, Cameroon, Chad, Ethiopia, Central African Republic, Gabon, and Democratic Republic of Congo all in Africa, while Cuba, Colombia, Haiti, Brazil, Jamaica and Venezuela are in Latin America (Table 7.1).

Over 95% of African yam production lies in the 'yam belt', which is Nigeria, Benin, Togo, Ghana and Côte d'Ivoire. So revered is the yam crop that the novelist Chinua Achebe (1958) called yams 'the king of crops'. Many annual festivals are devoted to yams, especially in West Africa, for example by the Igbo people and many other tribes in Nigeria (Asiedu and Sartie, 2010). During these festivals, yams that have been prepared in several ways are the main food served. In Côte d'Ivoire, funerals are sometimes postponed until after the local yam festival has been observed. In the 18th and 19th centuries some slave ships carried large quantities of yams for the voyage from Africa to the New World because of their much longer storage life than most other fresh root crops (Achebe, 1958). Yams continue to be a major source of food in West Africa, with the average yam consumption

DOI: 10.1079/9781789249279.0007

Table 7.1. Production and yield of yams by country. Modified from FAO Stats (2019).

	Mean yield tonnes ha^{-1} $year^{-1}$		Total yield thousand tonnes $annum^{-1}$	
Country	1961	2017	1961	2017
American Samoa	3.0	3.6	<1.0	1.0
Antigua and Barbuda	2.5	3.4	<1.0	0.2
Barbados	12.4	11.3	12.6	0.4
Belize	–	5.3	–	<1.0
Benin	8.9	14.7	614.2	3,133.7
Brazil	8.9	9.7	85.0	250.4
Burkina Faso	2.6	7.0	22.5	46.7
Burundi	5.7	6.9	5.7	4.0
Cameroon	4.2	11.3	150.0	648.4
Cayman Islands		3.3	–	<1.0
Central African Republic	5.0	8.1	100.0	476.7
Chad	7.6	9.6	1,10.0	493.8
Colombia	7.9	9.9	1,18.0	422.1
Comoros		5.4	–	4.6
Costa Rica		15.0	–	23.6
Côte d'Ivoire	7.7	5.8	1,150.0	7,148.0
Cuba		5.7	–	52.8
Democratic Republic of the Congo	7.5	4.3	60.0	87.2
Dominica	10.0	11.5	2.0	14.7
Dominican Republic	8.3	6.6	20.7	32.9
Ethiopia	3.6	2.9	197.0	1,400.0
Fiji	8.0	6.0	0.8	4.4
Gabon	5.9	6.4	50.0	222.9
Ghana	7.3	17.1	1,000.0	7,952.8
Grenada	2.6	1.9	0.5	0.6
Guadeloupe	13.0	14.2	19.5	3.4
Guyana	-	7.6	–	2.0
Haiti	4.0	7.9	80.0	439.3
Jamaica	14.3	16.5	42.9	144.3
Japan	16.9	21.7	54.0	145.0
Liberia	9.6	8.7	13.0	21.6
Mali	3.5	2.4	7.0	116.1
Martinique	10.6	1.1	19.0	0.2
Mauritania	6.9	6.3	1.8	2.8
New Caledonia	8.1	0.9	8.5	2.4
Nicaragua	–	13.4	–	1.0
Nigeria	7.8	8.1	3,500.0	47,942.7
Niue	10.0	1.1	0.1	0.2
Panama	7.5	4.8	10.5	8.0
Papua New Guinea	16.6	17.6	120.0	362.7
Philippines	5.0	5.9	35.0	14.4
Portugal	7.9	17.6	1.5	2.4
Puerto Rico	5.3	8.9	13.2	1.1
Rwanda	10.0	8.3	1.0	50.0
Samoa	15.0	4.6	0.6	8.6
Sao Tome and Principe	5.0	5.8	0.5	2.3
Solomon Islands	7.1	10.5	12.0	45.0
South Sudan		2.5	–	58.6
St Kitts and Nevis	2.5	9.1	0.4	<1.0
St Lucia	3.3	3.6	2.0	0.5

Continued

Table 7.1. Continued.

	Mean yield tonnes ha^{-1} $year^{-1}$		Total yield thousand tonnes $annum^{-1}$	
Country	1961	2017	1961	2017
St Vincent and the Grenadines	10.0	12.9	2.0	2.4
Sudan	3.0	21.1	120.0	181.9
Tanzania	7.1	6.4	5.0	11.0
Togo	10.0	9.1	300.0	826.6
Tonga	12.2	13.6	28.0	4.7
Trinidad and Tobago	15.7	3.6	1.9	<1.0
Venezuela	5.5	9.7	45.5	48.4
Wallis and Futuna Islands	11.3	9.8	0.5	0.5

per capita per day being highest in Benin (364 kcal), followed by Côte d'Ivoire (342 kcal), Ghana (296 kcal) and Nigeria (258 kcal) (Bassey, 2017).

The mean yield of yams worldwide has fluctuated over the last 50 years or so but was not much higher in 2017 than it was in 1961 (Table 7.2). In contrast to yams, three other major root crops, cassava, sweet potatoes and potatoes, have shown general progressive increases in mean yield between 1961 and 2017 (Tables 7.2, 7.3). The reasons for these differences are not clear, since considerable work has been done in identifying the factors that limit yields in all crop plants and all these three root crops suffer similar challenges in pest and disease control. Also, all three crops are produced at both the small-scale farmer level and at a large-scale production level, but perhaps potatoes are more commonly produced in the latter and yams in the former. Also, cassava is commonly produced in many countries on a large scale to meet processing needs. Total world production could also perhaps influence the priority given for research, since total production of potatoes was 388 million tonnes, cassava 292 million tonnes and yams only 73 million tonnes in 2017 (FAO Stats, 2019).

Table 7.2. World production of yams. FAO Stats (2019).

Year	World mean yield tonnes ha^{-1} $year^{-1}$	World area harvested million hectares	World total production million tonnes $annum^{-1}$
1961	7.23	1.15	8.32
1971	7.46	2.05	15.28
1981	6.69	1.80	12.02
1991	9.91	2.75	27.23
2001	9.74	4.11	40.08
2011	8.13	6.53	53.12
2017	8.53	8.56	73.02
2018	8.69	8.35	72.58

Table 7.3. World mean yield of selected root crops tonnes ha^{-1} $year^{-1}$. Modified from FAO Stats (2019).

Year	Potatoes	Cassava	Sweet potatoes	Yams
1961	12.2	7.4	7.3	7.2
1971	13.7	8.3	11.1	7.5
1981	14.3	9.2	12.4	6.7
1991	14.4	9.8	13.9	9.9
2001	15.7	10.6	14.3	9.7
2011	19.3	12.2	13.0	8.1
2017	20.1	11.1	12.3	8.5

Some Yam Producing Countries

Much of this section is based on data extracted from the United Nations FAO Stats (2020). Some of this data is labelled 'FAO data based on imputation methodology' while other data is labelled 'official data' or 'calculated data'. There can be confusion about what exactly should be included under 'yams'. For example, sweet potatoes (called 'yams' in parts of the USA) and taro (called 'cocoyams' in parts of West Africa) have separate listings. However, there may be some anomalies. For example, for Nicaragua, FAO lists yam production, over the last 40 to 50 years, as about 1,000 tonnes annum^{-1} with a mean yield of 13.1–13.4 tonnes ha^{-1} but this is probably not entirely *Dioscorea* spp., but is likely, at least in part, to be the elephant foot yam (*Amorphophallus paeoniifolius*) or even 'new' cocoyam (*Xanthosoma* spp.), which are both grown in Nicaragua. Taro – also called 'old' cocoyam (*Colocasia esculenta*) – production is listed separately and for Nicaragua annual production was 12,928 tonnes with a mean yield of 9.7 tonnes ha^{-1} in 2018.

American Samoa

American Samoa is a group of small Pacific islands close to Samoa. American Samoa consists of one main island with a few smaller islands and is an overseas territory of the USA. Taro appears to be a more important root crop and both production and yield of yams were reported to be small (Table 7.4). *D. bulbifera*, *D. pentaphylla*, *D. alata* and *D. nummularia* are listed as growing in the National Park of American Samoa, with *D. pentaphylla* listed as uncommon.

Antigua and Barbuda

Grant (1996) reported that crop production was considered a lucrative business in Antigua and Barbuda with most of the food crops consumed locally with minor exports of yams and sweet potatoes to the UK. Working germplasm collections of yams existed at one time, but no resources were available to retain collections. Both production and yield of yams in Antigua and Barbuda continue to be low (Table 7.5).

Barbados

Chandler *et al.* (1983) reported that the *Dioscorea* species grown in Barbados included 11 *D. alata* cultivars ('Ashmore', 'Barbados', 'Bottleneck', 'Coconut Lisbon', 'Harper', 'Moonshine', 'Oriental', 'St Vincent Red', 'Seal Top', 'Smooth Statia' and 'White Lisbon'), *D. trifida*, *D. rotundata*, *D. cayenensis*, *D. esculenta* and *D. bulbifera*. Kay (1987) also reported that in Barbados *D. alata* was cultivated, without staking, on a plantation scale. In Barbados yams are traditionally grown by small-scale farmers, with *D. alata* widely grown for local consumption. Yams are also important in the export trade from Barbados. However,

Table 7.4. Yam production (tonnes annum^{-1}) and mean yield (tonnes ha^{-1}) in American Samoa. Modified from FAO Stats (2020).

Year	Total production in tonnes	Mean yield in tonnes ha^{-1}
1961	60	3.0
1966	90	3.2
1971	54	2.7
1976	40	2.7
1981	120	3.0
1986	150	2.0
1991	134	4.6
1996	129	4.4
2001	913	3.5
2006	855	3.5
2011	919	3.5
2016	1,065	3.6
2017	1,083	3.6
2018	1,102	3.5

Table 7.5. Yam production (tonnes annum^{-1}) and mean yield (tonnes ha^{-1}) in Antigua and Barbuda. Modified from FAO Stats (2020).

Year	Total production in tonnes	Mean yield in tonnes ha^{-1}
1961	20	2.5
1966	50	3.3
1971	280	4.5
1976	57	4.8
1981	48	3.7
1986	30	3.8
1991	50	3.7
1996	50	3.7
2001	55	3.7
2006	85	3.5
2011	116	3.3
2016	149	3.2
2017	142	3.2
2018	145	3.2

Chandler *et al.* (1983) reported that being part of the food production effort during World War II, it was obligatory on all sugarcane plantations in Barbados to devote 35% of their arable land to food crops. After the war, this was reduced to 12% and later to 5%. As a result, the production of yams extended to large sugar estates from that time. With few exceptions, yam production appears profitable but margins were reported to be low on both small farms and large estates. Planting and harvesting of yams is normally by hand, but a mechanical planter has been used, especially for yams grown on a plantation scale (Jeffers, 1977). Yams grow well as a rotation crop with sugar cane since the land has residual nitrogen and potassium from the previous cane crop, but yields have been substantially increased by the additional application of 225 kg ha^{-1} of a 9:10:23 NPK fertilizer applied 2–3 months after growth commenced. However, yam production in Barbados has progressively declined in the past few years (Table 7.6).

Belize

Yams are not an important crop in Belize, which has very low production level, and mean yield was shown to vary considerably (Table 7.7). *D. alata* is an introduced species and *D. cayenensis* has an unconfirmed presence in Belize (CABI, 2019). Govaerts (2012) reported that *D. bulbifera* was grown in Belize and also that *D. bernoulliana* is native to Belize, Guatemala, Honduras and Mexico (EOL, 2019). A steroid-yielding wild yam from Belize was shown to be *D. belizensis* by Blunden *et al.* (1963). *D. belizensis* is locally called 'cocolmeca' or 'barba del viejo' and is used as an aphrodisiac (Arvigo and Balick, 1994; Rätsch and Müller-Ebeling, 2003).

Table 7.6. Yam production (tonnes $annum^{-1}$) and mean yield (tonnes ha^{-1}) in Barbados. Modified from FAO Stats (2020).

Year	Total production in tonnes	Mean yield in tonnes ha^{-1}
1961	12,600	12.4
1966	8,680	11.5
1971	14,349	12.5
1976	5,782	12.5
1981	4,000	18.1
1986	1,989	9.0
1991	1,989	9.0
1996	1,500	9.4
2001	1,200	13.3
2006	794	13.2
2011	243	12.0
2016	378	11.4
2017	374	11.3
2018	345	11.5

Benin

Cassava and yams are by far the most cultivated root crops in Benin. Yams are usually traded fresh and used by households to make pounded yam dishes. Some varieties are processed into chips that are used for milling into flour for the preparation of 'amala' or 'wassa-wassa' and other yam products (FAO Benin, 2019). Dumont and Vernier (2000) reported that in Benin the Bariba ethnic group has two systems for the domestication of yams. The first is that wild tubers are collected in the bush, locally called 'dika', and the second is a clonal selection from improved tubers from among those collected from the bush, which generally produces varieties very similar to those already used. This collection system can also produce new varieties and IITA are working on genetic improvement of *D. rotundata* and *D. alata* at several centres in Benin including Dassa.

Table 7.7. Yam production (tonnes $annum^{-1}$) and mean yield (tonnes ha^{-1}) in Belize. Modified from FAO Stats (2020).

Year	Total production in tonnes	Mean yield in tonnes ha^{-1}
1991	83	1.7
1996	48	12.0
2001	13	1.9
2006	97	10.8
2011	95	10.6
2016	29	9.7
2017	37	5.3
2018	37	5.3

Dansi *et al.* (1997, 1999) collected and characterized 560 accessions of cultivated *D. cayenensis-D. rotundata* complex throughout Benin. Ninety morphotypes were identified and further classified into 26 cultivar groups based on their morphological similarities. They also conducted a comparative study between the cultivated forms and their wild relatives for *D. abyssinica*, *D. praehensilis* and *D. burkilliana* in order to establish linkages between the identified morphotypes and wild species. Paterne *et al.* (2019) rejected the previous study by Bilgili (2013), who found that there would be a strong degradation of cultivar diversity in Benin by 2017–2018 with probably severe consequences for food security. Paterne *et al.* recorded 640 *Dioscorea* cultivars (subject to synonymys). They found that out of the existing diversity, only ten cultivars, which were morphologically distinct, were perceived as tolerant to nematode infestation and were absent in the markets surveyed as they were considered to have no market value. The 'Florido' cultivar of *D. alata* was aggressively promoted in Nigeria, Benin and Togo by different research and development agencies. The Sudan United Mission in Abakaliki multiplied and distributed 'Florido' in eastern Nigeria. The IITA, in collaboration with the NRCRI, multiplied and distributed 'Florido' in the southern geopolitical zone of Nigeria through the Green River Projects, and Dr Philip Vernier distributed truckloads of 'Florido' in Benin and Togo under the yam valorization project (Nweke, 2016). In Benin, tuber yield of yams varied between 13 and 15 tonnes ha^{-1} over 2010 to 2014. Production in 2016 was reported to be 3,041,245 tonnes. Previous production and yield of yams in Benin are given in Table 7.8.

Table 7.8. Yam production (tonnes annum^{-1}) and mean yield (tonnes ha^{-1}) in Benin. Modified from FAO Stats (2019).

Year	Total production in tonnes	Mean yield in tonnes ha-1
1961	614,172	8.9
1966	454,200	8.4
1971	527,500	10.7
1976	598,800	9.5
1981	665,941	8.4
1986	883,931	11.0
1991	1,177,541	11.4
1996	1,346,070	10.3
2001	1,700,982	10.9
2006	1,577,008	10.9
2011	2,734,861	15.0
2016	3,041,245	15.8
2017	3,133,367	15.8
2018	2,944,944	13.6

Brazil

There are many *Dioscorea* species native to Brazil and some of these are considered endangered. For example, Kirizawa *et al.* (2016) identified *D. altissima*, *D. bulbotricha*, *D. marginata*, *D. monadelpha*, *D. sanpaulensis*, *D. tauriglossum*, *D. torticaulis* and *D. trilinguis* as growing in the Reserva Biológica do Alto da Serra de Paranapiacaba, Santo André. Among them *D. sanpaulensis* and *D. trilinguis* were already listed as endangered, but they recommended that *D torticaulis* should also be listed. Yams are grown in the agricultural areas of the Vale do Ribeira, where four different yam species were collected and identified, of which the most abundant was *D. trifida* followed by *D. cayenensis*, *D. bulbifera* and *D. alata*. Farmers reported that other yam species had been cultivated in the past, but were no longer planted for various socioeconomic reasons including '(a) lack of secure land tenure; (b) environmental laws that have restricted farmers' ability to open new areas of the forest for farming; (c) the rural exodus or the myth of modernization of the urban centres and (d) the shortage of family labour' (Bressan *et al.*, 2007). *D. alata* is widely distributed in Brazil and is mostly cultivated in association with other crops in holdings of different sizes. Veasey *et al.* (2012) reported that slash and burn practices, although mostly disappearing, were still being used by many farmers for yam cultivation.

Monteiro and Peressin (2002) commented that the morphology of yam tubers showed considerable diversity in Brazil and the relationship between accessions and the local and commercial genotypes was difficult to determine. However, there was high intraspecific diversity in local accessions from different regions, with the highest for southeastern Brazil. Some *D. alata* genotypes derive from the cultivar 'Florida' that was originally introduced from Puerto Rico in the 1950s and was popular with farmers due to its uniform tuber shape and easy management. However, the culinary qualities of 'Florida' were less appreciated than the cultivar 'Mimoso', which had been extensively grown in the past but increasingly less so, mainly due to its susceptibility to anthracnose. Production and value of yams in Brazil are given in Table 7.9, which shows considerable increases in production while yield has remained relatively constant.

Siqueira (2011) reported a strong African influence on cultures in northeastern Brazil where some communities still maintain their traditional diet and use yams in local dishes. He reported that in recent years the media in Brazil explored several issues related to health and nutrition, which may stimulate yam consumption as an alternative starch source to traditional crops.

Burkina Faso

Yams are a major crop in Burkina Faso with *D. rotundata* being the main species grown.

Table 7.9. Yam production (tonnes annum^{-1}) and mean yield (tonnes ha^{-1}) in Brazil. Modified from FAO Stats (2020).

Year	Total production in tonnes	Mean yield in tonnes ha^{-1}
1961	85,000	8.9
1966	105,000	9.1
1971	128,000	9.0
1976	154,000	9.1
1981	184,000	9.0
1986	200,000	9.1
1991	215,000	9.1
1996	225,000	9.2
2001	235,000	9.0
2006	240,000	9.2
2011	246,651	9.6
2016	250,370	9.7
2017	250,400	9.7
2018	251,489	9.8

The local common name of yam is 'Nyu', and in the province of Passoré in Burkina Faso yams are grown under semi-arid conditions (700 mm rainfall year^{-1}) on hydromorphic soils, in rotation with other staple crops, using both organic and mineral fertilizers (Tiama *et al.*, 2016). Yields of yams in Côte d'Ivoire were reported to be considerably higher than in Burkina Faso. In managed trials, reported by Frossard (2018), tuber yields of *D. rotundata* were up to 20 tonnes ha^{-1} in Côte d'Ivoire but only 12 tonnes ha^{-1} in Burkina Faso. For *D. alata* tuber yields were up to 35 tonnes ha^{-1} in Côte d'Ivoire and only 5 tonnes ha^{-1} in Burkina Faso. It was concluded that in Burkina Faso low yields were due to difficult soil and climate conditions and the higher yields in Côte d'Ivoire were due to the use of 'clean seeds' of improved cultivars, nutrient inputs and more regular planting density.

Nweke (2016) reported that in several West African countries, including Burkina Faso, deficits were generated in yam production. In order to compensate for such deficits, yams are imported from Nigeria and Ghana. Nweke commented that in a survey in 2012, large quantities of yams were on display for sale in a market in Burkina Faso (mostly *D. rotundata*) and the sellers reported that the yams were all imported from Ghana and that yams from anywhere else faced low consumer acceptance. Projected estimates by Nweke predicted that aggregate yam consumption in Burkina Faso will increase to 70,000 tonnes by 2025 due to high population growth and increased yam consumption. Production and mean yield of yams in Burkina Faso are given in Table 7.10 showing considerable variation in both production and mean yield.

Burundi

Burundi is a small landlocked country in which 36% of the land is used for arable cultivation and about 90% of the population depend on agriculture for a living. Agriculture consists of subsistence farming, with only about 15% of the total production

Table 7.10. Yam production (tonnes annum^{-1}) and mean yield (tonnes ha^{-1}) in Burkina Faso. Modified from FAO Stats (2020).

Year	Total production (tonnes)	Mean yield in tonnes ha^{-1}
1961	22,530	2.5
1966	25,000	4.8
1971	40,000	6.7
1976	50,000	10.0
1981	77,420	9.8
1986	112,065	8.9
1991	36,500	6.1
1996	49,298	7.4
2001	70,669	6.9
2006	22,157	8.5
2011	99,730	7.5
2016	47,662	6.0
2017	46,765	7.0
2018	35,909	6.4

Table 7.11. Yam production (tonnes annum^{-1}) and mean yield (tonnes ha^{-1}) in Burundi. Modified from FAO Stats (2020).

Year	Total production (tonnes)	Mean yield in tonnes ha^{-1}
1961	5,700	5.7
1966	5,000	5.7
1971	5,900	5.4
1976	6,000	5.5
1981	6,000	5.5
1986	7,000	5.8
1991	7,800	5.6
1996	7,909	5.6
2001	9,924	5.8
2006	9,912	5.6
2011	9,898	6.9
2016	3,498	6.9
2017	4,041	6.9
2018	6,925	8.6

marketed. The main food crops are legumes, green bananas, cassava, sweet potatoes, potatoes, cereals and yams. *D. alata* is cultivated in Burundi, but also *D. quartiniana*, *D. cochleariapiculata* and *D. baya* are found in several African countries including Burundi (Contu, 2013; CJBG, 2018; GBIF, 2019). Yam production and yield in Burundi are given in Table 7.11.

Cameroon

Yam, cassava, plantains and bananas are the main food crops in Cameroon. Cameroon has the highest diversity of rainforest wild yam species in Africa. Dumont *et al.* (1994) reported that there were17 *Dioscorea* spp. specific to Cameroon. However, Ngo-Ngwe *et al.* (2014) reported that the main *Dioscorea* species in Cameroon were *D. alata*, *D. cayenensis/rotundata* complex and *D. dumetorum* and Agbo-Egbe and Trèche (1995) reported that the species most extensively grown and consumed in Cameroon were *D. alata*, *D. dumetorum*, *D. schimperiana*, *D. esculenta*, *D. bulbifera*, *D. cayenensis/rotundata* complex and *D. liebrechtsiana*. Oben *et al.* (2016) collected yams from nine sites in the southwestern region of Cameroon, which is a major yam producing zone, and found they were mainly the *D. rotundata* cultivars 'Ikom', 'Calabar' 'Agar', 'Igbo' and 'White'. Acquah and Evange (1994) studied the traditional yam marketing systems in Fako in the southwestern region of Cameroon and found some 97% of yams were transported by head porterage from the farm to the house and 85% by hired vehicles from the house to the markets. They also found that the average yam farm size was 0.43 ha with a range of 1 to 5 ha fragmented fields. Yam production in Fako was economically viable but required 'a huge amount of initial up-front capital', which very few of the small farmers could afford. Production and yield of yams in Cameroon are given in Table 7.12. Both show substantial increases.

Cayman Islands

The Caribbean Agricultural Research and Development Institute's (CARDI) Strategic Plan for 2018–2022 centres on improving production and productivity of cassava and sweet potatoes in the Cayman Islands, but yams are not mentioned. Production and yield of yams in the Cayman Islands are both very low (Table 7.13).

Central African Republic

The main crops grown in the Central African Republic are cassava, yams, millet, corn and bananas. Several *Dioscorea* species are native or found growing wild, including *D. abyssinica*, *D. baya*, *D. mangenotiana*, *D. dumetorum*, *D. semperflorens*, *D. bulbifera* and *D. steriscus*. Hladik *et al.* (1984), in their study of the density of wild edible yams in African forests, found that in the Central African Republic there were between 16 and 100 stems (presumably plants) of yams per hectare compared to 17 to 50 in Gabon and 10 to 100 in Cameroon. Yams harvested from the wild formed part of the diet of local tribes. Bahuchet (1988), working with Aka families in the Central African Republic, found no seasonal variation throughout the year in the frequency of wild yam tubers in their diet, although they were used in only 4% of their meals. Yam yields were generally very low with some indication that they may be improving (Table 7.14).

Table 7.12. Yam production (tonnes annum^{-1}) and mean yield (tonnes ha^{-1}) of yams in Cameroon. Modified from FAO Stats (2020).

Year	Total production (tonnes)	Mean yield in tonnes ha^{-1}
1961	150,000	4.2
1966	161,914	4.2
1971	268,060	4.4
1976	250,000	4.5
1981	210,000	3.8
1986	134,306	4.1
1991	120,000	5.0
1996	120,000	7.5
2001	264,166	9.4
2006	366,808	10.1
2011	517,069	10.1
2016	624,881	11.1
2017	648,407	11.3
2018	674,776	11.5

Table 7.13. Yam production (tonnes annum^{-1}) and mean yield (tonnes ha^{-1}) in the Cayman Islands. Modified from FAO Stats (2020).

Year	Total production in tonnes	Mean yield in tonnes ha^{-1}
1966	8	4.0
1971	10	5.0
1976	13	4.3
1981	10	3.3
1986	15	5.0
1991	25	4.1
1996	11	3.7
2001	19	3.5
2006	20	3.4
2011	29	3.4
2016	29	3.4
2017	29	3.4
2018	28	3.4

Chad

Chad's main crops were reported to be cereals (millet, sorghum, corn, rice and wheat), legumes (black-eyed peas, beans, peanuts and sesame), cassava, yams, cotton, sugar cane and tobacco (UNDP, 2012). The most cultivated species of yam in Chad are *D. rotundata*, *D. cayenensis* and *D. alata*. Also, wild yams are sometimes harvested in times of food shortage (Bergh *et al.*, 2012). There is some indication that yam production and yield in Chad are increasing, especially production (Table 7.15).

China

Coursey (1967) quoted references that reported that *Dioscorea* species were used as

Table 7.14. Yam production (tonnes annum^{-1}) and mean yield (tonnes ha^{-1}) in the Central African Republic. Modified from FAO Stats (2020).

Year	Total production (tonnes)	Mean yield in tonnes ha^{-1}
1961	100,000	5.0
1966	140,000	5.4
1971	125,000	5.0
1976	170,000	6.8
1981	160,000	6.4
1986	180,000	6.0
1991	244,437	6.7
1996	350,000	6.6
2001	366,841	6.8
2006	370,000	6.9
2011	444,918	7.8
2016	476,674	8.1
2017	476,651	8.1
2018	513,489	8.4

Table 7.15. Yam production (tonnes annum^{-1}) and mean yield (tonnes ha^{-1}) in Chad. Modified from FAO Stats (2020).

Production year	Total production (tonnes)	Mean yield in tonnes ha^{-1}
1961	110,000	7.5
1966	117,000	7.8
1976	135,000	7.6
1986	200,000	7.7
1996	237,938	8.3
2001	241,764	9.1
2006	275,000	9.5
2011	400,000	9.5
2016	502,314	9.6
2017	493,841	9.6
2018	484,700	9.6

early as 2,000 BCE for medical purposes in China and that *D. esculenta* was being used as a food in China in the 3rd century CE. Simoons (1990) reported that *D. alata*, *D. opposita* and *D. japonica* were widely grown in China. Other *Dioscorea* species native to China or that have been reported growing there include: *D. arachidna*, *D. batatas*, *D. cirrhosa*, *D. collettii*, *D. divaricate*, *D. decipiens*, *D. fasciculata*, *Dioscores fordii*, *D. garrettii*, *D. hispida*, *D. kamoonensis*, *D. menglaensis*, *D. nipponica*, *D. oppositifolia*, *D. quinqueloba*, *D. scortechinii*, *D. septemloba*, *D. tokoro*, *D. wallichii* and *D. zingiberensis*. However, the production of yams in China is not included in the FAO statistics although it has been estimated that China produced 5–6 million tonnes of yams annually (Xu and Xu, 1997). Yams are consumed widely as fresh vegetables (including in desserts and salads) and are also processed into flour, flakes, chips and dry-roasted slices in China's food industries. Yam tubers are also used in traditional Chinese medicine (Birkill, 1935; Yang and Chen, 1983; Wanasundera and Ravindran, 1994; Tu, 2002). *D. zingiberensis* is the dominant source for diosgenin production in China (Huang *et al.*, 2008b). In traditional Chinese and Western medicine, several species of yams are major sources of steroid precursors including *D. nipponica* as well as *D. zingiberensis*. Fresh Plaza (2017) reported that in China the main production areas for yam are the Henan, Hebei and Shandong Provinces. It was also reported that in 2017 yam production was down by 20–30% compared with previous years, but wholesale prices increased by about 30%. It was claimed that China was actively seeking foreign wholesalers to co-operate in marketing yams in South East Asia or Europe, although it was also reported that 'dry yam' imports from Nigeria into China were increasing. Fresh Plaza (2018) quotes Yao Aihua of the Cheng Lin Fruit Professional Cooperative:

> Our yams are from Xinji City, Shijiazhuang, Hebei Province, one of the main producing areas of Chinese yams. The main product is white yam. The products are mainly sold to the wholesale markets of first-tier and second-tier cities in China. Thanks to the limited production and use of white yam, in recent years, the demand in the domestic market has increased year by year. I am very optimistic about the future market prospects.

A considerable amount of research on yams is constantly being produced from Chinese institutions, for example Sun *et al.* (2012) developed a DNA barcoding system to identify *Dioscorea* species.

Colombia

Yam production in Colombia is mainly in the northern coastal region, but production there was badly affected by an epidemic of anthracnose in 1989. *D. alata*, *D. cayenensis* and *D. rotundata* were reported to be the most important yam species economically in Colombia (Montes *et al.*, 2009). Pinzón-Rico and Raz (2017) characterized yams in the Bogotá market over the period 2010 to 2013 and identified wild-harvested tubers of four native species, *D. coriacea*, *D. lehmannii*, *D. meridensis* and *D. polygonoides*, with *D. coriacea* being the most prevalent. Yam is not considered to be a high-priority crop so appropriate technological development in production and postharvest are limited. Also, losses during cultivation and harvesting are high and were reported to be between 20 and 30% postharvest because the methods and infrastructure used are rudimentary (Guzmán and Buitrago, 2000). Niño *et al.* (2006) evaluated intra-specific variability and relationships among accessions of *D. polygonoides* from the Gran Caldas region of Colombia. They grouped the accessions into different regions and found that similar groupings reflected the geographical origins of the accessions.

Production of yams appears to have increased in the last few years (Table 7.16) and exports have increased from zero in 2011 to 3,347 tonnes in 2016 when 78% of exports were to the USA and the balance (22%) to Puerto Rico.

Table 7.16. Yam production (tonnes annum^{-1}) and mean yield (tonnes ha^{-1}) in Colombia. Modified from FAO Stats (2020).

Year	Total production (tonnes)	Mean yield in tonnes ha^{-1}
1961	118,000	7.9
1966	139,000	9.3
1971	63,100	9.4
1976	105,300	10.2
1981	119,300	9.2
1986	67,300	7.0
1991	51,944	8.1
1996	205,607	11.0
2001	255,458	11.7
2006	260,685	10.6
2011	396,613	11.1
2016	407,817	9.8
2017	422,063	9.9
2018	419,267	10.5

Comoros

Production of yams in Comoros (Table 7.17) is mainly *D. alata*, which was introduced through Austronesian migrations (Blench, 2007).

Table 7.17. Yam production (tonnes annum^{-1}) and mean yield (tonnes ha^{-1}) in Comoros. Modified from FAO Stats (2020).

Year	Total production in tonnes	Mean yield in tonnes ha^{-1}
1991	3,130	5.1
1996	3,500	5.2
2001	3,813	5.3
2006	4,200	5.4
2011	4,509	5.4
2016	4,594	5.4
2017	4,629	5.4
2018	4,664	3.5

Congo

Congo is also called the Republic of the Congo-Brazzaville. Agriculture is mostly at the subsistence level with cassava as the main food crop. Sugar cane, tobacco, palm kernels, cacao, coffee, bananas, peanuts, plantains and yams are also grown (Leonard, 2005). In a survey by Moula *et al.* (2012) of households in Congo, farmers were found to be cultivating cassava (100%), vegetables (tomatoes, onion, peppers and eggplant) (100%), corn (84.4%), peanuts (81.8%), beans (57.1%), soybeans (54.5%), sweet potato (37.7%), yams (31.2%), sesame (26.0%), millet (13.0%), okra (10.4%) and coffee (5.2%). FAO Stats (2019) give separate figures for the Congo (Table 7.18), which are quite different to those they provide for the Democratic Republic of the Congo (see p. 103).

Table 7.18. Yam production (tonnes annum^{-1}) and mean yield (tonnes ha^{-1}) in the Congo. Modified from FAO Stats (2019).

Year	Total production (tonnes)	Mean yield in tonnes ha^{-1}
1961	30,000	3.5
1966	38,000	3.8
1971	48,000	4.0
1976	11,800	4.7
1981	11,000	4.7
1986	9,263	4.4
1991	10,967	3.7
1996	8,300	4.1
2001	8,000	3.9
2006	12,100	5.3
2011	14,397	4.6
2016	15,359	4.6
2017	15,586	4.6
2018	15,786	4.6

Costa Rica

Costa Rica was reported to be 28th in terms of world production of yams but is ranked ninth in yams exported. A native Costa Rican species is *D. lepida* (Hammel, 2000). Yams exported from Costa Rica include *D. japonica*, also called the 'jinenjo yam', Yamaimo and East Asian mountain yam (Tridge, 2019). *D. trifida* is grown and is locally called 'Blanca'. Solano (1996) reported that one of the limiting factors for the production of *D. trifida* in the Atlantic region of Costa Rica is the lack of availability of high-quality planting material. In 2001, Costa Rica was the major supplier of yams to the USA with a market share that had increased from 55% in 1997 to 62% in 2001. Costa Rica is also a major exporter to the UK (Fig.7.1). Over 80% of the cassava imported to Europe is supplied by Costa Rica. Taro and yautia (*Xanthosoma sagittifolium*) are imported into Europe through the Netherlands or directly to destinations in France and the UK, with Costa Rica being the leading supplier (CBI, 2019). Tuber yields, in most years, where records exist, were between 12 and 18 tonnes ha^{-1}. Production appears to have increased substantially (Table 7.19).

Côte d'Ivoire

In Côte d'Ivoire, the species grown were reported to be mainly *D. cayenensis-rotundata* complex but other species are grown including *D. alata*, *D. praehensilis*, *D. burkilliana*, *D. hirtiflora*, *D. minutiflora*, *D. dumetorum* and *D. bulbifera*. Sahoré *et al.* (2005) reported that yams were the most important food crop in Côte d'Ivoire, but yields were low due to the lack of good quality planting material. However, in 2009 it was estimated that there were 758,000 hectares of yams in Côte d'Ivoire with 75% of this area estimated to being planted with 'improved varieties' (ASTI, 2019). The most important constraints on production of yams were insect damage on both leaves and tubers, nematode attack on tubers, drought, soil poverty and wilting. Dansi *et al.* (2013) reported that the most sustainable, economically profitable and environmentally friendly solution to address the above constraints on yams was to use 'tolerant cultivars'. They calculated that the use of tolerant cultivars would address 79% of these constraints. Nweke (2016) reported that the *D. alata* cultivar 'Florido', also called 'Asana', 'Matches' and 'Sudan', introduced from Puerto Rico in the early 1970s, was initially distributed spontaneously among farmers because of its desirable qualities. However, its distribution was subsequently promoted by the INRA for a yam mechanization project. Mean yield of yams has fluctuated considerably over recent years with a current possible trend towards reductions in spite of total production progressively increasing (Table 7.20).

Fig. 7.1. Yam from a Huddersfield market in the UK, imported from Costa Rica in December 2019. The species was not identified by the importer, but they appear to be *D. alata*.

Table 7.19. Yam production (tonnes annum^{-1}) and mean yield (tonnes ha^{-1}) in Costa Rica. Modified from FAO Stats (2020).

Year	Total production (tonnes)	Mean yield in tonnes ha^{-1}
1989	4.500	12.9
1991	5,000	12.5
1996	14,749	18.0
2001	17,856	12.5
2006	49,639	12.1
2011	29,580	13.9
2016	23,625	15.0
2017	23,625	15.0
2018	22,920	15.0

Cuba

The main *Dioscorea* species grown in Cuba is *D. cayenensis*, locally called 'ñame de Cuba', or just' ñame'. It has been listed as a 'transformer species with a high possibility of becoming invasive' (Oviedo Prieto *et al.*, 2012). *D. alata* and *D. pentaphylla* are widely cultivated as a food crop and are both naturalized in Cuba (Govaerts and Wilkin, 2015). Raz and Pérez-Camacho (2016) described a new *Dioscorea* species, *D. pseudocleistogama* found in the Guamuhaya Massif in central Cuba. In Cuba, yam tubers are used for Afro-Cuban ritual foods and drinks (Hanelt, 2017). Extracts from yam tubers are also used for curing gastritis among Yoruba local groups (Kadiri *et al.*, 2014). Cassava has been cultivated in Cuba by Arawak Indians since pre-Colombian times and Cuba is the second largest producer of cassava in the Caribbean with production estimated at about 300,000 tonnes in 2001. Yams are a much less important crop and production and mean yield has fluctuated considerably over the years (Table 7.21).

Democratic Republic of the Congo

Also called DRC-Kinshasa. Before the introduction of cassava in the 16th century, yams were the most important root crop in Africa (Coursey, 1967). Yams are common in forest areas of West Africa and Bahuchet *et al.* (1991) studied the availability of wild plant foods such as yams in the western Congo basin. They hypothesized that these yams were present in large enough quantities to support hunter-gatherers in times past, including pygmy foraging peoples, but had become increasingly neglected in the wild with increasing availability of cultivated plant foods. Yams were reported to be increasingly cultivated in DRC and Headland (1987) reported that smallholder agriculture was traditionally based on 'shifting cultivation', primarily of cassava, yams, cocoyam, bananas and plantains. Farmers cleared an area of forest, cultivated for about 2 years, left it fallow for up to 20 years, depending on many factors, including soil conditions and land availability, and returned to clear and cultivate again.

Some two-thirds of the population of DRC are involved in agriculture with the main crops being cassava, yams, plantains, rice and maize but crop production has tended to stagnate since independence in 1960. Agriculture is handicapped by a poor internal transportation system that impedes the development of an effective urban food supply system. Rice is grown in the Niari valley and in the north around Djambala, but the diet is supplemented with yams,

Table 7.20. Yam production (tonnes annum^{-1}) and mean yield (tonnes ha^{-1}) in Côte d'Ivoire. Modified from FAO Stats (2019).

Year	Total production (tonnes)	Mean yield in tonnes ha^{-1}
1961	1,150,000	7.7
1966	1,320,000	7.7
1971	1,555,000	8.0
1976	2,032,000	8.3
1981	2,493,016	11.5
1986	2,822,861	10.9
1991	3,263,920	10.1
1996	3,705,000	9.8
2001	4,579,723	8.9
2006	5,568,989	8.5
2011	5,531,865	6.6
2016	5,952,685	5.9
2017	5,531,865	5.8
2018	7,252,570	5.5

Table 7.21. Yam production (tonnes annum^{-1}) and mean yield (tonnes ha^{-1}) in Cuba. Modified from FAO Stats (2020).

Year	Total production (tonnes)	Mean yield in tonnes ha^{-1}
1992	6,000	4.9
1996	3,244	1.6
2001	59,800	18.5
2006	116,000	4.3
2011	349,692	6.9
2012	366,182	6.1
2013	377,771	4.4
2014	63,305	5.7
2015	60,604	5.7
2016	50,476	5.7
2017	52,881	5.7
2018	87,468	5.7

taro, sweet potatoes, maize, peanuts and fruit. Some yam breeding is being done in DRC with the objective of producing early maturing yams since many of the types are not ready for harvest for years after planting. Kumar *et al.* (2013) reported that DRC was actively seeking to promote yam cultivation, specifically with the aim of increasing yam production eight-fold to account for 8% of its food production by 2021. Fufu is made in the DRC from white yams, sweet potatoes, plantains or cassava. FAO Stats (2019) give separate figures for the Democratic Republic of the Congo (Table 7.22) which are quite different to those they provide for the Congo (see p. 101).

Dominica

Yams are traditionally grown by small farmers in all the islands of the Commonwealth Caribbean (Chandler *et al.* 1983). Dominica is a small island where production of yams is relatively low. The CARDI identified that the quality of planting material was a constraint to yam production in Dominica and developed a training programme to demonstrate the mini-sett technique and techniques in trellising yams that used less poles because of issues of deforestation by farmers who go into the forest to collect posts. Dominica, among other Windward Islands, suffered badly from Hurricane Maria in September 2017 that destroyed its major source of income, banana exports. It was found that yams, dasheens (*Colocasia esculenta*) and sweet potatoes proved to be resistant to hurricane damage. CIAT (Centro Internacional de Agricultura Tropical) introduced clean *D. alata* planting material into small farm systems of Dominica and Clarke *et al.* (1986) tested 23 *Dioscorea* accessions (15 *D. alata*, 3 *D. trifida*, 2 *D. rotundata*, 1 *D. cayenensis*, 1 *D. esculenta* and 1 *D. bulbifera*) at the Botanic Gardens in Roseau. Anthracnose disease was observed on some of the *D. alata* accessions and two of the cultivars 'White Yam' and 'Kaplaou' were included in on-farm trials to evaluate imported selections for tolerance to anthracnose. Both yields and production in Dominica are generally high with a steady increase in production over the years (Table 7.23).

Dominican Republic

The Dominican Republic produces only small quantities of yams and yields are comparatively low. *D. cayenensis* grown in Dominican Republic was exported to the USA. In January 2011 a shipment to New

Table 7.22. Yam production (tonnes annum^{-1}) and mean yield (tonnes ha^{-1}) in the Democratic Republic of the Congo. Modified from FAO Stats (2019).

Year	Total production (tonnes)	Mean yield in tonnes ha^{-1}
1961	60,000	7.5
1966	101,600	7.5
1971	164,800	7.6
1976	183,400	7.6
1981	237,600	9.0
1986	262,700	7.5
1991	287,300	6.9
1996	90,315	6.7
2001	87,100	5.0
2006	85,940	4.4
2011	91,334	4.3
2016	93,355	4.3
2017	87,239	4.3
2018	90,917	4.3

Table 7.23. Yam production (tonnes annum^{-1}) and mean yield (tonnes ha^{-1}) in Dominica. Modified from FAO Stats (2019).

Year	Total production in tonnes	Mean yield in tonnes ha^{-1}
1961	2,000	10.0
1966	2,250	10.0
1971	2,500	10.0
1976	3,180	10.3
1981	5,280	14.7
1986	6,600	14.3
1991	6,131	15.0
1996	7,672	14.8
2001	7,500	15.0
2006	10,621	13.0
2011	13,000	11.9
2016	14,420	11.6
2017	14,704	11.5
2018	15,496	11.7

York in 40 lb cartons made US$ 56.00 per carton, which was higher than *D. cayenensis* coming from Jamaica (Ministry of Agriculture and Fisheries, 2013). Kermarrec *et al.* (1981) reported that the nematode *Scutellonema bradys*, which infects yams, occurs in some Caribbean islands including Cuba, Jamaica, the Dominican Republic, Haiti and Puerto Rico. Average yield of yams varied from 6.3 to 6.9 tonnes ha^{-1} over the period 2010 to 2017 (Table 7.24).

Ethiopia

In the southern and western parts of Ethiopia, cultivated yams (*D. rotundata* and *D. cayenensis*) and forest yams (*D. abyssinica* and *D. praehensilis*) together with *Ensete ventricosum* were reported to be the staple source of dietary starch (Mengesha *et al.*, 2013). In the Sheko district of Ethiopia, Belachew Garedew *et al.* (2017) reported that *D. abyssinica*, which is native to Central African countries including Eritrea and Ethiopia, together with *D. bulbifera* and *D. alata* were the main staple crops. They reported that yams were mostly grown (96.8%) as intercrops. Itefa Degefa Alemu and Baressa Anbessa (2017) studied the agronomic practices of the Abaya Woreda people of southern Ethiopia. They grew *D. alata*, which they called 'underground boyina', for their tubers and *D. bulbifera* var. *anthropogophagorum*, which they called 'aerial boyina', for their bulbils. Some 75% of farmers grew *D. alata* and 25% *D. bulbifera* var. *anthropogophagorum*. Other species found growing wild in Ethiopia include *D. cochleariapiculata*, which is native to Ethiopia and other parts of eastern Africa, *D. praehensilis* (Magwé-Tindo *et al.*, 2018), *D. quartiniana* (Contu, 2013) and *D. schimperiana* (Burkil, 1985–2004).

Tamiru *et al.* (2008) surveyed 339 farm households in the major yam growing regions of southern Ethiopia. They found a total of 37 named landraces of yam with up to six on individual farms. Two of these landraces, 'Bola-Boye' and 'Bunde-Buchi' belong to *D. bulbifera* and were apparently identified based on the shape and size of their bulbils. The remaining landraces were not identified but preliminary observations, based on morphological features, suggested that some of the landraces were *D. cayenensis/D. rotundata* complex. The decision made by growers of which landrace to grow was based on tolerance to drought, maturity time and market demand. Production of yams and yield were low, but there are indications that both are increasing (Table 7.25).

Table 7.24. Yam production (tonnes annum^{-1}) and mean yield (tonnes ha^{-1}) in Dominican Republic. Modified from FAO Stats (2020).

Year	Total production (tonnes)	Mean yield in tonnes ha^{-1}
1961	20,700	8.3
1966	24,942	8.0
1971	27,890	8.0
1976	29,161	8.1
1981	7,371	6.5
1986	5,273	4.9
1991	7,670	7.1
1996	14,429	5.9
2001	18.042	6.3
2006	25,056	6.4
2011	28,196	6.7
2016	30,173	6.7
2017	32,880	6.6
2018	31,418	6.5

Table 7.25. Yam production (tonnes annum^{-1}) and mean yield (tonnes ha^{-1}) given as Ethiopia PDR (1961–1991) and Ethiopia (1996–2017). Modified from FAO Stats (2020).

Year	Total Production (tonnes)	Mean yield in tonnes ha^{-1}
1961	197,000	3.6
1966	238,800	4.2
1971	260,000	4.3
1976	280,000	4.4
1981	278,000	4.3
1986	250,000	4.1
1991	261,000	4.1
1996	216,685	7.1
2001	279,571	7.8
2006	227,151	7.6
2011	315,242	7.9
2016	1,500,000	30.0
2017	1,400,000	29.1
2018	1,355,584	30.0

Fiji

Yams are known locally in Fiji as ‘uvi’. Starch grains that were identified to be from *D. esculenta* have been recovered from Fiji, dated to around 3,050 to 2,500 BCE (Blench, 2010). *D. alata* is the main species grown and major components of 19 varieties of *D. alata* from Fiji, on a dry weight basis, were 68.5% starch, 18.6% amylose, 8.03% protein, 4.25% minerals and 2.46% sugars (SPYN, 2003). There has been a general increase in yam production in Fiji but yield data provided by FAO appears to be very variable (Table 7.26).

Gabon

Ondo Ovono *et al.* (2015, 2016) commented that yam cultivation requires a loose, deep soil to allow permeability of air and water to tubers, which they claim might explain why *D. rotundata* is the main species cultivated in Gabon. However, *D. cayenensis* is also cultivated. Gabon’s Ministry of Agriculture estimated that there were around 70,000 farms – mostly subsistence smallholder farmers – between 1 and 2 hectares in size, primarily cultivating plantains, cassava, taro, yams and various vegetables. Forestry and farming are relatively under-developed, contributing just 5% of GDP, yet agriculture employs about 40% of the rural population. In 2004, Gabon produced about 230,000 tonnes of cassava, 155,000 tonnes of yams, 61,800 tonnes of other root and tuber crops, 270,000 tonnes of plantains, 35,410 tonnes of vegetables and 31,000 tonnes of corn. From an evaluation of propagation methods for *D. rotundata* it was recommended that mini-tubers weighing at least 100–300 g, planted slanting or vertically, gave higher yield than traditional propagation methods (IFAD, 2016). Ondo Ovono *et al.* (2016) also evaluated *D. rotundata* and recommended the production of seedlings from mini-tubers of at least 100–300 g, planted slanting or vertically. For *in vitro* culture, when micro-tubers were harvested and directly transferred on a new medium without hormones, the tubers rapidly sprouted. After 1 week, around 50% of the tubers had sprouted and after 8 weeks, all the tubers had sprouted (Ondo Ovono *et al.*, 2010). The presence of two dry seasons in Gabon influences agricultural practice, as it gives two cycles of crop culture, including of yams (Ondo Ovono *et al.*, 2015). Yam production in Gabon has shown progressive increases over recent years, but yield per hectare has been consistently low (Table 7.27).

Table 7.26. Yam production (tonnes annum^{-1}) and mean yield (tonnes ha^{-1}) in Fiji. Modified from FAO Stats (2020).

Year	Total production (tonnes)	Mean yield in tonnes ha^{-1}
1961	800	8.0
1966	800	7.0
1971	750	7.5
1976	157	6.0
1981	3,458	14.0
1986	2,980	5.4
1991	1,393	2.4
1996	4,401	10.3
2001	3,149	6.6
2006	2,572	2.9
2011	5,953	12.2
2016	3,512	6.8
2017	4,425	6.0
2018	4,447	15.0

Table 7.27. Yam production (tonnes annum^{-1}) and mean yield (tonnes ha^{-1}) in Gabon. Modified from FAO Stats (2020).

Year	Total production (tonnes)	Mean yield in tonnes ha^{-1}
1961	50,000	5.9
1966	35,000	5.0
1971	45,000	5.3
1976	61,550	5.1
1981	85,100	6.1
1986	90,000	6.0
1991	112,478	6.6
1996	135,000	6.8
2001	165,000c	7.5
2006	170,000	7.5
2011	190,000	6.7
2016	218,994	6.5
2017	222,891	6.4
2018	227,822	5.3

Ghana

Ghana is the second largest producer of yams in the world with mean yields progressively increasing over recent years (Table 7.28). Commercially, cultivation is mainly *D. rotundata* (26 varieties) (Fig. 7.2) and to a lesser extent *D. alata* (13 varieties) (Fig. 7.3), with a few farmers growing *D. cayenensis*, *D. bulbifera*, *D. dumetorum* and *D. esculenta* mainly for home use. In 1994 it was reported that 'Yam production in Ghana has been on the decline despite the increasing demand for local consumption and for export' (Tetteh and Saakwa, 1994), but it has subsequently increased steadily. However, in 2016 yam production was reported to be in stagnation (Osei-Adu *et al.*, 2016). It was found that farmers received training (43%) and advice (29%) from extension services but had limited access to credit (4.3%). Gross margins were quite favourable and, as a solution, strengthening the capital base of input dealers to make them more efficient was suggested.

Bancroft (2000), working in Ghana, reported that the major causes of postharvest losses of yams were weight loss due to evapo-transpiration, which is intensified by sprouting, rotting due to fungal and bacterial pathogens and insect infestation. Another major issue was that the quality of yams produced in Ghana is sometimes compromised due to poor road transport, infrastructure, harvesting practices and storage conditions. As with other crops, quality can affect the price. Tubers affected by pests, diseases, harvesting damage and overexposure to sunlight can, when combined, lead to a price discount of between 25 and 40% (Bancroft *et al.*, 1998). Ghana is the largest exporting country of yams in the world. Exports of yams from Ghana in 2019 was US$ 39.1 million giving it a world share of 22.1%.

Table 7.28. Yam production (tonnes annum^{-1}) and mean yield (tonnes ha^{-1}) in Ghana. Modified from FAO Stats (2020).

Year	Total production (tonnes)	Mean yield in tonnes ha^{-1}
1961	1,100,000	7.3
1966	874,000	10.0
1971	909,000	5.3
1976	575,000	5.8
1981	591,000	5.1
1986	1,048,100	5.9
1991	2,631,900	11.6
1996	2,274,790	12.8
2001	3,546,739	12.3
2006	4,288,000	13.2
2011	6,295,453	15.5
2016	7,440,354	17.0
2017	7,952,750	17.1
2018	7,858,209	17.5

Grenada

In Grenada yams are an important staple crop and Buckmire *et al.* (1986) stated that among other crops yams are mainly grown for subsistence in Grenada. It was observed by Grenada (2017) that many types of yam are seen for sale at roadside produce stands in the yam season (December to May) with *D. alata* being the most common species including genotypes 'Lisbon Yam' (tubers are smooth and potato-shaped with fine, dense white flesh lacking the grainier texture of their wild relatives), 'CushCush Yam' (purple tuber flesh) and 'Moonshine' (large tubers with a lot of 'legs'). Grenada (2017) also reported that they 'did not see this cultivar ['Lisbon Yam'] produce any bulbils in its first year' and they are usually propagated by cutting off the head of the yam and replanting it, with the rest of the root being consumed or sold. Other yam types mentioned were: 'Fancy Yam' (a sweet yam that bears in a big bunch of tubers that have a tender skin and can be steamed and eaten without peeling), 'Potato Yam', 'Sweet Yam' and 'Lesser Yam', all of which are *D. esculenta*. 'Atuta' is a bitter yam with yellow flesh and is 'very nice yam for soup'. Arnold Babwah and Associates (2016) reported that in Grenada the average prices per kg ($EC) paid by the buyer segments for fresh yams were: hotels 4.40, supermarkets 4.40 and the Marketing Board 4.40. Production and

Fig. 7.2. *D. rotundata* cultivars. 'Mankron Pona' (a), 'Labreko' (b), 'Pona' (c), 'Kukrupa' (d). Photographs courtesy of Nana Pepra-Ameyaw, Ghana.

Fig. 7.3. *D. alata* 'Afase Ahodenfo' in Ghana. Photograph courtesy of Nana Pepra-Ameyaw, Ghana.

yield of yams in Grenada are both very low (Table 7.29).

Guadeloupe

In Guadeloupe, three *Dioscorea* species are grown: *D. alata*, the most popular, *D. rotundata-cayenensis* and *D. trifida* (CIRAD, 2015). Three different ploidy levels were detected for *D. alata* using flow cytometry (Gamiette *et al.*, 1999). Yam breeders at CIRAD are primarily working on anthracnose tolerance, yields and tuber quality (shape, flesh, nonbrowning and taste) for which they are using 384 *D. alata* accessions. There have been many studies on diseases, ideotypes and consumer preferences and requirements for yams in Guadeloupe, for example, Penet *et al.* (2016). Lebot (2009) reported that *D. trifida* is cultivated widely in the French West Indies and is highly valued for its organoleptic qualities but its cultivation is declining because of its sensitivity to viruses, especially yam mosaic virus. A significant contribution to *D. trifida* breeding was made by INRA in the 1960s in Guadeloupe and various selections were obtained in 1971 with yields of approximately 30 tonnes ha^{-1} (not staked), but no progress has been made on the sensitivity of *D. trifida* to viruses (Degras, 1993). Mechanical planting of yams is very rare but has been developed in Guadeloupe for *D. alata* and *D. cayenensis-rotundata* (Joachim *et al.*, 2003). Yam production in Guadeloupe seems to be progressively declining (Table 7.30).

Table 7.29. Yam production (tonnes annum^{-1}) and mean yield (tonnes ha^{-1}) in Grenada. Modified from FAO Stats (2020).

Year	Total production in tonnes	Mean yield in tonnes ha^{-1}
1961	450	2.6
1966	490	2.6
1971	520	2.5
1976	465	2.5
1981	482	2.5
1986	272	2.5
1991	308	2.3
1996	370	2.3
2001	397	2.2
2006	468	2.0
2011	541	1.9
2016	591	1.9
2017	600	1.9
2018	632	2.0

Table 7.30. Yam production (tonnes annum^{-1}) and mean yield (tonnes ha^{-1}) in Guadeloupe. Modified from FAO Stats (2020).

Year	Total production (tonnes)	Mean yield in tonnes ha^{-1}
1961	19,500	13.0
1966	25,500	11.2
1971	22,500	15.0
1976	21,600	18.0
1981	9,400	7.8
1986	10,608	11.1
1991	12,895	10.8
1996	6,200	8.1
2001	10,764	11.1
2006	3,000	12.0
2011	4,675	13.0
2016	3,445	14.0
2017	3,364	14.2
2018	3,425	14.3

Guinea

Guinea's agriculture accounts for 24% of GDP and engages 84% of the active population. A common name for *D. rotundata* is 'Guinea yam' or 'white Guinea yam', but 'Guinea yam' and 'yellow Guinea yam' are also applied to *D. cayenensis* (Coursey, 1967; Kay, 1987). *D. alata* is also cultivated in Guinea.

In 1999, the principal crops of Guinea were cassava (812,000 tonnes), rice (750,000 tonnes), sweet potatoes (135,000 tonnes), yams (89,000 tonnes), corn (89,000 tonnes), plantains (429,000 tonnes), sugar cane (220,000 tonnes), citrus fruits (215,000 tonnes), bananas (150,000 tonnes), peanuts (174,000 tonnes), palm kernels (52,000 tonnes) and coconuts (18,000 tonnes). In 1999, coffee production was estimated at 21,000 tonnes, compared to 14,000 tonnes on average annually from 1979 to 1981. Yam production is comparatively substantial and there are indications that it is increasing (Table 7.31).

Table 7.31. Yam production (tonnes annum^{-1}) and mean yield (tonnes ha^{-1}) in Guinea. Modified from FAO Stats (2020).

Year	Total production (tonnes)	Mean yield in tonnes ha^{-1}
1961	50,000	8.3
1966	55,000	8.5
1971	60,000	9.2
1976	56,950	8.5
1981	58,000	8.5
1986	70,000	9.3
1991	95,000	10.0
1996	90,000	11.2
2001	90,000	11.6
2006	26,057	11.8
2011	102,686	11.1
2016	125,960	8.2
2017	121,474	9.0
2018	187,935	9.4

Guyana

Yams and cassava are the most widely grown root crops in Guyana. The main species of yam produced were reported to be *D. alata*, but *D. esculenta* and *D. cayenensis* are also grown (Guyana, 2004). The New Guyana Marketing Corporation (2004) reported that *D alata* was the main yam produced in Guyana, where plants normally produced a single tuber that is variable in size (often weighing from 5 to 10 kg) and shape, and have a brown to gold coloured skin with white or purplish flesh. Other species include *D. esculenta*, which produces a large number of small tubers that are indented and have purplish-brown skin and white flesh, and *D. cayenensis*, which produces tubers of variable size and shape, usually with pale yellow flesh.

Yam tubers are harvested once a year in Guyana and can be stored for several months inside properly ventilated, non-refrigerated covered storage structures. These types of structure include pits, trenches, silos and covered piles in the field. Ground pits along Guyana's coast are not appropriate due to the high level of the water table. *D. trifida* was reported to have been domesticated by Amerindians and appears to have originated on the borders of Brazil and Guyana, followed by a dispersion throughout the Caribbean islands (Ayensu and Coursey, 1972). Guyana (2004) reported that in Guyana, yams are harvested once a year and may suffer considerable postharvest losses due to weight loss, microbial decay, sprouting and insect attack. The quantity of yams produced in Guyana is comparatively small and FAO data appears to show considerable variation in mean yield (Table 7.32).

Table 7.32. Yam production (tonnes annum^{-1}) and mean yield (tonnes ha^{-1}) in Guyana (modified from FAO Stats, 2020). No data are provided by the FAO before 1997.

Year	Total production (tonnes)	Mean yield in tonnes ha^{-1}
1997	2,500	
1998	2,672	
1999	2,800	
2001	4,107	9.6
2006	1,300	9.3
2011	258	5.7
2016	1,466	7.7
2017	1,980	7.6
2018	1,580	20.5

Haiti

In terms of quantity of production, the most important crops grown in Haiti (ranked in order) were sugar cane, cassava, yams, bananas, sweet potatoes, plantains, maize, mangoes, guavas and rice. The main types of yam cultivated in Haiti were reported to be *D. rotundata* (cultivars 'Guinée' and 'Adigwé') that are more widespread in the southern production areas and *D. cayenensis* (cultivars 'Soussou', 'Batwel' and 'Bouba') that are more widespread in the northern production areas (MARNDR, 2017). Eitzinger *et al.* (2014) claimed that climate change will impact on Haiti's crop production. In areas below 500 m, dry beans and *D. trifida* will become less suitable, while *D. rotundata* and *D. cayenensis* will become more suitable. *Haiti News* (2014) reported that nearly 30,000 hectares of yams were cultivated in Haiti, with a high concentration in the Grand Anse and pockets of production in the valleys of Jacmel, Piacenza and Pilate. However, even with increasing production there is insufficient supply of yams, given the potential of yams and the investment made by peasant producers. The newspaper also goes on to quote Vernet Joseph: 'recent studies of characterization of agricultural sectors have shown the importance of yams, and identified as one of the sectors to promote. Demand for yam has increased sharply in both domestic and international markets'.

Torres (2001), working with 'Haitian Yellow Yam' (*D. cayenensis-rotundata*), reported that as much as 50% of the tubers graded for export may be lost due to postharvest damage including mechanical injuries during transport, diseases and weight loss. Torres concluded that these factors may limit exportable yam yields to 2,300–3,850 kg ha^{-1}. According to this analysis, 75–125 farmers, each with 700 m^2 of *D. cayenensis* production, may be needed to produce 20,000 kg (one reefer container load) of exportable yams. Torres investigated two farms. The farmers had either 3,086 or 3,846 mounds ha^{-1} with a yield per mound of 10.06 or 8.06 kg and 25 or 15% exportable yield (cylindrical shape between 35 and 45 cm long with no injures). CNSA (2013) gave the average yield of yams in Haiti as 12.0 tonnes ha^{-1}, which compared badly to the Dominican Republic that had average yields of 14.6 tonnes ha^{-1}. However, these figures are substantially higher than data from the FAO (Tables 7.24, 7.33).

Hawaii

According to Kumar *et al.* (2013), yams have been cultivated in Hawaii for centuries, although Professor Robert Paull from University of Hawaii at Manoa commented recently that he had 'never seen them for sale'. Besides *D. alata*, which is the most widespread, *D. bulbifera* and *D. pentaphylla* grow in Hawaii, of which *D. bulbifera* is the most widespread of the two (Wagner *et al.*, 1999). As early as the 4th century, Polynesians brought these three species to Hawaii (Bevacqua, 1994) and *D. pentaphylla* yam was brought at least once by southern Polynesians as recently as 800 years ago. In Hawaii, *D. alata*, *D. bulbifera* and *D. pentaphylla* mainly occur in habitats with a moderate or well-balanced supply of moisture. They are all common on abandoned agricultural and pasture lands dominated by *Schinus terebinthifolius* trees (Wagner *et al.*, 1999). Wagner reported that in Hawaii, *D. bulbifera* tubers are only eaten in times of famine and *D. alata* was one of the most commonly cultivated of the yam species. *D. pentaphylla* was described as persistent in moist areas of lower Hawaiian forests and often commonly occurred growing with various tree species including *Aleurites moluccana*, *Metrosideros polymorpha* and *Eugenia sandwicensis* (St John, 1954). *D. pentaphylla* bulbils are horseshoe-shaped and about 1 cm in diameter but were reported to be rare in Hawaii. *D. pentaphylla* tubers are produced singly, are irregular to elongated and egg-shaped, range from about 17cm to 1m long, weigh up to about 1.4 kg and may occur near the soil surface or more than 1 m below the surface (St John, 1954; Wagner *et al.*, 1999).

Table 7.33. Yam production (tonnes annum^{-1}) and mean yield (tonnes ha^{-1}) in Haiti. Modified from FAO Stats (2020).

Year	Total production (tonnes)	Mean yield in tonnes ha^{-1}
1961	80,000	4.0
1966	90,000	4.0
1971	96,000	4.0
1976	110,000	4.0
1981	115,000	3.3
1986	120,000	3.8
1991	190,000	5.3
1996	230,000	5.0
2001	197,000	5.6
2006	220,000	5.9
2011	340,000	7.4
2016	416,297	7.8
2017	439,329	7.9
2018	423,545	9.9

India

No data on total production of yams in India could be found, but Prain and Burkill (1936, 1938) reported that there were about 50 *Dioscorea* species in India, while Karthikeyan *et al.* (1989) listed 30 species and 27 varieties. However, only 29 species and two varieties have been recognized as the number of accepted taxa in India (Pagare *et al.* 2015). Goswami *et al.* (2013) investigated *Dioscorea* species in Assam and found that *D. alata* occurred in all districts. Other species they found in Assam were *D. trinervia*, *D. pubera*, *D. glabra*, *D. hamiltonii*, *D. decipiens*, *D. laurifolia*, *D. esculenta*, *D. bulbifera*, *D. pentaphylla*, *D. hispida* and *D. cumingii*. Kumar *et al.* (2017) reported that there were 13 *Dioscorea* species found in the Similipal Biosphere Reserve in India including *D. pentaphylla*, *D. hispida*, *D. alata*, *D. oppositifolia* and *D. pubera*, ten species of which are known to be 'bitter in taste or unpalatable when eaten raw'. They reported that local people make them edible using different traditional practices, mostly by soaking overnight in water (e.g. a stream) and subjecting them to successive boilings to remove the bitterness. The Andaman and Nicobar Islands are an Indian archipelago in the Bay of Bengal consisting of about 300 islands. Ghosh (2014) identified several *Dioscorea* species growing there including *D. alata*, *D. bulbifera*, *D. bellophylla*, *D. glabra*, *D. wallichii*, *D. oppositifolia*, *D. esculenta*, *D. pentaphylla*, *D. hispida* and *D. tomentosa*.

Jamaica

In Jamaica in the mid-1970s, Beckford *et al.* (2011) reported that *D. cayenensis* displaced *D. rotundata* as the main yam produced and accounted for close to 60% of total yam production and perhaps as much as 90% of all Jamaican yam exports (Tables 7.34, 7.35). *D. trifida*, locally called 'yampie' is also commonly cultivated in Jamaica and Jamaican Bitter Yam (*D. polygonoides*) is grown as a source of sapogenin. Jamaica is a leading yam exporter accounting for 16% of the global yam trade, but probably 100% of the *D. cayenensis* market. Jamaica exports roughly 5% of the yams it produces.

Table 7.34. Yam production (tonnes annum^{-1}) and mean yield (tonnes ha^{-1}) in Jamaica. Modified from FAO Stats (2020).

Year	Total production (tonnes)	Mean yield in tonnes ha^{-1}
1961	42,918	14.3
1966	63,486	13.1
1971	122,864	11.6
1976	119,566	10.8
1981	136,410	12.8
1986	165,633	12.9
1991	186,104	13.4
1996	253,371	16.9
2001	157,577	16.5
2006	123,005	16.6
2011	136,785	16.2
2016	156,103	17.1
2017	144,319	16.5
2018	148,675	16.8

Japan

Several species of *Dioscorea* are found in Japan including *D. japonica* and *D. opposita*, which are native to Japan, Korea and China. Sato (2012) discussed Japanese exports of yams (he refers to them as *D. batatas*, which is synonymous with *D. polystachya*) to Taiwan and reported that total yam production in 2009 reached 138,000 tonnes, of which Hokkaido, Aomori and Nagano accounted for 92.3% of all yam production in Japan that year. The total volume and value of yam exports has tended to increase, with the former increasing rapidly from 244.2 tonnes in 1990 to 5,499.5 tonnes in 2010. Particularly high growth rates were between 2000 and 2002. Two Japanese foods, 'yamaimo', a slimy, slippery food, is made from *D. japonica* and 'nagaimo' is made from *D. polystachya* that can be cooked, but are usually eaten raw after being finely julienned or grated and may be served in hot or chilled miso soup. Yams grown in Japan are said to have many beneficial qualities and are considered by some to be an aphrodisiac. *D. tokoro* also occurs in the wild in Japan, but has an extremely bitter taste and is not generally regarded as edible. Average yields of yams are relatively high and varied from 21.7 to 22.7 tonnes ha^{-1} over the period 2010 to 2014 (Table 7.36).

Kenya

The following *Dioscorea* species have been reported as growing in Kenya: *D. minutiflora*,

Table 7.35. Estimated production of *D. cayenensis* in Jamaica. Modified from Pike (2016).

Year	Production (tonnes)	Crop area harvested (ha)	Output (kg ha^{-1})
2010	89,944	4,838	18.6
2011	88,601	5,659	15.7
2012	100,325	5,800	17.3
2013	95,334	5,798	16.4
2014	95,654	5,694	16.8

D. dumetorum, *D. bulbifera D. cayenensis*, *D. alata*, *D. rotundata*, *D. mangenotitana*, *D. schimperiana* and *D. bulbifera* (Mwirigi *et al.*, 2009; Price *et al.*, 2016, 2017). Muthamia *et al.* (2014) reported that *D. odoratissima* was found in Kenya, but with a low distribution. Maundu *et al.* (1999) reported that there was a high degree of morphological polymorphism among the cultivated yams in Kenya. In the central highlands, where considerable yam production occurs around Mount Kenya, tubers are harvested leaving the mother tuber or part of the mother tuber in the soil for the next crop. However, this practice was reported to result in low yields and attacks by soil-borne pests and pathogens (Maina, 2008). Maundu *et al.* (1999) reported that yams are grown for food, mainly by elderly farmers in eastern, central, western and coastal regions of Kenya, with *D. minutiflora* the most commonly grown species; however, its use as food had declined rapidly in recent decades. It is now rarely cultivated, and is almost invariably maintained by elderly women as a matter of tradition on their small farms. However, yams are a preferred food security crop in the drier areas of Kenya. Many families in Kenya plant yams near the trees in the 'shamba' (farm), so that the vines can get support. Yam cultivars grown in Kenya include 'Baribate', 'Barimutwa', 'Kambo', 'Karukwaji', 'Kijara', 'Kimwere', 'Majara', 'Mbeu Iguru', 'Mbeu Mpuria', 'Mbeu Nkuru', 'Mbeu Ruguru', 'Mbeuku', 'Mbithi', 'Murijo', 'Murungwa', 'Mweru', 'Mwezi I', 'Mwimba Iguru', 'Nakirima', 'Nalo', 'Ndera Ngutu', 'Ndiathi', 'Ndiua Na Thi', 'Ngundu', 'Nkandau', 'Nkone', 'Ntigania', 'Ntokinyoni', 'Nyaara', 'Emondo', 'Embame' and 'Kihama' (Infonet, 2019). Mean and total yield of yams in Kenya are given in Table 7.37.

Table 7.36. Yam production (tonnes annum^{-1}) and mean yield (tonnes ha^{-1}) in Japan. Modified from FAO Stats (2020).

Year	Total production (tonnes)	Mean yield in tonnes ha^{-1}
1961	54,000	16.9
1966	73,500	18.4
1971	113,400	17.3
1976	121,200	17.6
1981	134,000	17.4
1986	152,300	18.0
1991	184,900	19.2
1996	170,200	19.5
2001	182,400	20.8
2006	192,200	22.5
2011	165,900	22.1
2016	145,700	20.5
2017	144,952	21.7
2018	164,201	22.4

Liberia

Staple food crops grown in Liberia are rice, cassava, plantain, eddoe, sweet potatoes and yams. Species of yam reported to be grown in Liberia include: *D. alata*, *D. rotundata*, *D. bulbifera* and *D. mangenotiana* (Govaerts, 2012). Cassava was introduced into coastal Liberia nearly 400 years ago and cassava production has increasingly supplanted yam cultivation and use. Mean yield of yams is within the range of other West African countries and production has shown steady increases over recent years with mean yields remaining relatively constant (Table 7.38).

Madagascar

Madagascar does not appear on the FAO statistics for yam production, although at

Table 7.37. Yam production (tonnes annum^{-1}) and mean yield (tonnes ha^{-1}) in Kenya (no data was provided before 1989). Modified from FAO Stats (2019).

Year	Total production (tonnes)	Mean yield in tonnes ha^{-1}
1989	9,520	8.1
1990	11,769	8.8
1991	11,200	8.0
1996	8,765	6.0
2001	7,898	7.9
2006	8,001	9.5
2011	9,635	9.1
2016	12,388	9.8
2017	10,217	9.8
2018	9,610	9.8

Table 7.38. Yam production (tonnes annum^{-1}) and mean yield (tonnes ha^{-1}) in Liberia. FAO data based on imputation methodology, modified from FAO Stats (2020).

Year	Total production (tonnes)	Mean yield in tonnes ha^{-1}
1961	13,000	9.6
1966	13,000	9.6
1971	13,500	9.6
1976	15,000	9.7
1981	15,500	9.2
1986	18,000	9.0
1991	16,144	8.5
1996	20,895	8.7
2001	20,367	8.7
2006	21,000	8.7
2011	19,500	8.8
2016	21,491	8.7
2017	21,567	8.7
2018	21,326	8.7

least 30 of the 40 *Dioscorea* species found there are edible and may be collected from the forest to provide a food source during the months when rice is unavailable or prohibitively expensive. Most yams are collected from the wild since cultivation is uncommon, but *D. alata* is cultivated on a small scale in home gardens (Cable, 2018). Twelve of the edible species identified in Madagascar are threatened with extinction including *D. bako* that has been classified as endangered and was reported by Harris (2018) as perhaps suffering from over-utilization by people collecting them from forests. Randriamboavonjy *et al.* (2013) found that wild yams had become rare in the Ambositra-Vondrozo forest corridor in Madagascar due to forest fragmentation and unsustainable collection of tubers for food. Also, knowledge of both cultivated and wild yams in this region is incomplete with no management plans for the conservation of yams in their natural habitat. Randriamboavonjy *et al.* (2013) carried out a yam cultivation demonstration project and found that most villagers in the area initiated and continued cultivation after the demonstration. They claimed that these demonstrations with farmers were achieving the objective of ensuring sufficient food production to protect the wild yams, especially in sites where wild yams are over-used.

In a study in southwestern Madagascar (Mahafaly region) Andriamparany *et al.* (2014) found that *D. alatipes*, *D. bemandry*, *D. fandra*, *D. ovinala*, *D. nako* and *D. soso* were collected from the wild as medicinal plants and were also important for food and income generation. They found that yams gathered from the wild were used to supplement cassava and maize, especially during periods of food shortage. As reported above they also found that recently the quantity of wild yam tubers harvested had strongly increased, putting many species in danger.

Mali

Yam is an important food crop in Mali and, as in other West African countries, *D. rotundata* is the predominant species, cultivated with *D. alata* to a lesser degree. In their studies of yams in West Africa Coyne *et al.* (2006) collected tubers from markets and roadside traders in Benin, Burkina Faso, Côte d'Ivoire, Ghana, Nigeria and Togo as well as Mali. They found that in Mali there was the lowest level of nematode infestation by *Scutellonema bradys* (0.28%), but Mali had the highest proportion of galled tubers due to *Meloidogyne* species with 14.4–19.6% on *D. rotundata*. *S. bradys* infection, based on visible symptoms, was more evident on *D. rotundata* (3.8%) than on *D. alata* (0.6%), although 5.2% of yams in the group, comprising unidentified *Dioscorea* species, had the greatest mean proportion of visually affected tubers. Both total production and mean yield of yams in Mali has shown a general increase over recent years (Table 7.39).

Martinique

In Martinique, *D. alata*, *D. cayenensis* and *D. rotundata* are the most commonly cultivated species. Yams are costly to produce and production in Martinique was reported to be in decline; imports of fresh yams have been increasing for several years. The *D. alata* cultivar 'AL56' was a favourite for consumers in Martinique amongst the six cultivars ('AL54', 'INRA15', 'Boutou', 'AL18', 'A17' and 'AL56') tested by Njoh Ellong *et al.* (2015). Martin (1973) reported that a cultivar of the *D. rotundata-cayenensis* complex grown in Martinique called 'Saint Prix' produced very large tubers. Total yam production has been shown to vary annually, but the trend is a general decline (Table 7.40).

Table 7.39. Yam production (tonnes annum^{-1}) and mean yield (tonnes ha^{-1}) in Mali . Data based on imputation methodology modified from FAO Stats (2020).

Year	Total production (tonnes)	Mean yield in tonnes ha^{-1}
1961	7,000	3.5
1966	8,500	4.3
1971	9.400	4.7
1976	11,000	3.9
1981	15,000	4.1
1986	11,264	2.4
1991	14,955	5.0
1996	13,000	3.7
2001	36,790	10.1
2006	77,750	25.0
2011	90,138	22.6
2016	118,863	24.4
2017	116,117	24.7
2018	90,976	17.3

Table 7.40. Yam production (tonnes annum^{-1}) and mean yield (tonnes ha^{-1}) in Martinique. Modified from FAO Stats (2020).

Year	Total production (tonnes)	Mean yield in tonnes ha^{-1}
1961	19,000	10.6
1966	20,000	10.5
1971	23,000	11.5
1976	15,750	7.9
1981	7,200	9.9
1986	13,200	9.8
1991	6,900	8.3
1996	4,000	8.9
2001	7,182	8.8
2006	1,800	9.0
2011	137	11.0
2016	231	11.5
2017	19,000	11.2

Mauritania

Most of the West African country of Mauritania is either desert or semiarid and the country is not agriculturally self-sufficient. Where the rainfall exceeds 430 mm a year (in less than 1% of Mauritania), millet and dates are the principal crops, supplemented by yams, sorghum, beans, maize and cotton. Production and mean yield of yams have remained relatively constant over recent years (Table 7.41).

New Caledonia

Traces of *D. esculenta, D. alata, D. bulbifera, D. nummularia* and *D. pentaphylla* have been identified from a cave in New Caledonia, dated to around 2,700 to 1,800 BCE (Horrocks and Nunn, 2007; Horrocks *et al.*, 2008). The economy of New Caledonia is dominated by the service sector and the nickel industry with agriculture representing less than 2%. However, family farming remains strong, and yam production is important in traditional Kanak society. Kanaks are the native people of New Caledonia, who comprise just under half of its total population. Yam cultivation systems

Table 7.41. Yam production (tonnes annum^{-1}) and mean yield (tonnes ha^{-1}) in Mauritania. Modified from FAO Stats (2019).

Year	Total production (tonnes)	Mean yield in tonnes ha^{-1}
1961	1,800	6.9
1966	2,000	6.9
1971	2,100	7.0
1976	1,900	5.6
1981	2,200	5.8
1986	2,600	6.2
1991	2,900	6.5
1996	2,582	6.3
2001	2,612	6.3
2006	2,800	6.2
2011	2,845	6.3
2016	2,797	6.3
2017	2,801	6.3
2018	2,965	6.3

have greatly changed since the 1950s, and yams are increasingly being grown for market sales as well as home consumption (Gaillard and Manner, 2010). In New Caledonia,

> the festival of the yam marks the beginning of the yam harvest, and is the most important event in the Kanak calendar. Usually held around mid-March, the yam is treated with the respect usually saved for an ancestor. When the yams are declared ready by the elders, they are pulled up and presented to the older clansmen and the chief, signalling the beginning of the harvest. The yams are blessed the following day by the priest, then distributed among the villagers. Out of respect, the yams are broken like bread rather than cut.
>
> (Jasons, 2019)

Mean yields have been reported to fluctuate, but are generally poor and the latest results show very low mean yields (Table 7.42).

Nicaragua

Nicaragua has been ranked 40th within the group of 42 countries in terms of yam production. However, they may not be all *Dioscorea* spp., and perhaps *Amorphophallus paeoniifolius* or *Xanthosoma* spp. are being referred to, although in a focus group in Guatemala, Barlagne *et al.* (2017) reported that the perception of yam varieties by local experts included *D. alata*, *D. esculenta*, *D. cayenensis*, *D. rotundata*, *D. bulbifera* and *D. trifida*. Total and mean yield shows little variation if FAO statistics are to be believed (Table 7.43).

Table 7.42. Yam production (tonnes annum^{-1}) and mean yield (tonnes ha^{-1}) in New Caledonia. Modified from FAO Stats (2020).

Year	Total production (tonnes)	Mean yield in tonnes ha^{-1}
1961	8,500	8.1
1966	10,000	9.1
1971	8,000	5.9
1976	10,000	6.5
1981	10,800	6.4
1986	11,000	6.5
1991	10,900	3.4
1996	7,561	5.7
2001	2,848	3.5
2006	4,848	1.5
2011	1,302	0.9
2016	2,217	0.8
2017	2,374	0.9
2018	2,362	1.0

Nigeria

Nigeria is by far the largest producer of yams in the world with production constantly increasing (Table 7.44). In addition to their food and medicinal uses in Nigeria, yams have cultural, religious and social meanings that often vary between specific ethnic groups and regional areas. Yams play an important role in African culture especially for the Igbos in Nigeria who celebrate what they call the 'New Yam Festival'. During this festival, yam is the main food served – either roasted, with palm oil, stewed or with a sauce.

The primary species for cultivation in Nigeria are *D. rotundata* and *D. cayenensis*. Both *D. rotundata* and *D. cayenensis* are considered indigenous and most important to the people of West Africa, except for the inhabitants of the Côte d'Ivoire, who consider *D. alata* more important. Other species in Nigeria include *D. bulbifera*, *D. preussii*, *D. praehensilis*, *D. sansibarensis*, *D. dumetorum* and *D. abyssinica*, which are found growing wild, mainly in the rain forests (Asiedu and Sartie 2010; Obidiegwu and

Table 7.43. Yam production (tonnes annum^{-1}) and mean yield (tonnes ha^{-1}) in Nicaragua. Modified from FAO Stats (2020).

Year	Total production in tonnes	Mean yield in tonnes ha^{-1}
1991	1,000	13.1
1996	1,000	13.2
2001	1,000	13.2
2006	1,000	13.3
2011	1,000	13.3
2016	1,000	13.4
2017	1,000	13.4
2018	1,000	13.4

Table 7.44. Yam production (tonnes annum^{-1}) and mean yield (tonnes ha^{-1}) in Nigeria. Modified from FAO Stats (2020).

Year	Total production (tonnes)	Mean yield in tonnes ha^{-1}
1961	3,500,000	7.7
1966	6,987,000	9.4
1971	9,766,000	7.9
1976	6,556,000	9.7
1981	5,212,000	5.6
1986	5,209,000	5.6
1991	16,956,000	10.3
1996	23,201,000	10.7
2001	26,232,000	9.8
2006	36,720,000	12.1
2011	33,134,172	7.4
2016	44,109,615	8.4
2017	47,942,712	8.1
2018	47,532,615	7.9

Table 7.45. Yam production (tonnes annum^{-1}) and mean yield (tonnes ha^{-1}) in Niue. Modified from FAO Stats (2020).

Year	Total production in tonnes	Mean yield in tonnes ha^{-1}
1961	100	10.0
1966	100	10.0
1971	100	10.0
1976	170	9.4
1981	200	2.0
1986	200	1.9
1991	91	0.7
1996	108	0.8
2001	118	0.9
2006	130	0.9
2011	145	1.0
2016	154	1.1
2017	155	1.1
2018	156	1.2

Akpabio, 2017). Yam contributes more than 200 dietary calories per day for over 60 million people in Nigeria and the country exported US$ 27.7 million worth of yam to the USA in 2011 (IITA, 2009; Bassey, 2017).

Niue

Niue is a small island nation in the South Pacific Ocean that had a population of only 1,624 in 2018. Crops are mainly cassava, taro and breadfruit, with bananas, papaya, coconuts and yams also cultivated. Yam production is small and mean yield of yams has decline over recent years (Table 7.45).

Pacific islands

In the South Pacific the two most important *Dioscorea* species are *D. alata* and *D. esculenta*. Both of these species originated in South East Asia and were carried by early voyagers into the western Pacific. Yams continue to make an important contribution to nutrition and food security in most Pacific islands including Fiji, Tonga, Vanuatu, Samoa and the Federated States of Micronesia. *D. nummularia*, *D. bulbifera*, *D. pentaphylla*, *D. rotundata* and *D. trifida* are also grown to some extent in the Pacific islands (Martin, 1974b, 1974c; University of Hawaii Manoa Library, 2018). St John (1954) reported that in Polynesia *D. alata* is grown on most of the high islands and can also persist in forests and thickets from former cultivation. Other species that St John (1954) reported included *D. bulbifera*, *D. pentaphylla*, *D. esculenta*, *D. batatas* and *D. nummularia*. Kenyon *et al.* (2008) screened *Dioscorea* species from seven South Pacific islands for *Badnavirus* infection and found positive readings for 26–33% of the samples of *D. bulbifera*, *D. nummularia*, *D. pentaphylla*, *D. alata*, *D. esculenta*, *D. rotundata* and *D. trifida* that were tested.

Panama

Coursey (1967) suggested that cultivation of *D. trifida* in Central America, including Panama, dates to prior to the arrival of Columbus in the western hemisphere and he suggested that other yam species were probably harvested from the wild at that time. Jiménez-Montero and Aguilar (2016) reported that the first documented mention of the genus *Dioscorea* in Panama was made by the naturalist Seeman in 1852, when he referred to a plant with the common name of 'Cabeza de Negro', used as 'an antisyphilitic by the Indians'. *D. cayenensis* was an introduced species, known locally as 'Ñame Chomo', and is cultivated, but also was reported to have escaped into the wild in the Panama Canal Zone (PROTA, 2017). *D. alata* was also reported to be an introduced species to Panama (Correa *et al.*, 2004). Jiménez-Montero and Aguilar (2016) noted that both *D. alata* and *D. trifida* are currently cultivated and enjoy a high domestic demand. PROTA (2017) reported that *D. bulbifera* had been collected in Belize, Costa Rica, Guatemala, Honduras, Mexico, Nicaragua and Panama and that *D. bulbifera* has also been a staple crop consumed by indigenous populations for a considerable time. They compared local cultivation practices with innovative technology techniques and concluded that innovative technical systems always produced much higher yields than those of traditional systems. They also reported that *D. bulbifera* has attributes that could enable it to become a plant resource that could contribute to the improvement of food security of peasant communities in Panama. *D. mexicana* was reported to be native to Panama as well as other Central American countries. Yam production and mean yield have both been shown to be variable over recent times and are given in Table 7.46.

Table 7.46. Yam production (tonnes annum^{-1}) and mean yield (tonnes ha^{-1}) in Panama. Modified from FAO Stats (2020).

Year	Total production (tonnes)	Mean yield in tonnes ha^{-1}
1961	10,500	7.5
1966	12,400	6.9
1971	14,500	7.6
1976	16,514	6.6
1981	20,376	6.6
1986	21,228	5.7
1991	13,646	3.4
1996	13,989	3.1
2001	22,564	8.4
2006	25,615	8.3
2011	22,402	4.0
2016	9,222	4.9
2017	7,974	4.8
2018	11,573	3.7

Papua New Guinea

Prior to the adoption of sweet potatoes in the New Guinea highlands, people depended on taro (*Colocasia esculenta*) as their main food crop, supplemented by bananas and yams (probably *D. alata*, *D. esculenta*, *D. nummularia*, *D. pentaphylla* and *D. opposite*). Bourke and Harwood (2009) reported that

> it is likely that people once ate wild yam with poor eating properties. Some of these species were domesticated and the quality of the tubers improved. There are indications that some species of yam, such as *Dioscorea pentaphylla* and *D. nummularia*, are very ancient crops in PNG, perhaps introduced from elsewhere but most likely domesticated in the New Guinea area. Tubers of *D. pentaphylla*, for example, have inferior eating qualities and yields appear to be poor, yet people still grow the occasional plant, probably for its cultural value ('Bilong Tumbuna') rather than for food. It is likely that superior varieties of the greater yam (*D. alata*) were developed and that greater yam became the most important species of yam before the introduction of lesser yam (*D. esculenta*). Lesser yam is likely to have been a later introduction into PNG. It is agronomically superior to other yam species in PNG, including greater yam, has fewer disease problems, a greater yield per plant and tubers that are more easily prepared for cooking than those of most yam species. More people grow greater yam than lesser yam, although greater yam is not usually an important food. *Dioscorea esculenta* is less important for ritual purposes than the other yam species, particularly greater yam. All of this suggests that lesser yam has been adopted because of its ability to provide food energy, while the other species have been retained for different reasons.

Bourke and Vlassak (2004) reported that sweet potatoes (*Ipomoea batatas*) accounted for 64% of production of staple food crops by weight and 63% by food energy. The contribution by weight for banana was 9.7%, cassava (*Manihot esculenta*) was 6.0%, yam (*D. alata* and *D. escuienta*) was 6.0%, Chinese taro (*Alocasia cucullata*) was 5.0%, taro (*Colocasia esculenta*) was 5.0%, coconut (*Cocos nucifera*) was 2.2% and sago (*Cycas revoluta*) was 1.8%. A major change in agriculture over the previous 40 years had been the increased significance of cassava, potatoes, sweet potatoes and Chinese taro compared to bananas, sago, taro and yams that had either decreased or was of similar magnitude to that of 40 years previously. Opara (2003) reported that yams accounted for 4.6% of the total dietary calorie intake in Papua New Guinea. It was quoted by FAO Trobriand Islands (2018) that the National Disaster Centre reported that the islands were prone to natural disasters including droughts and there was an incentive to increase yam production. A gift of yams from the Papua New Guinea government was made to the islands at that time. It was claimed that food security was at stake as yams and other food resources were becoming scarce, because yields had declined in many Trobriand gardens, resulting in increased levels of hunger, malnutrition, unhealthy lifestyles and poor livelihoods. Some outer islands and atolls could not produce yams at all. Both the production and yield of yams in Papua New Guinea, with the former progressively increasing and the latter varying only slightly, are presented in Table 7.47.

Table 7.47. Yam production (tonnes annum^{-1}) and mean yield (tonnes ha^{-1}) in Papua New Guinea. Modified from FAO Stats (2020).

Year	Total production (tonnes)	Mean yield in tonnes ha^{-1}
1961	120,000	16.6
1966	138,000	16.2
1971	155,000	16.3
1976	166,000	16.4
1981	185,000	16.8
1986	202,000	17.4
1991	215,000	17.6
1996	210,000	17.5
2001	260,000	17.3
2006	297,000	16.5
2011	340,000	17.4
2016	359,472	17.6
2017	362,672	17.6
2018	375,989	17.7

Philippines

In the Philippines, *D. alata* is considered a high value crop and is cultivated in small patches of land, often less than a hectare, particularly in regions such as the Ilocos, Southern Tagalog, Bicol, Central Visayas and Northern Mindanao. It is an upland crop and usually planted in well drained soils (Perlas *et al.*, 2010). The purple *D. alata*, also called 'ube yam', produces both tubers and bulbils and is eaten as sweetened desserts called 'ube halaya' and 'halo-halo' (CABI, 2017). *D. alata* is also used for the production of yam powder and flour and as a popular ingredient for ice cream, jam and other local delicacies. Annual production in the Philippines was 26,464 tonnes for the period 2000–2005, while that in Tugui it was 2,702 tonnes (Cornago *et al.*, 2011). Cornago *et al.* also evaluated five Philippine cultivars of *D. alata*: 'Daking', 'Kimabajo', 'Rapang-rapang', 'Sampero' and 'Shiket', and two cultivars of Tugui yam (*D. esculenta*), one 'Highland' and the other 'Lowland'. Their total phenolic content ranged from 69.9 to 421.8 mg GAE 100 g^{-1} (dry weight). Their half maximal effective concentration was over the range of 1.7–14.8, 6.2–31.7 and 17.5–35.1 mg mL^{-1} for radical scavenging activity, reducing power and iron chelating capacity, respectively. Total antioxidant activity, measured by the ferric thiocyanate method at 50 mg mL^{-1} was between 92.0–95.6% (Cornago *et al.*, 2011). *D. hispida* is also found in the Philippines. The Philippines is a substantial producer of yams, but reported yields are comparatively low (Table 7.48).

Puerto Rico

Yam production in Puerto Rico is located in the central mountainous zone, but Sotomayor-Ramírez *et al.* (2003) commented that there was interest in expanding production to the

Table 7.48. Yam production (tonnes annum^{-1}) and mean yield (tonnes ha^{-1}) in the Philippines. Modified from FAO Stats (2020).

Year	Total production (tonnes)	Mean yield in tonnes ha^{-1}
1961	35,000	5.0
1966	26,252	5.2
1971	22,318	4.5
1976	23,754	3.8
1981	18,194	2.3
1986	23,470	3.9
1991	26,093	4.9
1996	26,391	4.7
2001	24,287	4.8
2006	29,265	5.0
2011	17,844	6.0
2016	14,166	5.7
2017	14,376	5.9
2018	14,465	5.9

Table 7.49. Yam production (tonnes annum^{-1}) and mean yield (tonnes ha^{-1}) in Puerto Rico. Modified from FAO Stats (2020).

Year	Total production (tonnes)	Mean yield in tonnes ha^{-1}
1961	13,199	5.3
1966	13,608	5.3
1971	12,247	5.6
1976	14,152	8.8
1981	14,606	6.5
1986	12,020	5.7
1991	10,523	5.7
1996	4,445	14.1
2001	3,523	8.4
2006	2,600	8.8
2011	1,785	8.4
2016	1,485	9.0
2017	1,134	8.9
2018	1,582	6.4

southern semi-arid zone, where mechanization is possible, where irrigation is available and where soil pH is neutral to alkaline. Martin *et al.* (1975) selected a cultivar of *D. alata* 'Florido' in Puerto Rico that had a compact tuber size and shape. It was introduced into Côte d'Ivoire for a harvest mechanization project but the project was never implemented. However, 'Florido' was adopted widely by farmers and has spread to several West African countries. Yam production in Puerto Rico has declined substantially over recent years (Table 7.49) mainly through disease incidence, low yields and market price competition (Commonwealth of Puerto Rico, 2000). Gaztambide and Cibes (1975) previously reported that nutrient deficiencies during production of the *D. rotundata* cultivar 'Habanero' could be due to the deficiency of nitrogen, phosphorus, potassium, calcium and sulphur in the soil, but particularly potassium.

Rwanda

Yam is one of the premium tuber crops in Rwanda. It is third to potatoes and sweet potatoes in the order of production levels. The average price for a yam in Kigali was reported to be 500 Rwandese francs in comparison with 250 Rwandese francs for 'the other more consumed' tubers (Selina Wamucii, 2020). Yam production in Rwanda is progressively increasing, but mean yields have been variable (Table 7.50).

Table 7.50. Yam production (tonnes annum^{-1}) and mean yield (tonnes ha^{-1}) in Rwanda. Modified from FAO Stats (2020).

Year	Total production (tonnes)	Mean yield in tonnes ha^{-1}
1961	1,000	10.0
1966	1,300	8.1
1971	4,302	9.0
1976	5,622	5.5
1981	6,026	4.4
1986	6,530	6.3
1991	2,000	3.1
1996	7,065	3.7
2001	7,000	3.7
2006	12,928	4.9
2011	18,725	6.7
2016	45,000	7.9
2017	50,000	8.3
2018	53,394	9.3

Samoa

Samoa is an independent nation made up of two main islands, Upolu and Savaii and several smaller and uninhabited islands. Cable and Wilson (1984) reported that in Samoa *D. esculenta* and *D. alata* were the most important species, but *D. nummularia* was also grown. They reported that in Western Samoa four cultivars of *D. nummularia* were recognized: 'Palai Maoi', 'Palai Samoa', 'Palai Niu Kini' and 'Palai Maoi'. They found that the quality of 'Palai Niu Kini' was best when it just begins to sprout, but after that it became unpalatable. *D. alata* and *D. bulbifera* are locally called 'Soi' and 'Ufi' respectively. Yam production has been variable over the years, but mean yields have progressively reduced to comparatively very low levels (Table 7.51).

Table 7.51. Yam production (tonnes annum^{-1}) and mean yield (tonnes ha^{-1}) in Samoa. Modified from FAO Stats (2020).

Year	Total production (tonnes)	Mean yield in tonnes ha^{-1}
1961	6,000	15.0
1966	8,000	13.3
1971	9,000	14.3
1976	10,000	14.3
1981	11,000	13.8
1986	11,000	6.9
1991	11,500	4.6
1996	15,000	3.5
2001	2,653	4.7
2006	2,650	4.7
2011	4,961	4.6
2016	10,585	4.6
2017	8,644	4.6
2018	6,653	3.5

São Tomé and Príncipe

São Tomé and Príncipe is an island nation off the coast of West Africa, close to the equator. Total yam production has been progressively increasing over the years and mean yields have been low but relatively stable (Table 7.52).

Table 7.52. Yam production (tonnes annum^{-1}) and mean yield (tonnes ha^{-1}) in São Tomé and Príncipe. Modified from FAO Stats (2020).

Year	Total production in tonnes	Mean yield in tonnes ha^{-1}
1961	500	5.0
1966	600	6.0
1971	500	5.0
1976	500	5.0
1981	800	5.3
1986	1,186	5.9
1991	815	5.6
1996	1,000	5.3
2001	1,344	5.8
2006	1,666	5.8
2011	2,350	6.0
2016	2,220	5.9
2017	2,257	5.9
2018	2,294	5.9

Solomon Islands

Yams, breadfruit and taro are important staple food crops in the Solomon Islands, but since the introduction of sweet potatoes and cassava, the planting of yams, breadfruit and taro has reduced drastically partly because sweet potatoes and cassava are easier and cheaper to produce (Solomon Islands, 2014). Yams are a significant food crop, which accounted for 8.1% of the total dietary calorie intake in the Solomon Islands (Opara, 2003). Yam production remains substantial and has progressively increased over recent years and there is some indication of increases in mean yield (Table 7.53).

Table 7.53. Yam production (tonnes annum^{-1}) and mean yield (tonnes ha^{-1}) in the Solomon Islands. Modified from FAO Stats (2020).

Year	Total production (tonnes)	Mean yield in tonnes ha^{-1}
1961	12,000	7.1
1966	13,300	7.5
1971	14,500	7.3
1976	16,700	8.4
1981	17,700	7.4
1986	19,100	6.4
1991	21,900	6.8
1996	22,000	7.3
2001	29,000	7.9
2006	30,000	8.6
2011	35,000	8.5
2016	44,218	10.4
2017	44,940	10.5
2018	46,548	10.7

St Kitts and Nevis

Sugar cane was the main export crop of St Kitts and Nevis up until 2005 when the sugar industry was closed down. Since then, the travel and tourism sector has become the main economic activity but some small-scale production of crops, including rice, yams, bananas, sweet potatoes and cotton has continued. Yam production levels have become very low, although there has been some indication of improved mean yields (Table 7.54). Yam yields per hour of labour were similar on both small farms and estates. There were positive net returns to yams on small farms and estates in St. Kitts and Nevis. Both yam and sweet potato production were more costly on plantations/ estates than on small farms (Rankine, 1973).

St Lucia

Major cultivated crops in St Lucia are bananas, coconuts, cocoa, citrus, mangoes and root crops including cassava sweet potatoes and yams (three cultivars of *D. alata* 'Belep', 'Oriental DV' and 'Plimebite') (FAO Saint Lucia, 1996). Arnold Babwah and Associates (2016) reported that in St Lucia the average price per kg ($EC) paid by the buyer segments for fresh yams was: hotels 6.10, supermarkets 5.85, restaurants 6.40 and market vendors 5.80. Yams continue to be an important food crop in St Lucia, but mean yields are comparatively low and total production has continued to decline (Table 7.55).

St Vincent and the Grenadines

FAO (2008) reported that there were 26 yam varieties grown in St Vincent and the Grenadines, 20 of *D. alata*, three of *D. cayenensis* and one each of *D. rotundata*, *D. trifida* and *D. esculenta*. They had been mainly introduced, with a few indigenous cultivars being domesticated and utilized by a small number of farmers for both food and for medicinal purposes. The dominance of the banana export industry in St Vincent was reported by FAO (2008) as being replaced by dasheen, tannia, eddo, yam, cassava, sweet potato and plantains. These species, along with imported flour and rice, represent the major staple foods of the populace. Bananas are still the main crop although their importance has declined significantly in recent years. In 2018 it was reported in the local newspaper that St Vincent was

Table 7.54. Yam production (tonnes annum^{-1}) and mean yield (tonnes ha^{-1}) in St Kitts and Nevis. Modified from FAO Stats (2020).

Year	Total production (tonnes)	Mean yield in tonnes ha^{-1}
1961	380	2.5
1966	390	2.5
1971	400	5.0
1976	410	5.1
1981	430	5.1
1986	420	5.3
1991	300	5.0
1996	70	4.7
2001	5	2.5
2006	14	8.8
2011	37	9.3
2016	34	6.6
2017	25	9.1
2018	4	9.0

Table 7.55. Yam production (tonnes annum^{-1}) and mean yield (tonnes ha^{-1}) in St Lucia. Modified from FAO Stats (2020).

Year	Total production in tonnes	Mean yield in tonnes ha^{-1}
1961	2,032	3.3
1966	2,000	4.4
1971	3,450	4.7
1976	3,418	4.6
1981	3,650	3.8
1986	3,800	3.9
1991	4,100	3.9
1996	3,730	3.7
2001	1,177	3.7
2006	429	3.6
2011	419	3.6
2016	488	3.6
2017	489	3.6
2018	588	3.7

expecting to receive technical assistance from Ghana in relation to cocoa and yam production (*News 784*, 2018). Arnold Babwah and Associates (2016) reported that in St Vincent the average price in $EC per kg paid by the buyer segments for fresh yams was: hotels 5.10, supermarkets 4.45, restaurants 5.65 and hucksters 2.85. Mean yield has tended to increase over the years while total production has been relatively stable (Table 7.56).

Table 7.56. Yam production (tonnes annum^{-1}) and mean yield (tonnes ha^{-1}) in St Vincent. Official data or FAO data based on imputation methodology. Modified from FAO Stats (2020).

Year	Total production (tonnes)	Mean yield in tonnes ha^{-1}
1961	2,000	2.5
1966	2,286	2.5
1971	2,948	5.0
1976	1,980	5.1
1981	4,320	5.1
1986	2,000	5.3
1991	2,634	5.0
1996	1,273	4.7
2001	1,900	2.5
2006	2,250	8.8
2011	2,207	9.3
2016	2,366	6.6
2017	2,368	9.1
2018	2,584	13.4

South Sudan

South Sudan is a new country and was reported to have a population of 12,919,053 in 2018. Although there is a hotel in Juba called the 'Yam Hotel', little information is available on the type of yams grown in South Sudan, but yam production has been showing some indication of increasing (Table 7.57) and mean yields are consistently high.

Table 7.57. Yam production (tonnes annum^{-1}) and mean yield (tonnes ha^{-1}) in South Sudan. Modified from FAO Stats (2020).

Year	Total production (tonnes)	Mean yield in tonnes ha^{-1}
2012	43,000	25.3
2013	48,800	25.7
2014	51,000	24.7
2015	53,471	24.5
2016	57,351	25.0
2017	58,629	25.2
2018	61,391	25.2

Sudan

Landraces of *Dioscorea* species are known to have been grown by the inhabitants of the southern states of Sudan for many years (FAO Sudan, 1995). Corkill (1948) surveyed the distribution of the wild cluster yam, *D. dumetorum*, and found it growing in Africa south of the Sahara and the Sudan. They reported that eating it could cause various disorders that could, in the most severe conditions, lead to coma and death. *D. sansibarensis* (Obidiegwu and Akpabio 2017), *D. quartiniana* (Contu, 2013) and *D. bulbifera* (Govaerts, 2012) were also reported growing in Sudan. Substantial production remained relatively constant, but latterly has shown substantial increases as have mean yields (Table 7.58).

Table 7.58. Yam production (tonnes annum^{-1}) and mean yield (tonnes ha^{-1}) in 'former' Sudan (1961–2011) and Sudan (2016–2017). Official data or FAO data based on imputation methodology. Modified from FAO Stats (2020).

Year	Total production (tonnes)	Mean yield in tonnes ha^{-1}
1961	120,000	3.0
1966	150,000	3.0
1971	120,000	3.1
1976	115,000	2.9
1981	115,000	2.9
1986	116,000	2.8
1991	128,000	2.8
1996	133,000	2.6
2001	137,000	2.4
2006	137,000	2.4
2011	141,153	2.4
2016	177,940	21.1
2017	181,889	21.1
2018	187,058	21.6

Tanzania

Ruffo *et al.* (2002) reported that in the United Republic of Tanzania there are 14 *Dioscorea* species distributed in different areas of the country. Some of these are found in the wild and some are cultivated as food crops. Yams in Tanzania are called 'kiazi kikuu' in Swahili and are grown for sale, home consumption and food security. The main *Dioscorea* spp. grown in Tanzania were: *D. cayenensis, D. alata, D. abyssinica* and *D. rotundata* (Selina Wamucii, 2019). *D. sansibarensis*, also called Zanzibar yam, is native to Madagascar and to tropical Africa (from Tanzania, including Zanzibar, west to Guinea and south to Mozambique) (Govaerts *et al.*, 2007). Both yam and taro are cultivated in Tanzania by smallholder farmers, and over 25% of farmers are engaged in the 'Yam and Taro Project' from the University of Dar es Salaam led by Dr Gladness Temu. The aim of the project is 'to use tissue culture to improve existing varieties in order to increase their production and distribution' (Abdu, 2018). Ruffo *et al.* (2002) also described how yams are prepared in Tanzania, where tubers are peeled, cut into small pieces and soaked in water overnight to remove toxic substances before being cooked, or tubers are peeled then soaked in water for several days, washed, sliced into small pieces and dried in the sun. Dried slices may be pounded into flour and used for making porridge. In recent years yam production and mean yield have remained fairly constant (Table 7.59).

Table 7.59. Yam production (tonnes annum^{-1}) and mean yield (tonnes ha^{-1}) in Tanzania. Modified from FAO Stats (2020).

Year	Total production (tonnes)	Mean yield in tonnes ha^{-1}
1961	5,000	7.1
1966	6,000	6.7
1971	7,000	7.0
1976	7,800	6.5
1981	8,200	5.8
1986	8,800	5.5
1991	9,148	5.4
1996	10,128	5.6
2001	8,500	5.7
2006	8,200	5.7
2011	9,800	6.0
2016	10,808	6.3
2017	11,036	6.4
2018	11,210	6.6

Fig. 7.4. Harvesting *D. alata* tubers at a small-scale farm in Chachoengsao province in south-central Thailand, east of Bangkok, March 2020.

Thailand

The main species of *Dioscorea* cultivated in Thailand is *D. alata* (Fig. 7.4). Sagwansupyakom and Chantraprasong (1984) reported that in Thailand yams are not a traditional food and are often eaten as a sweet or as a part of a snack. During the Second World War some species were eaten as a rice supplement in villages in the north and northeast of Thailand. They reported that 38 species had been identified, which were distributed from sea level to 1,800 m (Table 7.60). The Sakai tribe have inhabited some forest areas of southern Thailand – specifically Trang, Phatthalung, Satun, Songkla, Yala and Narathiwat provinces – for more than 10,000 years (Maneenoon *et al.* 2008). The main sources of carbohydrate for the Sakai people are yams and Maneenoon *et al.* found 15 species of *Dioscorea* in the Banthad Range, 13 of which are edible. Eight of these (*D. calcicola, D. filiformis, D. glabra, D. orbiculata, D. pentaphylla, D. pyrifolia, D. stemonoides* and *D. wallichii*) are used as food by the Sakai. The remaining edible species,

D. daunea, *D. membranacea*, *D. piscatorum*, *D. prazeri* and an unidentified *Dioscorea* sp. were reported to be substitute species during famine since they are bitter and have an unpleasant taste. *D. hispida* is commonly used for making snack foods in Thailand. Joob and Wiwanitkit (2014) reported that *D. hispida* is used in a Thai dessert called 'khao neaw na kloy' (served with sweet sticky rice). They described the symptoms that can occur when some people eat this dish: itching can develop within 15 minutes and some cases have been reported where consumption has been be fatal. However, cooking can remove the toxicity and the above effects are usually due to poor cooking. No information was given by FAO Stats (2019) on yam production in Thailand.

Table 7.60. *Dioscorea* species identified in Thailand. Modified from Sagwansupyakom and Chantraprasong (1984).

Dioscorea species	Characteristics and uses
D. alata	Edible
D. arachidna	Watery
D. birmanica	Edible
D. brevipetiolata	–
D. bulbifera	Edible
D. calcicola	Rare
D. cirrhosa	–
D. collinsae	Edible
D. craibiana	Rare
D. cumingii	Rare
D. daemona	–
D. decipiens	Edible
D. depauperata	–
D. esculenta	Edible
D. filiformis	Edible
D. garrettii	Rare
D. glabra	Edible
D. graci	–
D. hamiltonii	Rare
D. hispida	Edible
D. inopinata	Rare
D. japonica	Rare
D. kratica	Edible
D. laurifolia	Medicinal properties
D. membranacea	–
D. nummularia	Edible
D. orbiculta	–
D. oryzetorum	Edible
D. panthaica	Rare
D. paradoxa	Edible
D. pentaphylla	Edible
D. pierrei	Edible
D. prazeri	Poisonous
D. pseudonitens	Rare
D. pseudotomentosa	Rare
D. stemenoides	–
D. velutipes	Rare
D. wallichii	Edible

Togo

In Togo, Dansi *et al.* (2013) identified 19 constraints that affect the production of *D. cayenensis*/*D. rotundata* complex of which the most important were insect damage on both leaves and tubers, nematode attack on tubers, drought, soil poverty and wilting. Criteria for varietal preference included high productivity, good quality of fufu, resistance to drought and adaptability to all types of soils. Over the years Togo has been the fifth or sixth major yam producer in the world and over the last 50 to 60 years has shown a progressive, though fluctuating, increase in production (Table 7.61).

Table 7.61. Yam production (tonnes annum^{-1}) and mean yield (tonnes ha^{-1}) in Togo. Official data or FAO data based on imputation methodology. Modified from FAO Stats (2020).

Year	Total production (tonnes)	Mean yield in tonnes ha^{-1}
1961	300,000	10.0
1966	280,000	9.3
1971	300,000	10.0
1976	416,539	9.3
1981	527,815	9.1
1986	409,386	9.9
1991	376,478	9.0
1996	604,732	8.7
2001	614,962	11.4
2006	621,055	10.3
2011	727,749	10.2
2016	813,985	8.9
2017	826,523	9.1
2018	858,783	9.2

Tonga

Yams are an important food crop in Tonga, and according to Opara (2003) accounted for over 20% of the total dietary calorie intake. One farmer, who grows food crops on his farm in Tongatapu, Tonga's largest island, was quoted as saying. 'I grow a particular yam called "Ufi Lose" which can survive cyclone weather' and 'the measure of a man's wealth is not in monetary value but by his root crops and pigs' (Tonga, 2019). Yam production was substantial but has reduced over recent years to comparatively low but consistent levels (Table 7.62).

Trinidad and Tobago

There was an enormous amount of research work on yams in Trinidad in the 1960s and 1970s, when the University of the West Indies at St Augustine in Trinidad had a major root crops research programme. However, since that time production has become very limited indeed and mean yield very low (Table 7.63). Production of other staple crops, including sweet potatoes, has also declined, with only a little over 50% market demand for cassava and 11% for plantain, both of which are grown locally. In addition, Eitzinger *et al.* (2015) commented that by 2050 Trinidad and Tobago will be expected to show a significant reduction in the area suitable for growing tomatoes and possibly cocoa. They recommend that crops such as cassava, sweet potatoes and yams are good diversification alternatives. Degras (1993) reported that *D. trifoliata* is native to South America, including Trinidad (Marsden, 1812). *D. convolvulacea* is native to Central America, also including Trinidad (Govaerts *et al.*, 2007). *D. trifida* is native to South America and the Caribbean islands and was taken to Trinidad by the Arawaks. In the town of Tunapuna in Trinidad and Tobago there is an annual yam festival and competition where enormous tubers are judged entirely on size with the biggest *D. alata* yam tuber winning. Mr D.G. 'Pat' Coursey was invited to be judge in 1969, when he was a visiting lecturer at the nearby Department of Crop Science at the University of the West Indies.

Uganda

It was estimated by DIIVA (2020) that there were 254,000 hectares of yams grown in Uganda of which 14% of this area was estimated to be grown with improved cultivars

Table 7.62. Yam production (tonnes annum^{-1}) and mean yield (tonnes ha^{-1}) in Tonga. Modified from FAO Stats (2020).

Year	Total production (tonnes)	Mean yield in tonnes ha^{-1}
1961	28,000	12.2
1966	30,000	13.0
1971	30,767	13.2
1976	34,000	13.9
1981	27,500	13.8
1986	21,000	14.0
1991	8,104	14.1
1996	6,385	15.0
2001	4,400	12.6
2006	4,600	12.8
2011	4,711	13.7
2016	4,600	13.6
2017	4,732	13.6
2018	4,502	13.6

Table 7.63. Yam production (tonnes annum^{-1}) and mean yield (tonnes ha^{-1}) in Trinidad and Tobago. Modified from FAO Stats (2020).

Year	Total production (tonnes)	Mean yield in tonnes ha^{-1}
1961	1,878	15.7
1966	2,629	15.6
1971	3,404	15.7
1976	3,200	16.8
1981	3,518	18.5
1986	186	12.4
1991	130	8.7
1996	135	10.4
2001	12	8.6
2006	12	6.0
2011	78	3.1
2016	23	2.9
2017	24	3.6
2018	29	2.1

in 2009. The major yam cultivars grown in Uganda were reported to be 'Kyetutumula' (*D. cayenensis*), 'Kikwa' (*D. burkilliana*), 'Makunyi' (*D. bulbisiana*), 'Masebe', 'Nandigoya', 'Ndaggu Nganda', 'Ndaggu Nziba', 'Balugu', 'Vigonzo', 'Buyu', 'Vitungi' and 'Mahombo' (all *D. alata*). Kagoda *et al.* (2010) also reported that the most commonly grown yam landraces in Uganda included 'Kyetutumula' (*D. cayenensis*), 'Kisebe' (*D. alata*), 'Nakasoma' (*D. alata*), 'Kaama' (*D. abyssinica*), 'Ndaggu Nganda' (*D. alata*), 'Ndaggu Nziba' (*D. alata*) and 'Balugu' (*D. cayenensis/D. rotundata*) from Nigeria (popularly referred to as 'Nigerian yams'), which have been supplied to farmers in different locations by IITA/NARO. There are also several introduced yam hybrid lines. Most Ugandan farmers (52%) grow yams in pits, which makes large-scale production impossible. The crop fits well into the different cropping systems as well as the social and cultural context of most Ugandan communities. Major yam growing areas in Uganda include Kalangala, Masaka, Gulu, Luwero, Mukono, Kapchorwa, Bundibugyo and Mbale (Ministry of Agriculture, Animal Industry and Fisheries, Entebbe, 1996).

Many farmers in Uganda sell yam tubers in both rural and urban markets for income generation (Ocitti p' Obwoya, 2001). Nematodes were reported to be the major constraint to yam production in Uganda with *Pratylenchus sudanensis* being a major threat to increasing and sustaining yam production (Mudiope *et al.*, 1998). Coyne *et al.*, (2005a) reported that even after the introduction of the IITA/NARO germplasm, the cultivar 'Kyetutumula' (*D. rotundata*) remained the most preferred, essentially due to its taste. The introduced germplasm 'Kyetutumula' was in general well liked, depending on the clone, but farmers were less willing to increase their investment in expansion of these introduced cultivars due to variability between clones, including taste.

Venezuela

Pérez *et al.* (2010) reported that in the east of the Bolivarian Republic of Venezuela, Guyana and Caribe states there is an important and abundant range of tuberous plants including potatoes, yams, cassava and taro. The tubers of *D. trifida*, locally called 'Mapuey', are a staple food for indigenous peoples from the Caribbean coast and Amazon regions of Venezuela and Pérez *et al.* analysed and described three varieties (Table 7.64). *D. bulbifera* and *D. alata* are commonly-grown introduced species (Hokche *et al.*, 2008) while *D. abysmophila* was reported by Govaerts *et al.* (2007) as being native to Venezuela and Dorr and Stergios (2003) reported finding a new *Dioscorea* species, *D. lisae*, in the Venezuelan Andes as well as *D. lehmanii*, *D. meridensis* and *D. coriacae*. Total yam production in Venezuela has fluctuated in recent years, but there is some indication of progressive improvement in mean yield of tubers (Table 7.65).

Wallis and Futuna Islands

D. pentaphylla is widely cultivated as a food crop on several Pacific islands including the Wallis and Futuna Islands (Morat and Veillon, 1985). *D. alata* and *D. bulbifera*, which were introduced, were also reported to be cultivated in the Wallis and Futuna Islands (Meyer, 2007; Govaerts, 2013). Total production of yam tubers is only small and has been consistent over recent years, but there is some indication of decreasing mean yields (Table 7.66).

Table 7.64. Properties of varieties of *D. trifida* in Venezuela. Modified from Pérez *et al.* (2010).

	'Amazonian White'	'Amazonian Purple'	'Amazonian Deep Purple'
Mean weight	243g	216g	112g
Moisture	69%	73%	75%
Crude protein	6.8%	4.7%	4.9%
Fatty material	0.30%	0.28%	0.03%
Starch	89.5%	91.6%	93.2%

Trade in Yams

CBI (2019) reported that yams are produced overwhelmingly as a traditional staple food crop, with only a small fraction exported as fresh products. Global exports of yams in 2019 was valued at US$ 177 million, with a recorded annual growth rate in exports of 3% from 2018 to 2019. Over the period 2015–2019 the total growth rate was 9%. GEPA (2017) reported that Ghana was among the top exporters of yams worldwide, reaching around $US 27.5 million in 2016, with the USA being the largest recipient, followed by the UK. Exports of yams from Ghana in 2019 amounted to US$ 39.1 million giving it a world share of 22.1%. In the UK, Ghana was reported to be by far the largest supplier of yams with over 80% market share. In the US market in 2016, Jamaica (33%) was the largest followed by Costa Rica (20%) and Ghana (15%). Top countries exporting to the USA were Jamaica ($24,435,927), Ghana ($11,329,474), Colombia ($9,224,105) and Costa Rica ($9,001,919) (Flexport, 2020).

Table 7.65. Yam production (tonnes annum^{-1}) and mean yield (tonnes ha^{-1}) in Venezuela. Official data or FAO data based on imputation methodology. Modified from FAO Stats (2019).

Year	Total production (tonnes)	Mean yield in tonnes ha^{-1}
1961	25,496	5.5
1966	44,049	4.9
1971	38,852	4.5
1976	30,438	4.7
1981	30,963	5.3
1986	38,290	5.7
1991	40,964	5.9
1996	57,088	8.2
2001	87,523	9.2
2006	87,243	8.6
2011	104,295	10.1
2016	47,698	9.6
2017	48,351	9.7
2018	47,692	9.8

Table 7.66. Yam production (tonnes annum^{-1}) and mean yield (tonnes ha^{-1}) in Wallis and Futuna Islands. Modified from FAO Stats (2020).

Year	Total production (tonnes)	Mean yield in tonnes ha^{-1}
1961	450	11.3
1966	450	11.3
1971	560	13.3
1976	560	13.3
1981	560	13.3
1986	500	11.9
1991	561	13.4
1996	533	12.7
2001	504	12.0
2006	530	11.8
2011	532	10.3
2016	524	9.9
2017	524	9.8
2018	502	8.7

Nigeria has a long-standing trade relationship with neighbouring countries including Niger, Benin, Cameroon, Chad and Equatorial Guinea, particular for yams and cassava that are the second largest category of products exported by Nigeria to these regional partners (Bergh *et al.*, 2012). For yams, taro and other similar roots (with the exception of cassava), no size classes have been defined. Usually, their average length and/or diameter and average weight is listed in the specifications. Also, the size requirement may vary depending on the market. For local markets in many producing countries yam tubers are simply stacked in the back of lorries for transport from the farm to the packhouse or wholesale and retail markets). In India *D. alata* tubers are commonly packed in 13 kg jute sacks. For export from West Africa and the Caribbean Islands, yams (mainly *D. rotundata* and to a lesser extent *D. trifida*) are usually transported by sea freight packed in fibreboard cartons (Fig. 7.5)

Fig. 7.5. Fully telescopic cartons containing 20 kg of *D. rotundata* tubers imported from Ghana, at a wholesale market in the UK, October 2020.

8

Some *Dioscorea* Species

Over 600 species of *Dioscorea* have been described (Chapter 2), but not all are included in this chapter.

D. abysmophila

Dioscorea abysmophila Maguire & Steyerm is native to Venezuela and was reported growing in the Amazonian region of São Gabriel, on the northern shore of the Rio Negro in Brazil (Plants of the World Online *Dioscorea abysmophila*, 2019). Kew (2020) reported that a specimen collected in 1945 indicated that it was from rocky terrain on a high mountain range.

D. abyssinica

Dioscorea abyssinica Hochst. ex Kunth is native to several African countries including Benin, Burkina Faso, Central African Republic, Côte d'Ivoire, Eritrea, Ethiopia, Ghana, Liberia, Mali, Nigeria, Senegal and Sudan. Its common names in Ethiopia are 'boyna' and 'dika' and in different parts of West Africa it is called 'dooya' or 'diabongua'. *D. abyssinica* is considered a savannah species but it also grows in rocky places in forests (Magwé-Tindo *et al.*, 2018). Belachew Garedew *et al.* (2017) reported that in the Sheko district of Ethiopia yams were identified as a main staple food where farmers were growing *D. abyssinica*, *D. alata* and *D. bulbifera*. Also, Obidiegwu and Akpabio (2017) commented that *D. abyssinica* made a secondary, but significant, contribution to the yam domestication process in Nigeria. In Benin, Djedatin *et al.* (2017) studied the domestication of wild yams by local farmers and commented that the choice of the genotypes they domesticated was made on the morphological characteristics of plants as judged by the farmers. Based on farmers' criteria, 140 tubers belonging to *D. praehensilis* and *D. abyssinica* were sampled and morphological characterization, using the International Plant Genetic Resources Institute descriptors, displayed differentiation between domesticable and non-domesticable genotypes, supporting the hypothesis of the existence of a strong genetic component in the ability to domesticate yam varieties. *D. abyssinica* is a non-spiny annual climber up to 3 or 4 m high with stems twining right-handed or scrambling over the ground. Tubers are produced deeply in the soil and are slender in shape, making harvesting difficult. In addition to the tubers, *D. abyssinica* produces small grey-brown bulbils that are easily dislodged.

DOI: 10.1079/9781789249279.0008

Composition and uses

Tubers are cooked and eaten in various ways. Tubers are commonly harvested from the wild but are sometimes cultivated, especially in East Africa. The proximate composition of *D. abyssinica* tubers on a dry weight basis was given as: 0.1% ash, 0.5% protein, 1% fat and 98.4% starch, with an amylose content of 29.7% (Swinkels, 1985; Gebre-Mariam and Schmidt, 1998). Aregahegn *et al.* (2013) also found that the major constituent of the tubers was starch which accounted for 80% on a dry weight basis. They also analysed the mineral content in the tubers and flour of *D. abyssinica* grown in different parts of Ethiopia and found that the mean concentrations ranged (in $\mu g\ g^{-1}$) as follows: 8,469–13,914 K, 133–405 Na, 172–448 Ca, 180–354 Mg, 28.3–144.5 Fe, 12.0–14.5 Mn, 12.3–44.5 Zn, 7.26–17.6 Cu, 1.91–8.68 Co, 0.86–3.41 Cr and 2.43–5.31 Ni.

D. acanthogene

Dioscorea acanthogene Rusby is native to Bolivia, west-central Brazil, Colombia, Paraguay and Peru. *D. acanthogene* occurs in open vegetation formations (dry diagonal) and forested biomes – for example in the Amazonia region (Couto *et al.*, 2018).

D. acerifolia

Dioscorea acerifolia Phil. (also referred to as *D. acerifolia* Uline ex Diels) has been reported growing in Argentina, China, Japan and Surinam (GBIF, 2019). Quattrocch (2016) lists tubers of *D. acerifolia* var. *rosthornii* Diels as being poisonous, but edible after cooking.

D. acuminata

Dioscorea acuminata Baker is synonymous with *D. bararum* H. Perrier and *D. variifolia* Bertero and is native to Madagascar. It is distributed in dry forests and is classified as endangered (it is only known from four botanical collecting sites on the central plateau). The habitat at the one of the known sites is threatened by destruction through a variety of factors including encroachment by *Pinus* sp., marble extraction and collection of plants for consumption. It has been found in elevations from 1,000 to 1,560 m. Stems are left twining and tubers grow vertically in the soil and are cylindric with a white skin (Wilkin *et al.*, 2008). Burkill and Perrier (1950) stated that tubers of *D. acuminata* were harvested from the wild and eaten, following cooking, and were of very good quality, but Wilkin *et al.* (2008) and Kennerley and Wilkin (2017) reported that there are no known direct uses of the tubers.

D. aculeata

Dioscorea aculeata L. is synonymous with *D. aculeata* Roxburgh, *D. esculenta* (Lour.) Burkill, *D. tugui* Blanco, *D. tiliifolia* Kunth, *D. spinosa* Roxburgh ex Hook.f., *D. papuana* Warb., *D. papillaris* Blanco, *D. fasciculata spinosa*, *D. fasciculata* Roxburgh, *D. esculenta tiliifolia*, *D. esculenta spinosa*, *D. esculenta fulvidotomentosa*, *D. esculenta fasciculate*, *D. aculeata spinosa*, *D. aculeata* Roxburgh, *Dioscorea aculeata* L., *Oncus esculentus* Lour. and *Oncorhiza esculentus* (Lour.) Pers (Kew, 2014). Common names include: afou, barhtlum, birch rind yam, Goa potato, Guinea yam, kaawi, purple yam, thagdi and ufi tau. It is native to eastern Asia and was reported to occur in central India (Arunachal Pradesh, Assam and Meghalaya provinces) as well as South East Asia and Africa (JSTOR, 2018). Hedrick (1919) wrote

> Tropical Asia. This yam is said to be a native of tropical, eastern Asia, and is cultivated in the Indian Archipelago, the Pacific islands and the West Indies. The root is of a sweetish taste and Dr Seemann regarded it as one of the finest esculent roots of the globe. It is cultivated in India and the tubers are dug, in the cold season, in the forests and sold in the bazaars.
> A variety cultivated at Caracas has a very delicious taste, though Lunan, 1814, at

> Jamaica, says this yam is slightly bitter. This yam is said by Seemann, at Viti (a Fijian island), never to flower or fruit.

The peel and flesh of tubers are white. There are up to 40 stalked individual tubers on each plant. In Africa the introduced clones were reported to be sterile. *D. aculeata* tubers have a sweet taste and are boiled and used as a vegetable. They are sparsely cultivated in coastal West Africa from Sierra Leone to Eastern Nigeria. They also have traditional medicinal uses (Abhyankar and Upadhyay, 2011).

D. alata

Dioscorea alata L. is synonymous with: *D. atropurpurea* Roxburgh, *D. colocasiifolia* Pax, *D. eburina* Lour., *D. eburnea* Lour., *D. globosa* Roxburgh, *D. javanica* Queva, *D. purpurea* Roxburgh, *D. rubella* Roxburgh, *D. sapinii* De Wild., *D. sativa* Del, *D. vulgaris* Miq., *Elephantodon eburnea* (Lour.) Salisb. and *Polynome alata* (L.) Salisb. Hedrick (1919) quoted Sturtevant's 1892 work *Sturtevant's Edible Plants of the World* in relation to *D. rubella* 'East Indies': 'This is a common but very excellent yam of India, as good perhaps as any in cultivation. The tuber is of great size, crimson-red on the outside and of a glistening white within'. Common names include: Asiatic yam, greater Asiatic yam, greater yam, Guyana arrowroot, khoai mỡ, Lisbon yam, purple yam, rasa valli kilangu, ratalu, ten-months yam, ube, ufi, violet yam, water yam, white yam and winged yam. Common names in India include sree roopa, sree keerthi and sree shilpa (TNAU, 2019).

Kay (1987) stated that *D. alata* is not known in the wild, but was developed from species originating in South East Asia and is now cultivated throughout the tropics. There is evidence that it was first cultivated in the Assam province of India or Myanmar (Ahmad and Javed, 2007). Burkill (1935) stated that it is not known in the wild and either *D. hamiltonii* or *D. persimilis* may be an ancestor. Hedrick (1919), quoting Sturtevant:

> *D. alata* Linn. white yam. Tropical Asia. This plant is cultivated in the tropics of the whole earth. Unger ... says the Indian Archipelago and the southern portions of the Indian continent is the starting point of this yam, thence it was carried first to the eastern coast of Africa, next to the west coast and thence to America, whence the names *yam* and *igname* are derived from the negroes. In the Negro dialect of Guinea, the word yam means 'to eat'. This is the species most generally cultivated in the Indian Archipelago, the small islands of the Pacific and the Indian continent. It is universally cultivated in the Carnatic region. There are several varieties in Jamaica, where it is called white yam.

Hedrick again quotes Sturtevant, who described *D. globose* as

> East Indies. This species is much cultivated in India as yielding the best kind of yam and is much esteemed both by Europeans and natives. Roxburgh says it is the most esteemed yam in Bengal, but Firminger ... thinks it not equal in quality to other varieties. In Burma, Mason says it is the best of the white-rooted kinds.

Some cultivars of *D. alata* have bulbils (Fig. 8.1). Tubers have brown skin and, commonly, white flesh, but flesh can be pink or deep reddish-purple – for example *D. alata* var. *atropurpurea* and *D. alata* var. *rubella* (Fig. 8.2). Tubers vary in size, usually up to about 50 cm long and up to 30 cm in diameter, and may be cylindrical, globular, branched, lobed, flattened or fan-shaped (Figs 8.3, 8.4) and normally weigh 5 to 10 kg, but can be as heavy as 60 kg (Prain & Burkill 1938; Coursey, 1967; Kay, 1987).

Composition and uses

D. alata is known to contain bioactive compounds such as dioscorine, diosgenin and water-soluble polysaccharides. Shah and Lele (2012) reported that the purple yam (*D. alata* var. *purpurae*) is an important source of diosgenin. The composition of raw *D. alata* tubers, in mg or g 100 g^{-1} edible portion, was: energy 135 kcal, moisture 63.4–73.8 g 100 g^{-1}, protein 1.9–3.2 g 100 g^{-1}, fat 0.1 g 100 g^{-1}, carbohydrate 31.8 g 100 g^{-1}, fibre 1.5 ± 0.3-2.4 g 100 g^{-1}, ash 0.7–1.0 g 100 g^{-1}, calcium 24-39 mg 100 g^{-1}, phosphorus 27-56 mg

Fig. 8.1. *D. alata* bulbils of bright purple variety and white variety. Photograph taken at Sarinya's yam farm in Chachoengsao province in the eastern region of Thailand, 2018.

Fig. 8.2. Mixed boiled tubers of *D. alata* (purple and white) and *D. esculenta* (yellow/brown) cut into pieces. Photograph taken at in Sarinya's yam farm in Chachoengsao province in the eastern region of Thailand, 2018.

100 g^{-1}, iron 0.2-1.7 mg 100 g^{-1}, β-carotene equivalent 10 mg 100 g^{-1}, thiamine 0.05 mg 100 g^{-1}, riboflavin 0.03 mg 100 g^{-1}, niacin 0.05 mg 100 g^{-1}, ascorbic acid 5–15 mg 100 g^{-1} (modified from FAO, 1965). Opara (1999) reported that the content of the edible part of *D. alata* per 100 g edible tuber was: water 65–76 g, carbohydrate 20–31 g, fibre 0.6–1.5 g and protein 1.9–2.3 g. Knoth (1993) gave the following analysis on a fresh weight basis: 65–73% moisture, 22–29% carbohydrates, 0.1–0.3% fats and 1.1–2.8% crude protein. Shajeela *et al.* (2011) gave the following analysis on a fresh weight basis for *D. alata* tubers: starch 49.13 ± 0.21 g 100g^{-1}, niacin 36.20 ± 0.24 mg 100g^{-1}, ascorbic acid 74.56 ± 1.21 mg 100g^{-1}. They also gave the following analysis for antinutritional factors: total free phenolics 0.68 ± 0.04 g 100 g^{-1}, tannins 0.41 ± 0.01 g 100 g^{-1}, hydrogen cyanide 0.17 ± 0.01 mg 100 g^{-1}, total oxalate 0.58 ± 0.03 g 100 g^{-1}, amylase inhibitor 6.21 AIU mg^{-1}, soluble starch trypsin inhibitor 3.65 ± 0.04 TIU mg^{-1} protein. In the apical and basal regions of the tuber, the starch content decreased while the sugars and α-amylase activity increased during storage (Muthukumarasamy and Panneerselvam, 2000). Kay (1987) reported that a typical analysis of the edible portion of *D. alata* tubers was: water 65 to 73%, protein 1.12 to 2.78%, fat 0.03 to 0.27%, carbohydrate 22 to 29%, fibre 0.65 to 1.4% and ash 0.67 to 2.06%. She also reported that the starch from white fleshed and purple fleshed varieties had similar typical composition of: moisture 13.6%, protein 0.14%, ash 0.22%, amylose 21.1%, reducing sugars 0.18%, pH 7.1 and an iodine value of 5.5. Ascorbic acid content ranges from 4.9 to 8.2 mg 100 g^{-1} of edible portion have been reported, while certain cultivars in the South Pacific have been found to contain 6 mg 100 g^{-1} of carotene. Three cyanidin glycosides have been isolated from *D. alata* var. *atropurpurea* and *D. alata* var. *rubella*. Baah *et al.* (2009) gave the mean chemical composition, as percentages, of 16 *D. alata* varieties from experimental plots of the yam breeding programme at the International Institute of Tropical Agriculture, Ibadan in Nigeria (Table 8.1).

In Ghana, Coursey and Aidoo (1966) measured the ascorbic acid levels of several *D. alata* varieties and found that they varied between 5.0–21.5 mg 100g^{-1}. Lubag *et al.* (2008) showed that purple fleshed *D. alata* contained a variety of acylated anthocyanins that exhibit higher levels of antioxidant activity than the corresponding nonacylated compounds. Moriya *et al.* (2015) isolated

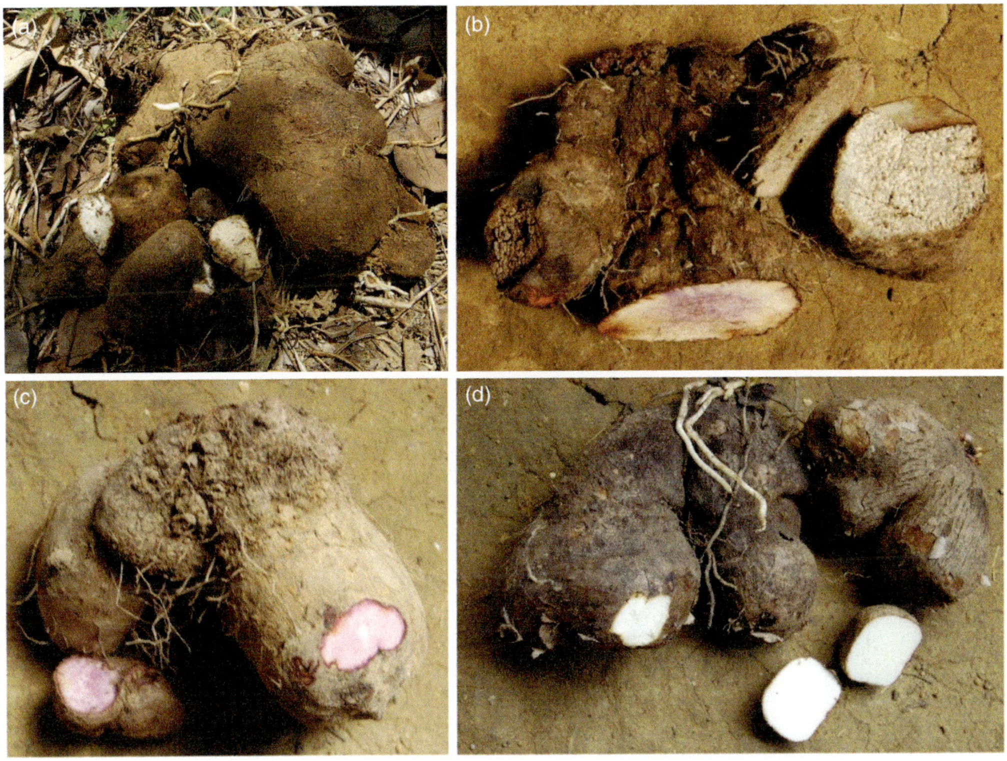

Fig. 8.3. *D. alata* tubers. Photographs taken at Sarinya's yam farm in Chachoengsao province in the eastern region of Thailand in 2018: (a) white variety; (b) purple variety; (c) bright purple variety; (d) all white variety.

Fig. 8.4. *D. alata* growing in Chachoengsao province in south-central Thailand. Called purple or blood yam, in Thai these three are called man lueat (มันเลือด), man lueat moo (มันเลือดหมู) and man lueat nok (มันเลือดนก) or man lueat kai (มันเลือดไก่).

eight anthocyanins from purple fleshed *D. alata*, measured their antioxidant activity and found that one of them, with a cyanidin skeleton, and two with sinapic acid units in their structure, exhibited the highest antioxidant activity of all the isolated anthocyanins. Rašper and Coursey (1967a,b) reported that the pigment occurring in the tubers of some *D. alata* varieties was anthocyanin, cyanidin-3,5-O-diglucoside.

In Côte d'Ivoire, Kouakou Dje *et al.* (2010) analysed tubers of *D. alata* and *D. cayenensis-rotundata* that had been stored for 6 months in a ventilated store where the

Table 8.1. Mean chemical composition (%) of 16 *D. alata* varieties from experimental plots of the yam breeding programme at the International Institute of Tropical Agriculture Ibadan, Nigeria (Baah *et al.* 2009).

	Fresh tuber		Flour				
	Moisture	Dry matter	Protein	Ash	Sugar	Starch	Total dietary fibre
Minimum	66.2	22.3	4.3	2.9	3.6	60.3	4.1
Maximum	77.7	33.8	8.7	4.1	11.0	74.4	11.0
Mean	71.8	28.2	5.9	3.5	5.8	68.4	6.8

conditions were 26.56 ± 3°C and 82 ± 5% RH. Their results showed that the total phenolic content varied in different parts of the tuber and decreased during storage. Protein content ranged from 7.90 to 7.94% and generally did not vary significantly during storage although a slight fall was observed in the distal part of one variety after 6 months. Lipids were low, only about 0.20% of the dry matter, and didn't vary significantly during storage. Shajeela *et al.* (2011) reported that *D. alata* tubers can contain antinutritional factors (Table 8.2).

Table 8.2. Antinutritional factors of tubers of *D. alata*. Modified from Shajeela *et al.* (2011).

Component	Content
Total free phenolics (g 100g^{-1})	0.68 ± 0.04
Tannins (g 100 g^{-1})	0.41 ± 0.01
Hydrogen cyanide (mg 100 g^{-1})	0.17 ± 0.01
Total oxalate (g 100 g^{-1})	0.58 ± 0.03
Amylase inhibitor (AIU mg soluble starch^{-1})	6.21
Trypsin inhibitor (TIU mg protein^{-1})	3.65 ± 0.04

Harvesting

D. alata normally matures 9 to 10 months after planting, but some varieties are harvested immature after only 6 months (Kay, 1987). Harvesting is usually by digging the tubers by hand. Experimental mechanical harvesters were developed and tested at the University of the West Indies in Trinidad and Tobago by the agronomist Dr Theo Ferguson and the engineer Dr Lewis Campbell in the late 1960s. These were successfully used to harvest yam tubers with minimum mechanical damage, but there seems to be no evidence of their subsequent use.

Curing

Gonzalez and de Rivera (1972) recommended 29 to 32°C and 90 to 95% RH for about 4 days directly after harvest. Ravindran and Wanasundera (1993) successfully cured *D. alata* tubers by exposing them to sunlight for 3 days after harvest in mean ambient conditions of 29°C and 84% RH. Wilson (1980) described a simple method of curing *D. alata* tubers. The tubers were first stacked on the ground in a lightly shaded area and covered with grass or mats. A canvas tarpaulin was then placed over the whole stack to cover the grass or mat, ensuring the tarpaulin did not touch the yams. As an alternative, a simple wooden frame could be built and the canvas draped like a tent over the piled yams. She indicated that plastic sheets should not be used for curing since they can make the yams too hot. If no canvas tarpaulin was available, she found that several layers of sacks or mats could be used.

Sprouting

The dormancy period for *D. alata* from the Caribbean or Nigeria was reported to be 14 to 16 weeks (Passam, 1982; Opara, 1999). Tubers of the *D. alata* cultivars 'Sree Keerthi' and 'Sree Roopa', stored at room temperature of 30 to 32°C and 80 to 85% RH in the dark sprouted 70 to 80 days after harvest (Muthukumarasamy and Panneerselvam, 2000). Ravindran and Wanasundera (1993) found that sprouting started after 60 to 90 days during storage in 24 to 28°C and 70 to 90% RH.

A range of chemicals, which have been shown successfully to prevent sprouting in potato (*Solanum tuberosum*) tubers, were studied on yams but they all proved ineffective (Cooke *et al.*, 1988). However, Passam (1982) showed that postharvest treatment of *D. alata* tubers with gibberellins could delay sprouting during storage. Tschannen (2003) reported that gibberellic acid (GA_3) applied to *D. alata* directly after harvest prolonged dormancy and reduced respiration rate at the time of sprouting. Balogun *et al.* (2006) reported that applying GA_3 to the foliage prior to harvest had some effects on subsequent sprouting. When applied to tubers soon after harvest, GA_3 extended dormancy by 13 weeks (Gerardin *et al.*, 1998). Tubers treated with GA_3 at the beginning of storage sprouted 4 weeks later than those not treated, the effect being less when treatments were given later; treatments given after 3 months of storage did not inhibit sprouting (Martin, 1977). Okagami and Tanno (1991) tested GA_3 and CCC (chlorocholine chloride) on the sprouting of bulbils and tubers of ten species of *Dioscorea* including *D. alata*. In general, GA_3 inhibited sprouting and CCC promoted sprouting, although their efficiency differed between species. In some species, however, diluted GA_3 (0.003-0.3 μM) promoted sprouting and diluted CCC (3-30 μM) inhibited sprouting. They explained the relationship between GA_3 or CCC concentrations and sprouting by assuming that the two counteractive reactions were activated by GA_3 in different degrees.

Refrigerated storage

Storage of *D. alata* tubers at either 3 or 12°C resulted in total physiological breakdown within 3 to 4 weeks (Czyhriniciw and Jaffe, 1951). Coursey (1961) reported that at 5°C they appeared to store well for 6 weeks but on removal to 25°C chilling injury symptoms occurred rapidly. Chilling injury symptoms included flesh browning and a susceptibility to microorganism infection. Opara (1999) reported that the optimum conditions for storage of *D. alata* were 15 to 17°C and 70% RH for 180 days for cured tubers and 150 days for non-cured tubers. However, Opara differentiated between different *D. alata* varieties and reported that the optimum conditions for storage of water yam or greater yam was 30°C for several months and for white yam or Guinea yam it was 16°C and 80% RH also for several months. Coursey (1961) recommended 12.5°C for 8 weeks for *D. alata* and Gonzales and de Rivera (1972) and Opara (1999) recommended 15 to 17°C and 70% RH for 180 days with about 11% weight loss for cured tubers and 150 days with about 12 to 25% weight loss for non-cured tubers.

Sorh *et al.* (2015) stored *D. alata* tubers grown in Côte d'Ivoire in ambient conditions (31.25 ± 3.00°C, 81.44 ± 12.25% RH) and found that moisture, protein, starch, magnesium and calcium decreased significantly ($p \leq 0.05$), whereas reducing sugars increased. Stored tubers had high phosphatase activity that decreased after 3 months storage (during natural dormancy) and subsequently remained stable for between 4 to 7 months. Contrasting this, low amylase and cellulase activities increased significantly. These enzymes are involved in the metabolic and physiological activities of yam tubers during storage. Sorh *et al.* (2015) also gave the mineral content as: Mg 266 ± 24.93, Ca 185 ± 18.14, P 41 ± 0.48, K 90 ± 0.01, Na 20 ± 0.01 and Fe 10 ± 1.41 mg/100g dry matter (Table 8.3).

D. alatipes

Dioscorea alatipes Burkill and H. Perrier is native to southwestern Madagascar. Schols *et al.* (2003) studied the pollen of various *Dioscorea* species and noted that the inflorescence of *D. alatipes* grows very low to the ground on leafless lower stems, unlike most other members of *Brachyandra*, which might be associated with ant pollination. Wilkin *et al.* (2008) described a new species

Table 8.3. Changes in proximate composition (g 100g^{-1} dry matter) of *D. alata* cultivar 'Brazo Yam' tubers during storage in Côte d'Ivoire in ambient conditions (Sorh *et al.*, 2015).

Storage in months	Moisture	Ash	Protein	Lipids	Reducing sugars	Total sugars	Starch
0	68.38 ± 1.43	1.78 ± 0.03	5.78 ± 0.36	0.76 ± 0.10	1.30 ±0.31	3.50 ± 0.10	87.70 ± 4.70
3	65.30 ± 2.00	1.73 ± 0.04	5.44 ± 0.14	0.68 ± 0.08	1.97 ± 0.07	2.40 ± 0.11	78.02 ± 3.08
7	60.00 ± 2.00	1.69 ± 0.04	5.44 ± 0.14	0.69 ± 0.10	2.50 ± 0.07	2.72 ± 0.41	58.51 ± 5.05

of *Dioscorea*, *D. bako* Wilkin, in western Madagascar that was similar to *D. alatipes*, and outlined the characteristics that differentiated the two species. *D. alatipes* was identified as a potential food resource in the Mahafaly region of Madagascar where 70% of the interviewed households were collecting wild yams for consumption (Andriamparany *et al.*, 2014).

D. althaeoides

Dioscorea althaeoides R. Knuth is native in a range from south-central China (Guizhou, Sichuan, Eastern Xizang and Yunnan provinces) to Thailand in mixed forests, on mountain slopes, in ravines and by roadsides, all at 1,400–3,200 m. Stems twine to the left and tubers are cylindric up to 1.5 cm in diameter, produced horizontally with little branching (Ding and Gilbert, 2010).

D. altissima

Dioscorea altissima Lamarck is indigenous to forested areas of Brazil, Bolivia, Peru, Central America (Panama) and the Caribbean islands. Its common name is dunguey. Its tubers are foraged and cooked for human consumption, and *D. altissima* is also cultivated in Brazil on a small scale for its edible tubers. This species has been introduced to Puerto Rico (Mansfeld's World Database of Agriculture and Horticultural Crops, 2019). Kirizawa *et al.* (2016) identified *D. altissima* as growing in the Reserva Biológica do Alto da Serra de Paranapiacaba, Santo André.

D. antaly

Dioscorea antaly Jum. and H. Perrier is found in the west and the east of Madagascar and the name, Antaly, is a town in southwestern Madagascar. Stems are left twining and tubers have a brown skin; their growth is in the horizontal direction with ramifications (Fig. 8.5). In times of scarcity, *D. antaly* tubers are used as food after soaking and cooking to remove bitterness and toxicity. Tubers contain 8-epidiosbulbin E, which had an LD_{50} of 0.86 mg mL^{-1} when tested on Japanese rice fish (*Oryzias latipes*) embryo larvae (Rakotobe *et al.*, 2010).

D. arachidna

Dioscorea arachidna Prain & Burkill is synonymous with *D. collinsae* Prain & Burkill and *D. filicaulis* Prain & Burkill. Its native range is Assam to south-central China and it also occurs in Thailand, northeast India, Cambodia, Laos and Vietnam in open and disturbed areas within dry and mixed deciduous forest and hill evergreen forest. It is found on various soils, including limestone, from near sea level to 1,250 m (Govaerts *et al.*, 2007). In 2019 there were 16 occurrences of recorded specimens of *D. arachidna* in the Global Biodiversity Information Facility (GBIF, 2019), which were: Thailand, ten sightings from 1930 to 1992; Vietnam, three sightings from 1914 to 1925; Cambodia, one sighting, 1965; India one sighting (not dated); and Afghanistan one sighting, not dated. Tubers are very small, up to 5 mm in diameter and 5–8 cm in length, cylindric to clavate, with the apex thickened globose to subglobose (Wilkin, *et al.*, 2009; Botany VM, 2018).

Fig. 8.5. *D. antaly* tuber. Reproduced with permission of Dr Marcella Corcoran.

D. aspersa

Dioscorea aspersa Prain & Burkill. Stems twine to the right and tubers are cylindric and have white flesh. They occur at 1,600–2,100 m in the forests and on the mountain slopes of western Guizhou and eastern Yunnan provinces of China (Flora of China, 2019).

D. bahiensis

Dioscorea bahiensis R. Knuth. As its specific name would suggest, it is native to Bahia in Brazil although Govaerts *et al.* (2007) gave its native range as 'cosmopolitan'.

D. bako

Dioscorea bako Wilkin is native to western Madagascar. Wilkin *et al.* (2008) described *D. bako* as a new species similar to *D. alatipes*, but with characteristics that differentiate the two. *D. bako* occurs in deciduous forest on sandy, loamy, lateritic or stony soils, sometimes in partially disturbed areas or light gaps, but is also encountered in deep shade and in wetter habitats from 20 to 262 m (Wilkin *et al.*, 2008). Harris (2018) classified *D. bako* as endangered as it may be suffering from over-utilization. Kennerley and Wilkin (2017) reported that *D. bako* is a tertiary wild relative of cultivated yams, *D. alata*, *D. bulbifera*, *D. cayenensis*, *D. dumetorum*, *D. esculenta* and *D. rotundata*. The people in the Menabe region of Madagascar informally reported that *D. bako* is rare and that the number of plants has reduced markedly in recent years through a high level of wild harvesting of tubers. Tubers are edible and, along with *D. maciba*, are the favoured yam of people of the Morondava region of Madagascar who harvest them from the wild (Wilkin *et al.*, 2008). The tubers are peeled, washed and cut into 12–20 pieces about 2–4 cm long. These pieces are boiled in water that is deep enough to cover them and served with either fish or honey as a main meal or alone as a snack. The larger pieces are called 'sambaiky' in Sakalava when cooked, whereas the smaller, which have a softer consistency, are called 'katokato' (Wilkin *et al.* 2008). It was reported to be an important crop when other food sources, especially rice, are in short supply (Burkill and Perrier, 1950).

D. balcanica

Dioscorea balcanica Košanin was discovered by Košanin in 1914, in northern Albania. Subsequently it has been found growing in a few more localities in Montenegro and southwestern Serbia (Košanin, 1929). *D. balcanica* is an endemic, endangered species and therefore has been put under protection. It grows in mixed deciduous shrubby forests of various habitats at 250 to 700 m in Albania and Kosovo. In the Koritniku and Gjallica mountains it grows in beech forests at 1,000 to 1,300 m (Shuka *et al.*, 2018).

D. banzhuana

Dioscorea banzhuana S.J. Pei & C.T. Ting. Stems twine to the left and tubers are dark brown, cylindrical, curved, irregularly branched and grow horizontally in open forests, scrub forests and on mountain slopes at 1,400–1,500 m in southeastern Yunnan province in China (Ding and Gilbert, 2010).

D. batatas

Dioscorea batatas Decne. is a synonym of *D. polystachya* Turcz. and is commonly called Chinese yam or cinnamon vine. However, both *D. opposita* and *D. oppositifolia* are also referred to as cinnamon vine and Chinese yam and *D. japonica* is also referred to as Chinese yam. *D. batatas* grows in forests in valleys and on the slopes of hills in central and north China and is cultivated from 100–2,500 m in China. *D. batatas* extract has been shown to exhibit medicinal properties. For example, mucilage from *D. batatas* tubers has been reported to exhibit antioxidant and angiotensin-converting-enzyme inhibitory activities that can reduce hypertension (Lee *et al.*, 2003). It was introduced into the USA in the 19th century for culinary and cultural uses and is now considered an invasive plant species in several states. It can form dense masses of vines that cover and kill native vegetation, including trees.

D. baya

Dioscorea baya De Wild, also called *D. baya* var. *baya* is found in Angola, Burundi, Cameroon, Central African Republic, Democratic Republic of the Congo, Gabon, Uganda, Zambia and Zaire (GBIF, 2019). It occurs in swamps and seasonally flooded sedge meadows, rainforests and in open areas. It persists in areas of disturbance, old termite mounds and forest regrowth in wet soils that are rich in organic matter, and also in secondary forests associated with the African corkwood tree or umbrella tree (*Musanga cecropioides*) (CJBG, 2018). It has been reported growing over the range of 200–1,500 m. Stems are right twining with large tubers covered by a woody plate, with woody roots from crowns and with prickles up to 1 cm long and no bulbils (Wilkin *et al.*, 2009; CJBG, 2018; Botany VM, 2018).

D. belizensis

Dioscorea belizensis Lundell is synonymous with *D. hondurensis* R. Knuth. It is native to Central America where it is locally called wild yam, cocolmeca and barba del viejo. Blunden *et al.* (1963) reported that the tubers of *D. belizensis* contain about 2% diosgenin (dry weight) and Arvigo and Balick (1994) reported that *D. belizensis* is used as an aphrodisiac. It is also a popular treatment for urinary tract infections, colds, bilious colic, rheumatism, arthritis and diabetes (Martin, 1969).

D. belophylla

Dioscorea belophylla (Prain) Haines is synonymous with *D. nummularia* var. *belophylla* Prain. Common names include: chupri alu, jangli matol, kadu genasu, sankhaluka, spear-leaved yam and turar. It is native to India and Pakistan where it is commonly found in the foothills, ascending to about 1,400 m. Tubers are small and sub-fusiform in shape (Fig. 8.6). Stems are glabrous, twining anticlockwise (Flowers of India, 2019). *D. belophylla* tubers are used in herbal medicine and contain bioactive compounds including saponins (18.46 mg 100^{-1}g), alkaloids (0.68 mg 100^{-1} g), flavonoids (8.84 mg 100^{-1} g), tannins (4.2 x 10^2 mg 100^{-1} g) and phenols (2.8 x 10^3 mg 100^{-1} g). The tubers are relatively rich in ascorbic acid, riboflavin and thiamine and were shown to contain 1.67 mg 100^{-1} g of ascorbic acid, 0.70 mg 100^{-1} g of thiamine and 0.43 mg 100^{-1} g of riboflavin (Poornima and Ravishankar, 2009). The tubers

Fig. 8.6. *D. belophylla*. Photograph Dr K. Abraham, CTCRI Emeritus, C-63, Gandhipuram, Sreekariam, Trivandrum 695017, India, reproduced with permission.

are also 'prized for culinary purposes' and sold in bazaars under the name of turar.

D. bemandry

Dioscorea bemandry Jum. and H. Perrier is native to Madagascar and occurs on sandy soils with low nutrient content. Andriamparany *et al.* (2015) reported that *D. bemandry* seems to perform better at low soil nutrient levels and low silt content and is found mainly in open forest habitats near roads. High harvest intensity and non-fragmented forests at lower altitudes were reported to be reducing the population. The tubers can be 40–80 cm long and 8–13 cm in diameter (Humbert, 1946–1984), are edible and sometimes gathered from the wild for local consumption (Andriamparany *et al.*, 2015). *D. bemandry* was identified as a potential food resource in the Mahafaly region of Madagascar where 70% of households interviewed were collecting wild yams (Andriamparany *et al.*, 2014).

D. bemarivensis

Dioscorea bemarivensis Jum. and H. Perrier is found mainly in dry forests in the Boina region of Madagascar. Stems twine to the left and bulbils are produced in leaf axils. Tubers grow down in the soil in a vertical direction and are ovoid with a brown detached epidermis (Botany VM, 2018).

D. benthamii

Dioscorea benthamii Prain & Burkill. Stems twine to the right, 'the rootstock is unknown', and it occurs in shrub forests, on mountain slopes, in valleys, along rivers and on roadsides at 300–900 m in Taiwan and western Fujian, Guangdong and Guangxi provinces of China (Ding and Gilbert, 2010).

D. bicolor

Dioscorea bicolor Prain & Burkill. Stems twine to the right, but the 'rootstock is unknown'. They occur in 'herb communities' and mountain valleys at 1,600–2,100 m in southwestern Sichuan and Northern Yunnan provinces of China (Ding and Gilbert, 2010).

D. biformifolia

Dioscorea biformifolia S.J. Pei & C.T. Ting. Stems twine to the left and tubers are cylindrical, irregularly branched, with the outside purplish black and a rough skin. They grow horizontally in the soil in scrub forests, forest margins and on mountain slopes at 600–1,800 m in Yunnan province of China (Ding and Gilbert, 2010).

D. birmanica

Dioscorea birmanica Prain & Burkill is native to China, occurring at about 400 m in northwestern Yunnan province, and is also known as lone dragon. These yams have been reported in Myanmar and Thailand and have also been found in India and other parts of South East Asia. They produce edible tubers with pink flesh that grow horizontally in the soil. Tubers are irregularly branched, bulky, with corky skin that may be cracked and tessellated. Bases of old roots become hard and somewhat thorn-like. Stems twine to the left (Ding and Gilbert, 2010). In Thailand, Jaiaree *et al.* (2010) investigated the cytotoxic activity of *D. birmanica* tuber extract against breast and ovarian cancer cells. Subsequently, Jaiarree *et al.* (2018) isolated its cytotoxic compound and identified it as prosapogenin A of dioscin, which showed high cytotoxic activity against ovarian and breast cancer cells, but none against human keratocyte cells.

D. brevipetiolata

Dioscorea brevipetiolata Prain & Burkill originated in South East Asia and was reported to be edible and not toxic, but not cultivated (Martin and Degras, 1978). In Thailand its common name is candle yam or man nok, but man nok is a general term for yams in Thailand. It was reported to be edible and also occurs in the Saravane region of Laos (Vidal, 1960). Tubers, one per growing season, were reported to be 10–15 cm long and 1.6–3 cm in diameter, cylindric to clavate, growing deeply vertically in the soil, with the withering tuber from the previous season on top (Fig. 8.7) (Wilkin *et al.*, 2009).

D. buchananii

Dioscorea buchananii Benth. was reported to be native to Angola, Democratic Republic of the Congo, Malawi, Mozambique, South Africa, Tanzania and the United Republic of Zimbabwe. It is widespread in Tanzania and southern and eastern Democratic Republic of the Congo (Kinshasa), Southern Mozambique, Zimbabwe, Angola and South Africa. In South Africa *D. buchananii* subsp. *undatiloba* was reported to be found exclusively in Limpopo, Mpumalanga and KwaZulu-Natal (Bachman, 2011; Wilkin and Muasya, 2015). Stems are left twining with a 'perennial subterranean tuber' (Wilkin *et al.*, 2009; Botany VM, 2018).

Fig. 8.7. *D. brevipetiolata* Prain & Burkill, candle yam. Photograph taken at Sarinya's yam farm in Chachoengsao province in the eastern region of Thailand, 2018.

D. buckleyana

Dioscorea buckleyana Wilkin was part of a 'mixed taxonomic concept' under the invalid name *D. sambiranensis* R. Knuth subsp. a*mbrensis* H. Perrier. Common names include ovialain, ovy and taravy. It appears to be restricted to limestone substrates, with at least six of the nine specimens reported by Wilkin *et al.* (2009) being from areas of calcareous rock. It was found in deciduous forests at altitudes from 50 to 300 m. Tubers are edible. It is similar to *D. pteropoda* H. Perrier, but Wilkin *et al.* (2009) describe the differences.

D. bulbifera

Dioscorea bulbifera L. occurs in a wide variety of forms. Many synonyms have appeared in the literature including: *D. anthropophagorum* A. Chev., *D. crispata* Roxburgh, *D. heterophylla* Roxburgh, *D. hoffa* Cordem., *D. hofika* Jum. and H. Perrier, D. *korrorensis* R. Knuth, *D. latifolia* Benth., *D. longipetiolata* Baudon, *D. oppositifolia* Campbell, *D. papilaris* Blanco, *D. perrieri* R. Knuth, *D. pulchella* Roxburgh, *D. rogersii* Prain & Burkill, *D. sativa* Thunb., *D. tamifolia* Salisbury, *D. tenuiflora* Schltdl., *D. violacea* Baudon, *D. tunga* Hamilton, *Helmia bulbifera* (L.) Kunth, *Polynome bulbifera* (L.) Salisb. *D. anthropophagorum* A. Chev, *D. bulbifera* var. *anthropophagorum* (A. Chev.) Summerh., *D. bulbifera* var. *crispata* (Roxb). Prain, *D. bulbifera* var. *elongata* (F.M. Bailey) Prain andamp Burkill, *D. bulbifera* var. *pulchella* (Roxb.) Prain, *D. bulbifera* var. *sativa* Prain, *D. crispata* Roxburgh, *D. heterophylla* Roxburgh, *D. hoffa* Cordem., *D. hofika* Jum. and H. Perrier, *D. korrorensis* R. Knuth, *D. latifolia* Benth., *D. longipetiolata* Baudon, *D. perrieri* R. Knuth, *D. pulchella* Roxburgh, *D. rogersii* Prain & Burkill, *D. sativa* L., *D. sativa* var. *domestica* Makino, *D. sativa* var. *elongata* F.M. Bailey, *D. sativa* var. *rotunda* F.M. Bailey, *D. sylvestris* De Wild, *D. tamifolia* Salisb., *D. tenuiflora* Schltdl., *D. violacea* Baudon, *Helmia bulbifera* (L.) Kunth and *Polynome bulbifera* (L.) Salisb.

D. bulbifera is now common throughout the tropics, and it occurs in the wild both in Africa and Asia. Burkill (1911) commented that *D. bulbifera* now occurs widely throughout the world, even on the most remote islands of the Pacific and that the wild forms are bitter and often poisonous. Chevalier (1913, 1936) argued that *D. bulbifera* was taken into domestication from the wild independently in both Africa and Asia. However, there is some disagreement as to whether the original source was in South East Asia or whether there was also a centre of origin in Africa (Kay, 1987; Hammer, 1998). Hladik and Dounias (1993) suggested that even the apparently ancient forms of *D. bulbifera* may ultimately be of Indian origin in an unknown prehistoric era. However, Terauchi *et al.* (1991) found that chloroplast genomes of 15 accessions of *D. bulbifera* and *D. opposita* could be classified into nine types and their phylogenetic relationships showed that African and Asian chloroplast genomes diverged from each other at the earliest point in time.

Common names of *D. bulbifera* include: aerial yam, air potato, air yam, batata-de-rama, bitter yam, bonay, bonda, bondanza, bonday, bulbil bearing yam, cara-de-aire, cará-de-árvore, cara-de-espinho, cara-de-sapateiro, cará-do-ar, cará-moela, cheeky yam, danda yam, gunda, hoei-oepas, hoi, huang du, igname bois, igname bulbifere, igname massokor, igname pousse en l'air, kaachil, kaile, kaile manu, knollen-yam, masoco, massokop, monday, ñame blanco, ñame bobo, ñame cimarrón, ñame congo, ñame criollo, ñame de monte, ñame del aire, otaheite potato, papa aérea, papa caribe, papa del aire, papa voladora, potato yam, sarau and turkey liver yam. Obidiegwu and Akpabio (2017) commented that *D. bulbifera* made a secondary but significant contribution to the domestication process of *Dioscorea* in Nigeria. Martin and Degras (1974) tested and evaluated several cultivars in Puerto Rico, showing considerable variation in their performance (Table 8.4)

D. bulbifera is a strongly climbing vine with smooth (non-spiny) stems up to 8 mm in diameter with edible bulbils (Fig. 8.8) in the leaf axils above-ground and tubers (Fig. 8.9) below ground, although tubers are sometimes absent (Coursey, 1967). Martin and Degras (1974) commented that since the bulbils were the principal food source, early man selected plants on that basis and many African and Asian varieties, for this reason, do not produce tubers. When produced, the tubers are hard, bitter and unpalatable (Purseglove 1972). Wilkin and Thapyai (2009) and Botany VM (2018) reported that tubers are solitary, up to 20 cm long and 5–15 cm in diameter, variable in shape, usually globose to cylindric, sometimes digitate or irregularly lobed, shallowly buried, annually replaced and, as reported above, sometimes absent. The bulbils are the main storage organs and are

Table 8.4. Yield of *D. bulbifera* in a trial in Puerto Rico. Modified from Martin and Degras (1974).

Cultivar/Variety	Country	Bulbil yield in tonnes hectare^{-1}	Tuber yield in tonnes hectare^{-1}
'Sharp Angled'	Puerto Rico	4.9	0.0
'Smooth Angled'	Puerto Rico	2.0	1.1
'Round'	Puerto Rico	9.7	18.7
'Sativa'	India	6.9	13.4
'Poison'	Hawaii	10.6	22.1
'Thuma'	New Caledonia	19.5	2.5
'Olode'	Nigeria	3.0	0.3
Wild	Côte d'Ivoire	1.7	1.0
Cultivated	Côte d'Ivoire	1.9	3.7
Wild	Sierra Leone	0.05	0.15
Cultivated	Sierra Leone	8.0	3.5

Fig. 8.8. Bulbils of *D. bulbifera*. Photograph, Dr K. Abraham, CTCRI Emeritus, C-63, Gandhipuram, Sreekariam, Trivandrum 695017, India, reproduced with permission.

Fig. 8.9. *D. bulbifera* tubers beginning to sprout. Photograph: Dr K. Abraham, CTCRI Emeritus, C-63, Gandhipuram, Sreekariam, Trivandrum 695017, India, reproduced with permission.

eaten. They can weigh up to 2 kg, but are more commonly around 500 g. Bulbils are grey or brown in colour with white or yellow mucilaginous flesh. Flowers are white or pinkish tinged, diminutive and sessile and their fruits are capsules, three-winged, and about 2.5 cm long (Acevedo-Rodríguez and Strong, 2005). *D. bulbifera* follows a repeated annual cycle of growth and dormancy that corresponds to the rainy and dry seasons within its native range (Coursey, 1967).

Composition and uses

The bulbils are normally cooked and eaten in a manner similar to other starchy root crops, though many African forms require detoxification by soaking in water or prolonged boiling, and discarding the water, before they are safe to consume. However, it has been reported that certain varieties are very succulent and may be eaten raw. Achy *et al.* (2016) pointed out that the high level of antinutritional factors (total phenol 558 ± 3.46 mg $100g^{-1}$, oxalate 320 ± 2.65 mg $100g^{-1}$, phytate 469.33 ± 2.08 mg $100g^{-1}$ 'could pose a serious problem of public health. Therefore, it would be wise to cook bulbils before eating'. The flavour is reported to be inferior to that of most common yams and some are bitter

(Purseglove 1972). Kay (1987) reported that the proximate composition of the bulbils, in terms of the fresh weight, was: 63–67% water, 1.12–1.5% protein, 0.04% fat, 27–33% carbohydrate, 0.7–0.73% fibre and 1.08–1.51% ash. Teponno *et al.* (2008) isolated 2 clerodane diterpenoids, bafoudiosbulbins F (1) and G (2), together with bafoudiosbulbins A-C, 3,5,40-trihydroxy-30-methoxybibenzyl and kaempferol from the bulbils of *D. bulbifera* var *sativa*. Bafoudiosbulbins and 2 clerodane diterpenoids are common in a variety of plant species and have both been reported to have shown anti-bacterial, anti-fungal and, to some extent, anti-insect activity. Achy *et al.* (2016) gave the proximate composition of flour made from the *D. bulbifera* bulbils of the cultivar 'Yellow' (on a dry weight basis) as 6.22 ± 0.87% moisture content, 2.36 ± 0.87% fat, 8.12 ± 0.02% protein, 3.44 ± 0.05% ash, 0.91 ± 0.08% cellulose, 79.86 ± 0.09% carbohydrate and 373.16 kcal $100g^{-1}$ energy. They found that the most predominant mineral was potassium (847mg $100g^{-1}$) and the major organic acids were oxalic acid (486 ± 0.03 mg $100g^{-1}$) and citric acid (365.4 ± 0.5 mg $100g^{-1}$).

Tubers are edible and are cooked and used as a vegetable. Knoth (1993) gave the following analysis for tubers on a fresh weight basis: 63–67% moisture, 27–33% carbohydrates, 0.1% fats and 1.1–1.5% crude protein. Proximate composition of *D. bulbifera* tubers in g per 100 g fresh weight was: 3.4 protein, 27.0 carbohydrate, 1.1 fat, 67.0 moisture, 0.5 fibre and 132.0 kcal energy (Rajyalakshmi and Geervani, 1994). Opara (1999) reported that the content of the edible part of tubers per 100 g edible tuber portions was 71–79 g water, 1.4–1.5 g protein, 18–26 g carbohydrate and 0.9–1.2 g fibre. Data from South East Asia gave: 62.6–66.9% water, 27–33% starch, 0.08–0.10% reducing sugars, 0.25–0.35% non-reducing sugars, 0.4% fat and 1.12–1.50% crude protein (Coursey, 1967). Kay (1987) reported the fresh weight composition of the tubers as 69.1% water, 0.89% protein, 0.1% fat, 26.5% carbohydrate and 6.74% fibre. Shajeela *et al.* (2011) gave the following analysis for *D. bulbifera* var. *vera* for 100g fresh weight of tuber: starch 38.10 ± 0.1 g, 7 niacin 33.74 ± 0.21 g, ascorbic acid 91.04 ± 0.86 g. Martin (1974c) found that β-carotene was not detected in *D. bulbifera* and the yellow pigments were xanthophylls. Shajeela *et al.* (2011) gave the following analysis for antinutritional factors for 100g fresh weight: total free phenolics 2.20 ± 0.01 g, tannins 1.48 ± 0.10 g, hydrogen cyanide 0.19 ± 0.01 mg, total oxalate 0.78 ± 0.01 g, amylase inhibitor 1.36 AIU mg^{-1} soluble starch and trypsin inhibitor 1.21 ± 0.01 TIU mg^{-1} protein.

Bhandari *et al.* (2003) gave the following analysis of *D. bulbifera*, *D. versicolor*, *D. deltoidea* and *D. triphylla* tubers on a fresh weight basis: crude protein 1.6–3.1, ash 0.5–1.2, crude fat 0.2-0.3, crude fibre 0.6–1.5%, K 250–560 mg 100 g^{-1}, Na 4.15–17.8 mg 100 g^{-1}, P 33.1–61.6 mg 100 g^{-1}, Ca 14.3–46.9 mg 100 g^{-1}, Mg 18.3–27.3 mg 100 g^{-1}, Cu 0.10–0.21 mg 100 g^{-1}, Fe 0.39–2.92 mg 100 g^{-1}, Mn 0.14–0.35 mg 100 g^{-1} and Zn 0.22-0.53 mg 100 g^{-1}. In India Arinathan *et al.* (2009) gave the composition specifically for *D. bulbifera* var. *vera* (Tables 8.5, 8.6) and Shajeela *et al.* (2011) gave their anti-nutritional composition (Table 8.7).

Table 8.5. Proximate composition of *D. bulbifera* var. *vera* tubers (g $100g^{-1}$). Modified from Arinathan *et al.* (2009).

Component	Level
Moisture	68.70 ± 0.25
Crude protein	5.16 ± 0.23
Crude lipid	9.13 ± 0.18
Crude fibre	1.23 ± 0.06
Ash	2.91 ± 0.30
Nitrogen free extract %	81.57
Calorific value (kJ $100g^{-1}$)	1792.59
Starch g $100g^{-1}$	18.10 ± 0.07
Niacin mg $100g^{-1}$	23.69 ± 0.35
Ascorbic acid mg $100g^{-1}$	106.52 ± 0.11

Table 8.6. Mineral composition of *D. bulbifera* var. *vera* tubers (mg 100 g^{-1}). Modified from Arinathan *et al.* (2009).

Na	K	Ca	Mg	P	Zn	Mn	Fe	Cu
66.78 ± 0.44	1600.31 ± 1.48	238.15 ± 0.09	441.17 ± 0.08	134.14 ± 0.53	1.30 ± 0.01	11.60 ± 0.12	4.90 ± 0.01	2.74 ± 0.01

Table 8.7. Antinutritional factors of tubers of *D. bulbifera* var *vera*. Modified from Shajeela *et al.* (2011).

Component	Level
Total free phenolics (g 100g^{-1})	2.20 ± 0.01
Tannins (g 100g^{-1})	1.48 ± 0.10
Hydrogen cyanide (mg 100g^{-1})	0.19± 0.01
Total oxalate (g 100g^{-1})	0.78 ± 0.01
Amylase inhibitor AIU mg soluble starch^{-1}	1.36
Trypsin inhibitor TIU mg protein^{-1}	1.21 ± 0.01

Considerable differences in the composition between bulbils and tubers were reported by FAO (1965) (Table 8.8). However, when Celestine and David (2015) compared the tubers and bulbils of *D. bulbifera* they found there were no significant differences (p = 0.05) between the tubers and bulbils in their content of crude fats, fibre, ash and carbohydrates and their calorific values while crude protein was significantly (p = 0.05) higher in the bulbils. The levels of K, Na, Ca, Mg, Fe, Zn, Cu, P, oxalate, tannins and phenols were all significantly (p > 0.05) higher in the tubers than the bulbils, while the bulbils had higher levels of alkaloids, hydrogen cyanide, saponins and flavonoids.

In Australia, Webster *et al.* (1984) reported that *D. bulbifera* var *rotunda* tubers had a bitter taste, but were not toxic. They found that a major bitter component was diosbulbin D (0.07 mg g^{-1}) and the traditional aboriginal methods of food preparation were very efficient in removing diosbulbin D, thus making them palatable. However, they found that boiling or baking did not remove diosbulbin D, but leaching reduced levels by about 86%. They concluded that boiling or baking softened and opened up the tissue aiding subsequent leaching.

Choudhary *et al.* (2009) reported that *D. bulbifera* was used as a medicinal plant in Rajasthan in India. The tuber is also used as contraceptive by tribal women with dried tuber powder (10g) given once in a day for 5 days after menses (Swarnkar and Katewa, 2008).An amount of 8-epidiosbulbin E acetate extracted from *D. bulbifera* has also been shown to have potential as a plasmid curing compound against multidrug resistant bacteria (Shriram *et al.*, 2008). In India a folk medicine is made from the tuber, where it is made into a paste and used as a cure for snakebite. In Jamaica it is used for treatment of scorpion and centipede stings, in Indonesia a fish poison is made from the bulbils of toxic varieties and in Africa poisonous varieties may be planted among safe varieties to discourage thieves (Kay, 1987). *D. bulbifera* grows wild and its bitter tubers are used by tribal people in Bangladesh for treatment of leprosy and tumours (Murray *et al.*, 1984).

Harvesting

Kay (1987) reported that it may take 2 years from planting to harvesting for *D. bulbifera* and that harvest maturity is judged to be when the vines begin to wither. At this point the bulbils may fall off naturally or they can easily be dislodged with a twist. Immature bulbils may be harvested 3 to 4 months after planting and harvesting may continue for the life of the plant. When the bulbils are ready, they can be picked like fruit. Tubers are normally harvested when the vine dies back, which is after some 15 to 24 months. Forks or other digging tool are used to harvest the tubers.

Table 8.8. Composition of *D. bulbifera* mg or g 100 g^{-1} raw edible portion. Modified from FAO (1965).

	Energy kcal	Moisture	Protein	Fat	Total carbohydrate	Fibre	Ash	Calcium	Phosphorus	Iron
Tuber	112	71.0	1.5	0.1	25.6	0.9	0.9	69	29	
Bulbil	78	79.4 ± 77.8–81.0	1.4 ± 1.3–1.4	0.2 ± 0.1–0.3	16.8	1.2	1.0 ± 0.7–1.2	0.7–1.2	58 ± 45–70	2.0

Storage

When bulbils or tubers are left in the shade in tropical ambient conditions harvest wounds heal quickly, which makes them more resistant to subsequent fungal infections (Kay, 1987). The tubers may be left for several months in the soil until needed. Okagami and Tanno (1991) reported that bulbils can also last a year or more on the ground waiting for suitable environmental conditions to sprout. In Nigeria they were reported to have a dormancy period of 14 to 16 weeks (Opara, 1999) and it has been suggested that bulbils require a chilling period before they sprout. Kay (1987) indicates that both tubers and bulbils can be stored under dry cool conditions away from sunlight, which appears to give a moderate storage life.

Invasiveness

D. bulbifera has been described as one of the most aggressive weeds ever introduced into the USA (Florida Exotic Pest Plant Council, 2008). USDA-NRCS (2012) classified *D. bulbifera* as a 'noxious weed' in Alabama and Florida and as an 'invasive species' in Cuba, the Bahamas, Puerto Rico and the Pacific islands, including Hawaii, Fiji, French Polynesia, Niue and Palau.

D. bulbotricha

Dioscorea bulbotricha Hand.-Mazz. is native to Brazil and has been found growing wild in the Reserva Biológica do Alto da Serra de Paranapiacaba, Santo André, where it was considered endangered by Kirizawa *et al.* (2016).

D. burkilliana

Dioscorea burkilliana J. Miège is a West Africa species. In the year 2000, Sato (2006) found *D. burkilliana* growing wild in a forest area of southeastern Cameroon. Additionally, JSTOR (2019) reported it as occurring only in the bush and forest zone of Sierra Leone and Côte d'Ivoire. It may also occur in the wild in Nigeria. *D. burkilliana* is cultivated as a food crop (Sahoré *et al.*, 2005). In Côte d'Ivoire, Alexis and Amani (2013) extracted starch from *D. burkilliana* tubers and fractionated it according to its sensitivity to acid hydrolysis. They found that only 5% was easily hydrolysable. Stems twine right-handed from a perennial fibrous tuber protected by a 'woody rootstock'. *D. burkilliana* was considered very similar to *D. minutiflora* and *D. smilacifolia*, but it was recognized in Côte d'Ivoire as a different species (Miège, 1958). Information is lacking regarding its uses. Tubers may be edible, but probab require special preparation.

D. burkillii

Dioscorea burkillii R. Knuth. The Plant List does not consider *D. burkillii* to be the currently accepted name (Plant List *D. burkillii*, 2019). Stems twine to the left, tubers are cylindric and grow vertically in the soil in secondary scrub forests and forest margins at 2,000–3,000 m in northern Guizhou, western

Sichuan and Yunnan provinces of China (Ding and Gilbert, 2010).

D. cachipuertensis

Dioscorea cachipuertensis Ayala is native to Peru (Govaerts *et al.*, 2007), where a plant was reported to be found growing at 250 m in the Peruvian Andes.

D. calcicola

Dioscorea calcicola Prain & Burkill, . Its common name in Thailand is Yan Lueat. It is a member of *Enantiophyllum* and has been reported growing in limestone soils in Thailand. It has small, narrow leaves and wiry stems in order to reduce evapotranspiration, with the narrowest leaves being either alternate or opposite (Thapyai *et al.*, 2004). Maneenoon *et al.* (2008) found *D. calcicola* was among the plants that provided the main sources of carbohydrate for the Sakai people in Thailand. In phylogenomic studies of *Dioscorea*, Soto Gomez *et al.* (2019) suggested that *D. calcicola* may have utility in future breeding programmes for *D. alata*. Proximate composition of *D. calcicola* tubers, in g per 100 g fresh weight, was: 0.94 protein, 34.35 carbohydrate, 0 fat, 61.79 moisture, 0.83 ash, 0.38 fibre and 148.00 kcal energy (Maneenoon *et al.*, 2008).

D. caucasica

Dioscorea caucasica Lipsky. Its common name, Caucasus yam root, reflects its native range, which is Transcaucasia where it is found growing in part shade (deciduous forest edges), at 900–1,600 m above sea level. *D. balcanica* Košanin and *D. caucasica* are morphologically very similar. Stems are right-twining and both are dioecious with female plants being larger than the males. The most conspicuous difference between the two species is the leaf size and shape. In *D. caucasica* all leaves are of equal size, longer than wide, while in contrast, the lower leaves in *D. balcanica* are larger, approximately of equal length and width, while the upper leaves are smaller, but more elongated and acute (Košanin, 1929). Deliu *et al.* (1992) and Ćulafić *et al.* (1994) found that there was a negative correlation between carotenoid and diosgenin biosynthesis in both *D. caucasica* and *D. balcanica*. They also showed that the intensity of diosgenin biosynthesis was correlated with cell proliferation, being maximum in the exponential phase and minimum at its beginning. Diosgenin content for *D. caucasica* tubers was given by Martin (1969) as 1.5%.

D. cayenensis

Dioscorea cayenensis Lamarck has also been referred to as *D. oppositifolia* L. and in older literature many taxonomists considered *D. cayenensis* to be a variety of *D. rotundata* and others refer to it variously as *D. rotundata-cayenensis* complex. However, Okeke (2004) investigated the taxonomic position of *D. cayenensis*, *D. rotundata* and *D. pruinosa* and found diagnostic characteristics that were more than enough to warrant the recognition of each as a separate species. This view was supported by hybridization experiments, which indicated that members of the group were almost completely inter-sterile. There is considerable variation in both species but even so the overall differences appear to support Okeke's findings (Table 8.9).

Common names for *D. cayenensis* include: attoto yam, balugu, congo amarillo, cut and come again yam, dye yam, gname jaune, Guinea yam, hard yam, igname de Guinée, inhame da Guiné, niame, twelve months yam, yellow guinea yam and yellow yam. It occurs in Africa from Senegal to Ethiopia and Uganda, and probably arose in cultivation resulting from hybridization in the section *Enantiophyllum* of *Dioscorea*. Domestication of plants from the wild still continues (e.g. in Benin) where local farmers collect wild yam plants and cross them with cultivated ones. It is a staple food in a belt from Côte d'Ivoire to Cameroon and it is also planted in Central and East Africa, the Caribbean islands, Brazil and the Philippines. It has recently been introduced into Papua New Guinea and Oceania (PROTA4U, 2018). Tubers are borne singly and can weigh up to 10 kg, but more usually they are within the range of 500 g to 2 kg. They have brown skin and yellow flesh. It is interesting to note that the price of *D. cayenensis* tubers imported from Jamaica and sold in the same UK shop in January 2018 was £4.55 per kg and in October 2020 it had risen to £8.00 per kg. (Fig. 8.10).

Composition and uses

The tubers are used almost exclusively for human consumption with their peel often used as animal feed. Unpeeled tubers may be boiled, roasted or baked; peeled tubers may be boiled or fried. A traditional method of preparation in West Africa is to pound the boiled peeled tubers to produce a thick dough (pounded yam). Pounded yam is consumed by rolling it into small balls which are then dipped into a sauce and often just swallowed (usually without chewing). Occasionally, the peeled tuber is cut into small chips, dried and then milled to produce yam flour. Chips may also be fermented before further processing. At meal time, the flour is stirred in boiling water and kneaded to produce a paste, which can be consumed in the same way as pounded yam. Yam flour is also prepared. Industrial processes have resulted in the production of yam flakes which can be reconstituted in hot water to give a product similar to pounded yam or mashed yam. Instant yam flakes can be made from suitable cultivars, though *D. cayenensis* yams are not always favoured because of their yellow colour. In West Africa, *D. cayenensis* has a high

Table 8.9. Comparison of characteristics of *D. cayenensis* and *D. rotundata*. Martin and Sadik (1977).

Characteristic	*D. rotundata*	*D. cayenensis*
Tuber flesh colour	White	Yellow
Leaf shape	Narrowly ovate	Broadly ovate
Climatic preference	Intermediate rainfall	High rainfall
Planting to harvest	7 to 8 months	10 to 12 months
Number of harvests	2	1
Possible time of harvest	Limited to late summer	Almost all year round

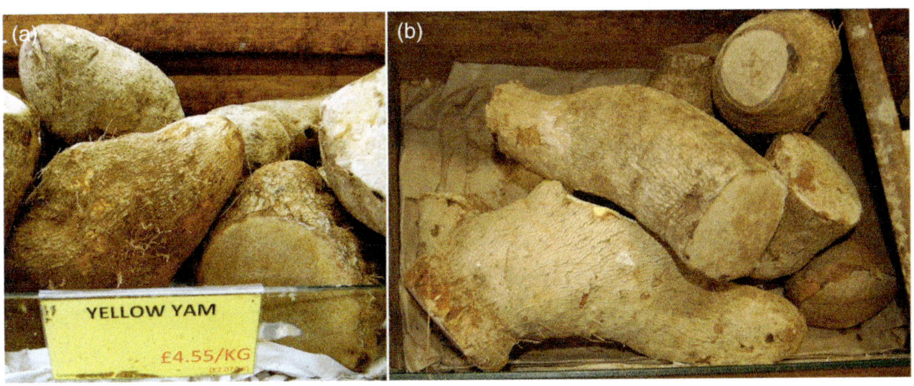

Fig. 8.10. *D. cayenensis* tubers imported from Jamaica by sea on sale in Huddersfield, UK, in January 2018 (a), and October 2020 (b).

sociocultural value attached to its production and use. It is a prime object for traditional religious observances, social gift exchange and cultural festivals (PROTA4U, 2018).

FAO (1965) gave a general composition for tubers of *D. cayenensis* and *D. rotundata* (Table 8.10). Opara (1999) reported that the content of the edible part of *D. cayenensis* tubers was: water 80 g, protein 1.5 g, carbohydrate 16g and fibre 0.6g in 100 g. Knoth (1993) gave the following analysis of *D. cayenensis* tubers on a fresh weight basis: 58 to 80% moisture, 15 to 23% carbohydrates, 0.1 to 0.2% fats and 1.1 to 2.0% crude protein. The approximate composition of the edible portion of mature *D. cayenensis* tubers was quoted by Kay (1987) as: energy 439 kJ 100 g^{-1}, water 58–33%, protein 1.02–1.99%, fat 0.05–0.12%, carbohydrate 15–23%, fibre 0.35–0.79% and ash 0.53–2.56%. Higher moisture contents occurred in immature tubers.

In Ghana, Coursey and Aidoo (1966) measured the ascorbic acid levels in several *D. cayenensis* local varieties and found that they varied between 4.5 and 8.2 mg 100^{-1}, which was generally slightly lower than in *D. rotundata* at 6.5–11.6 mg 100 g^{-1}. Differences in the nature of the starch are reported: *D. rotundata* granules were larger (10–70 m) but those of *D. cayenensis* were smaller (3–25 m). It was reported in PROTA4U (2018) that on a fresh weight basis 100 g of *D cayenensis* tubers were composed of: 58–80 g water, 15–23 g carbohydrate, 1–2 g crude protein, 0.05–0.12 g lipids, 0.35–0.79 g crude cellulose and 0.68–2.56 g ash. The starch granules of *D. cayenensis* are generally smaller than those of *D. rotundata* and triangular in shape. Starch granule size generally increases from the top of the tuber downwards, and from the subcutaneous region towards the centre. The protein content, though generally low, is highest closer to the skin, high in aspartic and glutamic acids and low in tryptophan and cystine. Some cultivars have significant amounts of vitamin C and thiamine. Freshly cut *D. cayenensis* tubers can cause skin irritation due to the presence of raphides (needle-shaped crystals of calcium oxalate), which are destroyed when the tubers are cooked.

Adepoju *et al.* (2018) found that raw *D. cayenensis* tubers contained (per 100g edible portion) 66.79g moisture, 2.62g crude protein, 0.27g lipid, 0.17g fibre, 0.63g ash, 29.69g carbohydrates, 262.30mg potassium, 61.53mg magnesium, 0.79mg iron, 0.39mg zinc.They yielded 108.26 kcal energy with insignificant vitamin content. Tubers were processed into various local foods, which significantly (p = 0.05) improved macronutrients and energy content with significant (p = 0.05) reductions in all antinutritional products. Abubakar and Gana (2017) showed tubers of *D. cayenensis* contained significantly higher (p < 0.05) magnesium (16.3 ± 1.52 mg/100g), phytate (31.2 ± 0.40 mg) and lower fibre (0.9 ± 0.03 g/100g) than *D. rotundata* or *D. alata*. In Côte d'Ivoire Kouakou Dje *et al.* (2010) reported that tubers of *D. alata* and *D. cayenensis-rotundata* were stored for 6 months in a ventilated store where the conditions were 26.56 ± 3°C and 82 ± 5% RH. Their total phenolic content varied in different parts of the tuber and decreased during storage. Protein content ranged from 7.90 to 7.94% and generally did not vary significantly (p = 0.05) during storage, although a slight fall was observed in the distal part of one variety after six months. Lipids were low, only about 0.20% of the dry matter, and did not vary significantly (p = 0.05) during storage.

Sautour *et al.* (2004) isolated three compounds from *D. cayenensis* tubers and tested their antifungal activity against the human pathogenic yeasts *Candida albicans*, *C. glabrata* and *C. tropicalis* (MICs of 12.5,

Table 8.10. Composition of *D cayenensis, D. rotundata* in g 100g^{-1} edible portion. FAO (1965).

Energy kcal	Moisture	Protein	Fat	Carbohydrate	Fibre	Ash	Calcium	Phosphorus	Iron
71	80.8	1.5	0.1	16.4	0.6	1.2	36	17	5.2
–	70.5–85.9	1.4–1.6	–	–	0.3–0.7	0.8–1.4	30–48	6–39	4.9–5.4

12.5 and 25 μg mL^{-1}, respectively). The compounds were dioscin that exhibited antifungal activity, diosgenin 3-*O*-α-*L*-rhamnopyranosyl-(1→4)-α-*L*-rhamnopyranosyl-(1→4)-[α-*L*-rhamnopyranosyl-(1→2)]-β-*D*-glucopyranoside that showed weak antifungal activity and 26-*O*-β-*D*-glucopyranosyl-22-methoxy-3β, 26-dihydroxy-25(*R*)-furost-5-en-3-*O*-α-*L*-rhamnopyranosyl-(1→4)-α-*L*-rhamnopyranosyl-(1→4)-[α-*L*-rhamnopyranosyl-(1→2)]-β-*D*-glucopyranoside that showed no antifungal activity (Kay, 1987).

Curing

Exposure of the tubers to 35 to 40°C and 95 to 100% RH for 24 hours was found to initiate the curing process and also controlled storage rots (Table 5.4). This treatment could well replace chemical fungicides for postharvest disease control. See the section on *D. rotundata* for details of the process (pp. 64–86) (Thompson *et al.*, 1977). Nnodu and Nwankiti (1986) recommended curing at 26°C and 92% RH for 11 to 15 days for *D. cayenensis*. Weight loss during storage of *D. cayenensis* was considerably reduced when tubers had been cured directly after harvest (Fig. 5.6). For non-cured tubers it was found that 78% of the tuber surface was covered with fungi after 105 days storage in ambient conditions in Jamaica, compared with zero for those that had been cured. Curing also reduced the necrotic tissue in tubers from 55% to 22% (Thompson, 1972b).

Storage

In ventilated storage in Jamaican ambient conditions of 24 to 31°C and 52 to 68% RH, tubers lost 41% in weight after 4 months (Thompson *et al.*, 1973a). It was concluded that the storage period of 4 months was too long and should probably be confined to a maximum of 1 month. Refrigerated storage recommendations for *D. cayenensis* tubers were as follows:

13°C and 95% RH for less than 4 months with 29% weight loss and some development of internal necrosis (Thompson *et al.*, 1973a).

13°C and 95% RH for less than 4 months (Opara, 1999).

16°C and 80% RH for 60 days (Tindall, 1983; Opara, 1999).

Dormancy and sprouting

D. cayenensis has a short dormant period and tubers do not store well even under optimum conditions. However, Passam (1982) and Opara (1999) gave the dormancy period of *D. cayenensis* in Nigerian ambient conditions as 4 to 8 weeks. At tropical temperatures *D. cayenensis* tubers sprouted some 28 days after harvesting (Coursey, 1961). The most popular variety of *D. cayenensis* in Jamaica is called 'Roundleaf' and these began to sprout some 90 days after harvest compared with another variety, 'Common', that sprouted after 50 days. This study was at ambient temperatures of 25 to 34°C and 64 to 92% RH (Thompson 1972b).

Tschannen (2003) reported that GA_3 applied to *D. cayenensis* directly after harvest prolonged dormancy and reduced respiration rate at the time of sprouting. A proprietary potato sprout suppressant containing CIPC and IPC in a dust formulation had no effects on sprouting, but low temperature storage could delay sprouting. No sprouting occurred in any *D. cayenensis* tubers at 13°C during 5 months of storage, but there was chilling injury at temperatures below 15°C when stored for over a month (Thompson *et al.*, 1973a), so this method was not appropriate.

Diseases

A number of storage rots are known to affect *D. cayenensis*, *D. rotundata* and *D. alata* (see Postharvest Fungal Diseases, p. 76) including: *Botryodiplodia theobromae*, *Rhizopus*, *Fusarium*, *Hendersonula*, *Macrophomina* and *Penicillium* species (Okafor, 1966). In Jamaica, *Penicillium sclerotigenum* was a particular problem on *D. cayenensis*, developing on tubers during export transportation, particularly on their cut surfaces (Fig. 6.2). Tubers were dipped in benomyl before packing as a standard practice to control all fungi, including *P. sclerotigenum*, but resistant strains developed. However, these strains were successfully controlled by replacing the benomyl dip with an imazalil dip (Table 8.11).

D. chingii

Dioscorea chingii Prain & Burkill. Stems twine to the left and tubers are brown-black in colour, cylindrical, up to 5 cm in diameter, irregularly branched and grow horizontally in scrub forests and on mountain slopes. They have been found growing at 1,200–1,800 m in Guangxi and Yunnan provinces of China and also in Vietnam (Ding and Gilbert, 2010).

D. chouardii

Dioscorea chouardii Gaussen is synonymous with *Borderea chouardii* (Gaussen) Gaussen and Heslot and is native to Spain. It is endemic to the Pyrenees mountains where it grows in vertical limestone crags and overhangs facing north. *D. chouardii* is unusual and has been recorded as occurring only on two adjacent vertical cliffs and is thus considered a threatened species. García *et al.* (2012) contended that the plants may be very old: '... some individuals exceed 300 years in lifespan and its presence in a climatically very stable habitat, inaccessible to large herbivores. Such a combination of traits and habitat properties may explain the persistence of this relict species'. They produce dark grey-brown, cylindrical or globose tubers some 8 cm long, slim and becoming wider towards the base (Segarra-Moragues *et al.*, 2005; Botany VM, 2018).

Table 8.11. Effects of benomyl and imazalil on the growth (in cm^2) of a benomyl tolerant strain of *Penicillium sclerotigenum* on *D. cayenensis* tubers during storage at 20°C. Modified from Plumbley *et al.* (1984).

	Storage time in days				
Treatment	0	7	14	21	28
Control	0	1.10	2.20	3.19	3.66
Benomyl (500 ppm)	0	0.83	1.87	2.05	3.49
Benomyl (1000 ppm)	0	0.60	1.34	2.35	2.99
Imazalil (500 ppm)	0	0	0	0	0
LSD (p = 0.05)			0.56		

D. cirrhosa

Dioscorea cirrhosa Lour. Tubers are globose, ovoid, gourd-shaped oblong or cylindric. The tuber skin is dark brown with light brown or red. The flesh is reddish, drying purplish black. Stems twine to the right. *D. cirrhosa* is cultivated in southern China, Taiwan and northern Vietnam for its edible tubers (Fern, 2014). Tubers are harvested when about 3 years old, usually in the dry season when the flesh has a high tannin content. Particular care should be taken to ensure that tubers are not broken or bruised as this can affect tannin content. *D. cirrhosa* tubers were formerly traded internationally as an important source of tannins (tubers contain 6–13% tannin) for use as a preservative and dye, but currently *D. cirrhosa* is only of local importance. If tubers are to be used as dyes, they should be protected against desiccation because they lose much of their colouring properties when desiccated. For dyeing and tanning purposes, tubers are peeled and the flesh rasped. About 3 litres of water is added to 1 kg of rasped flesh, which is then filtered. It is used locally by dipping the cloth in a hot or cold solution and then drying the cloth in the sun. This process is often repeated several times, until the desired reddish-brown colour is attained. The dye rapidly loses its activity, and best results are obtained with fresh solutions. At the beginning of the 20th century dye from *D. cirrhosa* was an important export from northern Vietnam, with a maximum of 8,000 tonnes per year to Hong Kong at its peak. Since 1930 exports have rapidly diminished because of the increasing use of synthetic dyes, but excessive exploitation of the wild populations may also have caused the market to decline, as happened at the end of the 19th century in Hong Kong (Asian Plants, 2019).

D. cochleariapiculata

Dioscorea cochleariapiculata De Wild is found in Burundi, Democratic Republic of the Congo, Malawi, Mozambique, Namibia, Tanzania, Zambia and Zimbabwe where it grows at 400–1,900 m in sandy soils bordering rivers, lakes, forest margins and woodlands associated with baobab (*Adansonia digitata*) and frankincense trees (*Boswellia sacra*), and termite hills. *D. cochleariapiculata* is rarely found in heavily disturbed places (CJBG, 2018). The vines are left twining and the tubers are globose and about 10 cm in diameter growing near the soil surface. Tubers are commonly solitary or in clusters of up to five tubers (Wilkin, 1999; CJBG, 2018). Tubers can be eaten as a famine food, but thorough washing and drying is necessary to remove toxic alkaloids. Extracts of the tubers were reported to have been used as an arrow poison in Mozambique (CJBG, 2018).

D. collettii

Dioscorea collettii Hook.f. is widely distributed in southwestern China in mixed forests and on mountain slopes, at altitudes between 200–3,200 m. It is also found in India, Laos, Myanmar, Thailand and Vietnam (Ding *et al.*, 2000). Thapyai *et al.* (2005d) reported that *D. collettii* was newly recorded from Doi Chiang Dao in Chiang Mai province of Thailand. Ding *et al.* (2000) divided *D. collettii* into *D. collettii* var. *collettii* and *D. collettii* var. *hypoglauca*. Li *et al.* (2016) reported that tubers of *D. collettii* are called 'yellow ginger'. Stems twine to the left and tubers are irregularly branched and lobed, up to 17 cm long and up to 2 cm in diameter. Tubers are produced close to the soil surface, with a periderm 'not hard and lacking rigid and spine-like roots' (Wilkin and Thapyai, 2009; Botany VM, 2018). Acidified extract from tubers is used in traditional Chinese medicine as a treatment for certain cancers. Yang and Chen (1983) isolated the following steroidal compounds from acid treated wild *D. collettii* tubers from Sichuan province in China: diosgenin, yamogenin, β-sitosterol, Δ (3,5)-deoxytigogenin, isonarthogenin, Δ (3,5)-deoxyneotigogenin, diosgenin palmitate, yamogenin palmitate and yamogenin-β-D-glucoside. Jin *et al.* (2017) isolated 29 compounds from *D. collettii* tubers, including six steroid saponins, 13 monocyclic phenols, two flavonoids, three sterols and five cyclodipeptides. Also, Li *et al.* (2016) reported that some medicines containing extracts of tubers of *D. collettii* as their main ingredient were being sold in China. They reported that tubers of *D. collettii* were 'often used in the treatment of chyloid stranguria, gonorrhea, leucorrhagia, rheumatism arthralgia pain, joints disable [*sic*] and lumbar knee pain syndrome'. Yi *et al.* (2014) sampled yams from different parts of China, measured their diosgenin content and found that in *D. collettii* it was 13.19 ± 0.26 mg g^{-1}.

D. composita

Dioscorea composita Hems is native to Belize, Costa Rica, El Salvador, Guatemala, Honduras and Mexico. Its common names include Atlantic yam, barbasco, batata silvestre, black yam, China root, colic root, composita yam and devil's bones. *D. composita* is a plant of the moist tropics and subtropics that grows best in areas where annual daytime temperatures are within the range 25–30°C, but can tolerate temperatures in the range of 15–38°C. It will die at 9°C or lower. The optimum mean annual rainfall is in the range 1,200–1,500 mm, but *D. composita* can tolerate 1,000–2,100 mm. The plants prefer a position in full sun, but are tolerant of light shade. Optimum soil conditions are light to medium soil of moderate fertility and a pH in the range 6–6.5 (tolerating 5.8–7.3) (Fern, 2014). *D. composita* can be propagated from true seeds and Viana and Felippe (1990) found that seed germination was delayed in freshly harvested seeds but this dormancy was reduced in seeds stored for about 9 months. In seeds stored for 3 years, germination occurred within 2–3 weeks, but their viability decreased after 4 or 5 years of storage (Cruzado *et al.*, 1964). Tubers reach their maximum size in about 4–5 years after sowing (Ecocrop, 1993–2007).

D. composita is grown commercially largely for the production of diosgenin. It can have a much higher content of diosgenin than *D. mexicana*, which was the species previously used (Coursey, 1967). Martin (1969) reported that the diosgenin content of *D. composita* varied between 0–13% depending on variety. Russell Marker was an American who collected wild yams in Mexico. He found that *D. composita* 'contained a nearly pure sapogenin (a subtype called diosgenin). This particular wild yam was called *barbasco* by the indigenous population, and it became the industry's choice for the raw material in hormone synthesis' (Dach, 2013; Anonymous, 2017). Yang and Liu (2005) isolated 3 new diarylpropanes from the dried tubers of *D. composita*, which they claimed may have anti-malarial and anti-fungal functions. These diarylpropanes were: 1, 3-bis-(2-hydroxy-4-methoxyphenyl) propane, 1, 3-bis-(2, 4-dihydroxyphenyl) propane and 1-(2′-hydroxy-4′-O-β-D-glucopyranosyphenyl)-3- (2″,4″-dihydroxyphenyl) propane).

D. convolvulacea

Dioscorea convolvulacea Cham. and Schltdl, also *D. convolvulacea* subsp. *convolvulacea* is synonymous with *D. hispida* Dennst. It is native to Central America (Mexico and Panama), where its common name is gadong, and Trinidad where it is eaten after cooking (Govaerts *et al.*, 2007). Tubers are dish-like in shape, slightly palmate and up to 17 cm in diameter.

D. cotinifolia

Dioscorea cotinifolia Kunth is synonymous with *D. malifolia* Baker and is found in Botswana, Mozambique, South Africa (with a wide distribution from the Eastern Cape near Port Elizabeth to the northern border of Limpopo) and into southern Mozambique. Viruel *et al.* (2016) reported that *D. cotinifolia* is the only *Dioscorea* species in the *Enantiophyllum* section found in sub-Saharan Africa. It grows in open dry forests, forest margins, among scrubby vegetation and rocky places from 0–400 m altitude (Conservatoire et Jardin Botaniques de la Ville de Genève, 2019). Stems are left twining and plants produce several tubers that are ovoid to pyriform in shape and up to 5 cm long, usually on shortish stout rhizomes, eventually forming a caudex (Wilkin and Timberlake, 2009; Botany VM, 2018). Tubers were reported to be regarded as edible in southern Mozambique (Hills and Wilkin, 2017).

D. communis

Dioscorea communis L. Caddick and Wilkin is synonymous with *Tamus communis* L. Common names include: Bryony, black bindweed, black Bryony, lady's seal and norça. It is widespread in Europe from Ireland to the Canary Islands, North Africa and Western Asia, east to Iran and Crimea. It is the only member of *Dioscoreaceae* to be found in the UK. It is a climbing perennial plant with weak stems twining clockwise and almost cylindrical tubers up to 3 cm in diameter and up to 8 cm long with scattered, thin root fibres. The peel is blackish-brown with whitish flesh that produces a slimy paste when peeled. The stems die back in winter but the rhizomes are perennial. *D. communis* is normally propagated by seed and grows best in moist, well drained fertile soil (Heywood *et al.*, 1980; Botany VM, 2018). The large, fleshy tuber is high in starch, black on the outside, very acrid and extremely poisonous and should not be eaten, at least when raw, but Coursey (1967) reported that 'during prehistoric and possibly medieval times this tuber was certainly used as a food in Europe'. Young shoots are eaten in southern France, Spain,

Portugal, Italy and Croatia after cooking in boiling water with the water removed and replaced during cooking to remove toxins. Parts of the plant contain calcium oxalate crystals, histamines, saponins and a glycoside in both the berries and rhizomes. *D. communis* has been used as a poultice for bruises and inflamed joints, but it can cause painful blisters and calcium oxalate and histamines in the tubers and berries may contribute to skin irritation and contact dermatitis. Kovács *et al.* (2007) reported that the cytotoxic activity of the tubers was due to the presence of 7-hydroxy-2,3,4,8-tetramethoxyphenanthrene, 2,3,4-trimethoxy-7,8-methylenedioxyphenanthrene, 3-hydroxy-2,4,-dimethoxy-7,8-methylenedioxyphenanthrene, 2-hydroxy-3,5,7-trimethoxyphenanthrene and 2-hydroxy-3,5,7-trimethoxy-9,10-dihydrophenanthrene.

D. coriacea

Dioscorea coriacea Humb. and Bonpl. ex Willd. is synonymous with *D. caucensis* R. Knuth, *D. frutescens* Rusby, *D. pennellii* R. Knuth, *D. pennellii* var. *pilosula* and *Helmia coriacea* (Humb. and Bonpl. ex Willd.) Kunth. It is native to western South America from Colombia to Venezuela at an altitude of 1,200–3,900 m (Useful Plants of Boyacá Project, 2019). In Colombia, Raz *et al.* (2013) reported that *D. coriacea*, *D. lehmanni* and *D. polygonoides* all contained sapogenins, with the highest yields occurring in *D. coriacea*. Pinzón-Rico and Raz (2017) characterized yams in the Bogotá market during 2010, 2012 and 2013, using semi-structured interviews with vendors. They identified wild harvested tubers of four native species: *D. coriacea*, *D. lehmannii*, *D. meridensis* and *D. polygonoides*. *D. coriacea* was the most prevalent, but tubers of all four species were sold under the name 'zarzaparrilla'. All vendors interviewed recommended *Dioscorea* species for treating 'ailments of the blood', as a 'blood cleanser', and for lowering triglycerides and cholesterol (the most often-cited specific uses.) Total estimates of sales of wild yams over the study period was small and averaged only 488 kg per week.

D. craibiana

Dioscorea craibiana Prain & Burkill. Its common names in Vietnam are từ craib and củ nần. Sagwansupyakom and Chantraprasong (1984) reported that *D. craibiana* occurred in the wild in Thailand but it was rare. Thapyai *et al.* (2005a) commented that *D. craibiana* appears to be endemic to Thailand. They reported that the specimen from Vietnam, cited in previous studies, was in fact *D. arachidna* Prain & Burkill. Tubers were reported to be incompletely known, probably annually replaced, slender, spreading, cylindric to clavate, probably with thickened apices, epidermis thin and papery, pale brown with whitish flesh (Thapyai *et al.*, 2005a).

D. cubensis

Dioscorea cubensis R. Knuth, also referred to as *D. cubijensis* R. Knuth, which is not an accepted name. It is native to Cuba and is found in pine woods at between 100 and 800 m. Tubers are borne singly and are about 3 x 3.5 cm in size. No bulbils were reported (Flora de la República de Cuba, 2019).

D. cumingii

Dioscorea cumingii Prain & Burkill. Co's Digital Flora of the Philippines (2019) lists four varieties: *D. cumingii* var. *cumingii*, *D. cumingii* var. *inaequifolia*, *D. cumingii* var. *polyphylla* and *D. cumingii* var. *ramosii*. Common names include: lima-lima, kasi and par (Chung, 2001). It is native to the Philippines, Taiwan, Indonesia and possibly the Yunnan province of China (Flora of China, 2019). The usual habitat consists of areas with considerable rainfall close to mountains, in thickets and forests at low and medium elevation up to 1,400 m altitude (Chung, 2001). Burkill (1951) reported that it is used as a food in the Philippines to a limited extent, but it may sometimes be confused with *D. pentaphylla*. Sagwansupyakom and Chantraprasong (1984) described it as rare. Flora of China (2019) described the rootstock as unknown.

D. deltoidea

Dioscorea deltoidea Wall. ex Griseb. is synonymous with *D. nepalensis* Sweet ex Bern. and *Tamus nepalensis* Jacq. Common names include: baniatakari, bhyakur tarul, harvish, janj, jung kinch, nepal yam and shingli-mingli. It occurs in the Himalayas up to an altitude of 3,500 m (Coursey, 1967, 1968), and is grown in the East Indies, where it produces 'edible roots' that occur both wild and cultivated in the Indian Archipelago. Ding and Gilbert (2010) reported that it occurs in broad-leaved forests, scrub forests at 2,000–3,100 m in Sichuan, Xizang and Yunnan provinces of China as well as Bhutan, India, Myanmar, Nepal, Sikkim, Thailand and Vietnam. Experimental cultivation has been attempted in the Kenyan highlands and in tropical climates in Bangalore in India, where it showed no dormant period and had a yield of 1.25 kg plant^{-1} (Coursey, 1967). Tubers are up to 1.5 cm in diameter and grow horizontally in the soil (Ding and Gilbert, 2010).

D. deltoidea is a good source of steroidal sapogenins, but it had become depleted in the wild due to heavy collection in its native habitat in the highlands in India (Coursey, 1967). Bhandari *et al.* (2003) reported that wild yams (*D. deltoidea*, *D. bulbifera*, *D. versicolor* and *D. triphylla*) make a significant contribution to the diets of tribal people in Nepal. They gave the following analysis on a fresh weight basis for the four species: crude protein 1.6–3.1, ash 0.5–1.2, crude fat 0.2–0.3, crude fibre 0.6–1.5%, K 250–560 mg 100 g^{-1}, Na 4.15–17.8 mg 100 g^{-1}, P 33.1–61.6 mg 100 g^{-1}, Ca 14.3–46.9 mg 100 g^{-1}, Mg 18.3–27.3 mg 100 g^{-1}, Cu 0.10–0.21 mg 100 g^{-1}, Fe 0.39–2.92 mg 100 g^{-1}, Mn 0.14–0.35 mg 100 g^{-1} and Zn 0.22–0.53 mg 100 g^{-1}. The tubers of *D. deltoidea* were reported to contain about 4.8% diosgenin. The 'juice of the tuber' is used in the treatment of various disorders including those of the genitals, asthma, arthritis, roundworm, constipation and also for contraception. Rokem *et al.* (1985) described how diosgenin could be obtained from cell cultures of *D. deltoidea* and its production could be increased by growth regulators, nutrient medium and biotic elicitors.

D. demourae

Dioscorea demourae Uline ex R. Knuth is also referred to as *D. de-mourae* Uline ex R. Knuth. Its native range is Argentina (Chaco, Corrientes, Misiones), where it is commonly called cará, and Brazil (Minas Gerais, Paraná, Rio de Janeiro, Rio Grande do Sul, Santa Catarina, São Paulo, Distrito Federal Espinhaço and Bahia) (Pedralli, 2002, Pedralli, *et al.*, 2002; USDA, 2019) where *D. demourae* was reported to be one of the 27 *Dioscorea* species growing there.

D. divaricata

Dioscorea divaricata Decne. is synonymous with *D. divaricate* Blanco., *D. opposite* Thunb., *D. oppositifolia* L. *D. foxworthyi* Prain & Burkill, *D. oxyphylla* R. Knuth, *D. soror* Prain & Burkill, *D. polystachya* Turcz. There is some confusion over the correct name for this species. (Govaerts *et al.*, 2007). Common names include: buloy, pakit, kiroi, dulian and bakliakang. It is closely related to, and much resembles, *D. nummularia* Lamk (Groen *et al.*, 2016). It is native to the Philippines and temperate areas of eastern Asia (Groen *et al.*, 2016) and grows wild on the slopes of hills in China. It was reported that it can be invasive in both 'open to shady areas' in eastern USA (Govaerts *et al.*, 2007). It is a climbing twining perennial with flowers that have a pleasant scent of cinnamon. It produces tubers up to about 1 m long (Groen *et al.*, 2016). Tubers store well for long periods and can also be left in the ground and harvested as required in the winter. However, Groen *et al.* (2016) reported that *D. divaricata* is rarely cultivated because the tubers grow too deeply in the soil, though Tongco *et al.* (2014) reported that *D. divaricata* is cultivated in China for its edible tubers, which contain about 20% starch, 75% water, 0.1% vitamin B1, 10–15 mg vitamin C. Values from proximate analysis of boiled tubers were: moisture 69.3–76.4%, crude protein 2.0–2.4%, crude fat 0.2–0.6%, crude fibre 0.9–2.1% and ash 0.6–0.7%. Tubers have a floury texture with a very pleasant flavour and can be boiled, baked, fried, mashed, grated and added to soups. *D. divaricata* tubers are eaten by the Magbukún Ayta community in the Philippines (Tongco *et al.*, 2014). In West Africa, Diby *et al.* (2009) reported that it is often found in subsistence conditions where yields are low, varying between 9 and 10 tonnes ha^{-1} in low fertility soils compared with a potential yield of 51 tonnes ha^{-1} for *D. alata* and 27 for tonnes ha^{-1} for *D. rotundata*. The starch extracted from the tubers can be used as a substitute for arrowroot, though it is not as good at binding as the starch from *D. japonica*.

D. daunea

Dioscorea daunea Prain & Burkill. It was reported to be growing in Thailand (Wilkin and Thapyai, 2009) and was reported to be distributed from western Indo-China to Peninsula Malaysia, although Purdue University (2019), on their location map, show its location only Thailand. Also, Saikia and Sarma (2015) reported that it is grown in India where it is confined to a narrow range of habitats in both Arunachal Pradesh and Assam where it is a forest species found in hilly evergreen forests, especially in undisturbed areas. Schols *et al.* (2001, 2003), in their study of pollen morphology of *Dioscorea* reported that pollen of *D. daunea* had the second largest orbicules of the *Dioscorea* species included in their survey. The vertically oriented tubers are cylindrical to clavate and produced in groups of two or three. They are up to 10 cm long and up to 3 cm in diameter and replaced annually (Saikia and Sarma, 2015). Tubers are eaten in northeast Thailand (Purdue University, 2019). Proximate composition of *D. daunea* tubers in g per 100 g fresh weight were: 2.32 protein, 2.32 carbohydrate, 0 fat, 65.09 moisture, 1.02 ash, 0.64 fibre and 133 kcal energy (Maneenoon *et al.*, 2008).

D. decipiens

Dioscorea decipiens Hook.f. Wilkin and Thapyai (2009) and Botany VM (2018) described *D. decipiens* as being found in evergreen broad-leaved forests, scrub and on mountain slopes from 500 to 2,200 m. Its native range is south-central China (Yunnan) to Indo-China (Laos, Myanmar, Thailand, Vietnam) and Bangladesh. Ding and Gilbert (2010) described two varieties. These were: *D. decipiens* var. *decipiens* that occurs in the environment described above (evergreen forests, scrub, mountain slopes), in China (Yunnan province), Laos, Myanmar and Thailand and *D. decipiens* var. *glabrescens* that occurs in forests and mountain slopes at 1,100–1,500 m in western Yunnan province, China. Stems twine to the right and tubers grow vertically in the soil and are cylindric to clavate, brown in colour and become wrinkled if they dry out. The flesh is white or light yellow, drying slightly purplish brown. Tubers, one per growing season, often with a withering tuber of the previous season on top, are usually 5–20 cm long but up to 30 cm by 0.5–1.5 cm in diameter. *D. decipiens* also produces bulbils (Fig. 8.11), which can be used for propagation Botany VM, 2018).

D. depauperata

Dioscorea depauperata Prain & Burkill is native to Indo-China and was identified growing in Thailand where it is locally called sa man or sa kaeo (Sagwansupyakom and Chantraprasong, 1984). Wilkin and Thapyai (2009) described the tubers as deeply vertically buried in the soil, cylindrical with a woody crown growing to about

Fig. 8.11. *D. decipiens* bulbils growing in Sarinya's yam farm in Chachoengsao province in the eastern region of Thailand in 2018.

13 to 15 cm long and up to about 1 cm in diameter. Plants were found to be restricted to growing in open vegetation on limestone soils at 100–500 m. Ding and Gilbert (2010) described tubers as brown with white or light-yellow flesh drying slightly purplish brown, cylindric and 'drying wrinkled'.

D. doryphora

Dioscorea doryphora Hance (also referred to as *D. doryophora*) is called the 'Halberd' field potato in Taiwan (Huang, 2003). It has been reported to be synonymous with *D. polystachya* or *D. opposite*, but Peng *et al.* (2016) concluded that, based on leaf and tuber shapes and molecular data, that *D. doryphora* might be a separate species. It is grown wildly by farmers in Taiwan where the tuber is in great demand as a food and for medical uses because of its diosgenin content. Vanisree *et al.* (2004) and Gupta *et al.* (2012) described a cell suspension culture using *D. doryophora* for the production of diosgenin. Callus tissue was induced from the surface of both micro-tuber and stem nodes, and the explants were placed on solid growth media. These explants formed callus tissue in 2–6 weeks, which were cut from the explants and sub-cultured further. They showed that callus tissue can be inoculated in a liquid medium, and through continuous shaking, cell suspension cultures were obtained. Finally, dispersed cell suspension cultures were obtained from both micro-tuber-derived and stem-node-derived callus in liquid culture medium supplemented with 2,4-dichlorophenoxy acetic acid (0.1 mg L^{-1}) plus sucrose (3%) by incubation in a rotator shaker at 120 rpm. Although 6% sucrose was found to be optimum for the growth of cell suspension culture, it was observed that cells cultured in 3% sucrose containing medium produced a higher level of diosgenin. Analysis showed that both micro-tubers and stem node-derived cultured suspensions contained diosgenin. The diosgenin contents were analysed from the different plant parts (leaf, node, root, seed and stem) and a higher level of diosgenin was obtained from root tissue. An approximately similar level of diosgenin was obtained in field grown and *in vitro* grown tubers. Diosgenin content as high as 3.3% per gram dry weight was obtained from micro-tuber derived cell suspension cultures, but only 0.3% diosgenin from the stem node-derived callus cell suspension.

D. dregeana

Dioscorea dregeana (Kunth) T. Durand and Schinz, commonly called African wild yam, dakwa, ingcolo, isidakwa, ilabatheka, udakwa and wildejam. It occurs naturally in woodlands, forests and bush clumps in the eastern parts of South Africa particularly in Limpopo province, Mpumalanga, Swaziland, Kwa Zulu-Natal and Eastern Cape (Waizel, 2009; Mothogoane, 2013). Tubers are ovoid to spheroidal in shape, solitary and lobed or a main tuber may be produced with several smaller subsidiary tubers in a dense cluster near the soil surface that can persist for several years. Tubers are pale brown with white flesh and up to 7 cm in diameter (Mothogoane, 2013; Botany VM, 2018).

Previous work has shown that extracts from *D. dregeana* tubers could have medical applications. It was reported by Mothogoane (2013) that '*Dioscorea dregeana* is commonly sold on [*sic*] the *muthi*' markets in South Africa. It may be used in different ways for different purposes. The Zulus use it as a sedative in the treatment of epilepsy, hysteria, insomnia and acute psychosis. The Zulu word 'isidikwa' means drunkard, referring to the reported effects that it may have. It is also used topically for scabies.

In ancient times, it was used as a general anaesthetic to enable fractures of the limb to be manipulated and stabilized by traditional bone-setters. The fresh tuber is generally taken orally as a weak decoction, with an adequate dose resulting in sleep within 20–30 minutes. *D. dregeana* is sometimes combined with *Boophone disticha* (century plant or tumbleweed) for the purpose of divination. However, human deaths have been reported after the use of the plant as famine food or as medicine. *D. dregeana* is reported to make a person 'mad drunk', and it has been used in poison bait to kill monkeys. This *Dioscorea*, due to its toxicity, is often planted to eradicate moles, often among crops, especially root and tuber plants, such as amabhatata (*Ipomoea batatas*) and amadumbe (*Colocasia esculenta*)' (Herbal Fire Botanicals, 2018). As acetylcholinesterase inhibitors are important in the therapeutic strategy for controlling Alzheimer's disease, Murray *et al.* (2013) described searches for new molecules with anti-acetylcholinesterase activity and isolated the alkaloid crinamine from *D. dregeana*. Mulholland *et al.* (2002) previously isolated crinamine from the tubers of *D. dregeana* and Likhitwitayawuid *et al.* (1993) had reported that crinamine, present in *Crinum asiaticum* and *C. amabile*, had anti-malarial properties. Kelmanson *et al.* (2000) showed that extracts of *D. dregeana* were active against *Pseudomonas aeruginosa* and in general, methanolic extracts exhibited higher activity than aqueous and ethyl acetate extracts.

D. dumetorum

Dioscorea dumetorum (Kunth) Pax. is synonymous with *D. triphylla* L. var. *dumetorum* (Kunth) R. Knuth and *Helmia dumetorum* Kunth. Common names include: African bitter yam, bitter yam, cluster yam, esuru, ikamba, ñame amargo, ono, quinim, three-leaved yam, trifoliate yam and unquinim. This African species is closely related to the Asian species *D. hispida* Dennst., and is found wild throughout tropical Africa in woodland, often persisting in cleared areas, between 15° north and 15° south. It is cultivated in West Africa, especially in Nigeria (Coursey 1967, Purseglove, 1972). Obidiegwu and Akpabio (2017) commented that *D. dumetorum* made a secondary but significant contribution to the yam domestication process in Nigeria. According to Chevalier (1936) cultivated forms of *D. dumetorum* are not known west of the Benin Republic. The most important area of their cultivation appears to be southeast Nigeria; among the Bondjo people on the Ubangi River. *D. dumetorum* seems to be deeply embedded in the culture of certain areas of West Africa (Blench and Dendo, 2004). Ardener (1954), discussing Kpe-speakers of coastal Cameroon, comments that *D. dumetorum* 'is the most ritually embedded cultigen, and *Colocasia esculenta* and *Dioscorea alata* seem to be late-comers'. It is extremely widespread in Nigeria, cultivated throughout the Niger Delta and eastwards to the border of Cameroon, on the Jos Plateau, as well as in southern Zaria and in the montane regions of Adamawa.

Tubers are deeply lobed, spheroid or ovoid in shape and may be solitary or there can be several in a tight cluster (Flora of Zimbabwe, 2019). Purseglove (1972) reported that in tropical Africa, wild *D. dumetorum* are widely eaten in times of food shortage, although, as one of their common names implies, they have a bitter taste. Tubers are used as boiled vegetables but should first be thoroughly soaked in water to remove the toxic constituents. However, in its wild form, *D. dumetorum* is highly poisonous, but there is no report of alkaloids in cultivated varieties (Food-Info.net, 2019). Toxicity is due to a large amount of dihydrodioscorine in tubers, and death has been recorded for humans in Nigeria, Gabon and East Africa after consuming D. *dumetorum* (Corkill 1948; Bevan *et al.*, 1956). *D. dumetorum* was reported to be used in some areas to make arrow poison (Maundu, *et al.*, 1999). Kay (1987) reported that tubers are detoxified by slicing, soaking and boiling, frequently with the addition of salt. The slices may be subsequently dried. She also reported that *D. dumetorum* is 'becoming a preferred yam in Cameroon'. Kay also reported that it is commonly used for pharmaceutical or medicinal purposes in Africa and Asia. In southeastern Nigeria, *D. dumetorum* is used in treatment of diabetic patients and as a herbal treatment for various ailments including the treatment of malaria. Tubers contain 75–84% starch dry weight with small amounts of proteins, lipids and vitamins and is rich in minerals. The protein also showed a positive amino acid balance, which is important for human nutrition (Afoakwa and Sefa-Dedeh, 2001; Ecocrop, 2018).

The composition of the edible portion of raw *D. dumetorum* tubers was given as: energy 124 kcal, protein 3.2 g 100 g^{-1}, fat 0.1 g 100 g^{-1}, carbohydrate 28.3 g 100 g^{-1}, ash 1.1 g 100 g^{-1}, calcium 52 mg 100 g^{-1}, phosphorus 45 mg 100 g^{-1} (FAO, 1965). Siadjeu *et al.* (2020) reported that *D. dumetorum* tubers are protein-rich (9.6%) with a fairly balanced essential amino acids composition (Alozie *et al.*, 2009). The provitamin A and carotenoid contents of the tubers of deep yellow genotypes are equivalent to those of yellow corn maize lines that had been selected for increased concentrations of provitamin A (Ferede *et al.*, 2010). The deep yellow yam tubers are used in antidiabetic treatments in Nigeria, probably due to the presence of dioscoretine, which is a bioactive compound with hypoglycaemic properties (Nimenibo-Uadia *et al.*, 2017). Opara (1999) reported that the dormancy period of *D. dumetorum* from Nigeria was 14 to 16 weeks. Drying of sliced tubers is used as a method of storage.

Harvesting, handling and storage

Harvesting is usually 8 to 10 months after planting and carried out by hand. Kay (1987) commented that *D. dumetorum* tubers are easily harvested by hand, and can also be mechanically harvested where it is grown on a large scale. The tubers do not store well, a high proportion becoming hard and inedible within 4 weeks after harvest (FAO, 1965). Siadjeu *et al.* (2020) also reported that *D. dumetorum* suffers from postharvest hardening of tubers that can start within 24 hours of harvesting and renders the tubers inedible. Afoakwa and Sefa-Dedeh (2002) found that postharvest hardening of *D. dumetorum* tubers was characterized by thickened cell walls and middle lamella in the cells. They found several treatments and storage methods that 'limited the contact of the tubers with the environment'. These were found to slow the hardening process. Tubers stored under ambient and 4°C showed a rapid reduction in moisture and starch content and an increase in the total alcohol-soluble sugars and reducing sugars after 72 hours of storage. A similar trend was observed by Afoakwa and Sefa-Dedeh (2001) for the cultivar 'Jakiri'. They found that after 110 days of storage starch content decreased by approximately 3.5–4.5 g 100 g^{-1} while sugar and fibre levels increased. Drying sliced tubers is also used as a method of storage. Kay (1987) also reported that the insects *Araecerus fasciculatus* and *Lepidobregma minuscula* have been identified in Nigeria as infesting stored *D. dumetorum* tubers.

D. elephantipes

Dioscorea elephantipes (L'Héritier) Engler is native to southern Africa. Its common names include: elephant's foot, Hottentots' bread, pane degli ottentotti, turtleback, tortoise plant and tortoise back plant. It appears to have a wide range of habitats including weathered rock, dry stony slopes and in karroid scrubland (semi-desert areas of South Africa). The plant is adapted to growing in areas with seasonal rainfall, by going dormant in the dry season and can tolerate temperatures over 40°C down to –4°C. Kew Science (2019) reported that *D. elephantipes* occurs at 150 to 1,200 m altitude, in thorny and succulent karroid bush, with *Carissa ferox*, *Gymnosporia* species, *Rhus* spp., *Aloe* spp. and *Crassula* spp. Tubers were reported to be of hypocotylar origin and are perennial, epigeal and pear-shaped when young (Sparshott, 1935). In mature plants tubers are shaped like a beehive, base subterranean (about one-quarter of the whole), with wiry roots up to 5 mm thick; the upper three-quarters of tuber is exposed above ground, to a total height up to 60 cm (a specimen recorded by Sir William Hooker in the *Guide to Glasgow Botanical Gardens* was 90 cm diameter, 210 cm high and weighed about 700 lbs). The surface is greyish brown with numerous four- to seven-sided tortoise-like plates of thick cork. The plates increase in number by splitting from the crown, and increase in size by peripheral growth, showing definite growth lines to 8 cm across and up to 3 cm thick at the centre. The crown of one or more growing points is at the top of the tuber, somewhat immersed, marked by stiff, brown, erect, deltoid, bracts up to 10 mm long, 15 mm broad at the base. The inner flesh is hard, yellowish-white, opaque and brittle (Archibald, 1967; Botany VM, 2018). The *Encyclopaedia of Succulents* (2013–2019) describes the enlarged, exposed tuberous base as a water-storing organ, called a caudex (tuber) about 60–90 cm in diameter, but sometimes growing to over 3 m and covered in tough armour-like plates. It is woody-looking, firm in texture but succulent inside and covered on the outside with thick greyish-brown bark that becomes deeply cracked. Tubers contain very high levels of saponins. In the past the tubers were eaten, after processing to remove toxic compounds, by indigenous people in South Africa. Due to the large amount of effort required to obtain edible material, such tubers were normally only consumed in times of famine (Kew Science, 2019).

D. esculenta

Dioscorea esculenta (Lour.) Burk., including *D. esculenta* var. *spinosa*, *D. esculenta* var. *fasciculate* and synonymous with *D. aculeata* L., *D. fasciculata* Roxburgh, *D. papillaris* Blanco, *D. atropurpurea*, *D. purpurea*, *D. sativa*, *D. vulgaris*, *D. javanica* and *D. papuana* Warb. The classification of *D. esculenta* used to be based on the presence or absence of prickles on the tubers but this method is no longer recognised (Coursey 1968; Kay, 1987). Common names include: Asiatic yam, Chinese yam, couche-couche douce, greater yam, igname, igname de Chine, igname des blancs, kachil, kangar, karen potato, katula, khanulu, kizahangu, lesser Asiatic yam, lesser yam, name, ñame azucar, obbi, oewi, oowi kelapa, ovy, ratula, sakourou, sweet yam, ten months yam, ubi kodi, ubi kemali, uwi, water yam, white yam and winged yam. Common names in India include sree latha and sree kala (TNAU, 2019). Its origin was reported to be within the area of India, Vietnam, Papua New Guinea and the Philippines and it is now grown widely throughout the tropics (Kay, 1987). It is cultivated as a minor tuber crop in the Indian subcontinent, South East Asia and New Guinea and has been introduced to China, Madagascar and the Caribbean islands. It was introduced to Madagascar and the Comoros by Austronesians, from where it spread to the East African coast. *D. esculenta* is also a dominant crop in parts of Oceania. However, it did not reach to the furthest islands in Polynesia, being absent in Hawaii and New Zealand. Starch grains identified to be from *D. esculenta* have been recovered from archaeological sites of the Lapita culture in Viti Levu in Fiji, dated to around 3,050 to 2,500 BCE (Sykes, 2003; Horrocks and Nunn, 2007; Blench, 2010).

The vines twine anti-clockwise. The average tuber weight was reported to be in the range of 250 g to 1 kg but in Papua New Guinea some varieties produce tubers weighing up to 3 kg (Kay, 1987) and tubers weighing up to 60 kg have been recorded. As with other *Dioscorea* species tubers are formed as swellings at the end of stolons that are borne in clusters, with up to 20 tubers per plant. Tubers are oval, resembling long, narrow sweet potatoes but can also be spindle shaped or branched. In the West Indies the tubers usually weigh up to 200 g and are up to 10–20 cm long and 5–8 cm in diameter. The flesh is yellow or white and the shape is usually digitate, ovate or globose in cultivated plants, but often cylindric to irregularly lobed in wild forms. The plants are produced shallowly in the soil with the crown bearing thicker roots and sometimes several rigid thorny roots especially in wild plants (Wilkin and Thapyai 2009; Botany VM, 2018).

Fig. 8.12. *D. esculenta* cultivated at Phitsanulok, Thailand. Reproduced with permission of Dr Chirdsak Thapyai.

Production

D. esculenta grows well in moist, light and medium acid, neutral or basic soils but grows best in well-drained soil. It can grow in shaded or open areas. Enyi (1972a, 1972b) reported that larger seed tubers out-yielded smaller ones both in the total yield and ware tuber yields. Total and ware tuber yields increased with the decrease in spacing distance. Total dry matter production per plant increased with the increases in both seed size and spacing distance but the dry matter produced per unit area decreased with the increase in spacing distance. *D. esculenta* can be grown in the tropics at elevations up to 1,000 m, although some varieties and cultivars have been reported to thrive at elevations up to about 2,500 m. The common latitudinal range is 23°N to 20°S. It can grow in both dry and humid environments. Enyi (1972a, 1972b) reported that a daylength of less than 12 hours was required for tuberization, but daylengths longer than 12 hours were necessary for adequate vine development. Reported yields were between 7 and 25 tonnes ha^{-1}. Cultivated tubers may be harvested 220–300 days after planting (Fig. 8.12).

Composition and uses

Tubers can be boiled (Fig. 8.2), baked, roasted, fried or used raw as a salad vegetable. *D. esculenta* has a sweet and pleasant flavour similar to sweet potatoes or chestnuts. It was a staple food in southern China from the 2nd and 3rd centuries CE (Hedrick, 1919). Opara (1999) gave the following analysis per 100 g of the tuber fresh weight as water 70–74 g, carbohydrate 24–25 g, protein 1.5–3.5 g and fibre 0.8 g. Kay (1987) reported the composition of the edible part as: water 67–81%, protein 1.29–1.87%, fat 0.04–0.29%, carbohydrate 17–25%, fibre 0.18–1.51% and ash 0.5–1.24%. The carbohydrate was mainly starch but with 7–11% of sugars. Shajeela *et al.* (2011) reported the following per 100g fresh weight: starch 62.40 ± 0.44 g, niacin 41.36 ± 0.35 g, ascorbic acid 84.06 ± 0.24 g. Knoth (1993) gave the following analysis on a fresh weight basis: 67–81% moisture, 17–25% carbohydrates, 0.1–0.3% fats and 1.3–1.9% crude protein. Shajeela *et al.* (2011)

gave an analysis for antinutritional factors in *D. esculenta* tubers (Table 8.12). FAO (1965) also gave composition of *D. esculenta* tubers (Table 8.13).

Curing

Ravindran and Wanasundera (1993) successfully cured *D. esculenta* tubers by exposing them for 3 days after harvest to mean ambient conditions of 29°C and 84% RH. Martin (1974a) recommended 26–28°C and high humidity for 5–7 days. Wilson (1980) described a simple method of curing *D. esculenta*. The tubers were first stacked on the ground in a lightly shaded area and covered with grass or mats. Then a canvas tarpaulin was placed over the whole stack to cover the grass or mats, ensuring the tarpaulin did not touch the yams. As an alternative, a simple wooden frame could be built and the canvas draped like a tent over the piled yams. She indicated that plastic sheets should not be used for curing since they can make the yams too hot. If there is no canvas tarpaulin, then several layers of sacks or mats could be used.

Harvesting, storage and sprouting

Harvesting is usually 6–10 months after planting when the vines begin to turn yellow. Commercial potato harvesting machines can be used for *D. esculenta* when they are planted on ridges. Storage in ventilated barns at about 25°C for up to 60 days has been recommended (Tindall, 1983; Opara 1999). Uninjured tubers can be stored for 4 months or more in ambient tropical temperatures in well-ventilated stores.

The dormancy period of *D. esculenta* was reported to be 4–8 weeks for the Caribbean and 12–18 weeks for Nigeria/West Africa (Passam, 1982; Opara, 1999). In their studies Ravindran and Wanasundera (1993) reported that tubers were stored in thin layers on the floor of a well-ventilated room in ambient conditions of 24–28°C and 70–90% RH for 150 days. They started to sprout after 120 days. Crude protein, starch and ascorbic acid content decreased significantly during the first 60 days of storage, but the decrease was gradual thereafter. Losses of some minerals and a significant reduction in levels of oxalates were also noted during storage. Panneerselvam *et al.* (2007) stored fully matured *D. esculenta* tubers in wooden boxes at 28±2°C and 65–75% RH in the dark for 70 days. They found that starch degradation and the enzymes involved (α-amylases and starch phosphorylase) showed a lower level of activity during early periods of dormancy, while sugar content and enzymes of carbohydrate metabolism increased rapidly during sprouting. The key enzymes of glycolysis, tricarboxylic acid cycle and the pentose phosphate pathway (aldolase, succinic dehydrogenase, malic dehydrogenase, glucose-6-phosphate dehydrogenase and 6-phosphogluconate dehydrogenase) increased even before the visible appearance of sprouting and their activities were at their maximum during sprouting. Tubers of the cultivars 'Sree Priya' and 'Sree Subra' stored at room temperature of 30–32°C and 80–85% RH in the dark sprouted 70–80 days after harvest. In the apical and basal regions, the starch content

Table 8.12. Antinutritional factors of tubers of *D. esculenta*. Modified from Shajeela *et al.* (2011).

Component	Level
Total free phenolics ($g100g^{-1}$)	0.79 ± 0.07
Tannins ($g100g^{-1}$)	0.20 ± 0.01
Hydrogen cyanide ($mg100g^{-1}$)	0.21 ± 0.03
Total oxalate ($g100g^{-1}$)	0.33 ± 0.02
Amylase inhibitor (AIU mg soluble starch^{-1})	7.80
Trypsin inhibitor (TIU mg protein^{-1})	1.92 ± 0.07

Table 8.13. Composition of raw *D. esculenta* tubers mg or g 100 g^{-1} edible portion. Modified from FAO (1965).

Energy	Moisture	Protein	Fat	Carbohydrate	Fibre	Ash	Ca	P
112 kcal	70.2	3.5	0.1	25.2	0.5	1.0	62 mg 100 g^{-1}	53 mg 100 g^{-1}

decreased while the sugars and α-amylase activity increased (Muthukumarasamy and Panneerselvam, 2000). Passam (1982) showed that postharvest treatment with gibberellins could delay sprouting, but no records were found of its commercial use and postharvest uses of gibberellins are not permitted in all countries. The effect of maleic hydrazide in controlling sprouting in yam showed that soaking the tubers in 1000 μl L^{-1} solutions for ten hours before storage reduced the rate of sprouting by 16% (Ramanujam and Nair, 1982).

D. esquirolii

Dioscorea esquirolii Prain & Burkill. Stems twine to the left. Tubers were reported to be unknown. *D. esquirolii* plants have been found in forests and on mountain slopes at 600–1,500 m in central and western Guangxi, southern Guizhou and northwestern and southeastern Yunnan provinces in China (Ding and Gilbert, 2010).

D. exalata

Dioscorea exalata C.T. Ting & M.C. Chang. Stem twine to the right and tubers are brown with white flesh and ovoid or cylindrical in shape. They occur in mixed forests, on mountain slopes, in valleys, along rivers and by roadsides from near sea level to 1,200 m in the Fujian, Guangdong, Guangxi, southern Hunan and southern Zhejiang provinces of China (Ding and Gilbert, 2010). Govaerts *et al.* (2007) reported that they also occur in Indo-China.

D. fasciculata

Dioscorea fasciculata Roxb. was reported to be synonymous with *D. esculenta* (Lour.) Burkill (The Plant List, 2018). According to Hedrick (1919):

> Tropical eastern Asia. Indian subcontinent, southeast Asia, New Guinea. Cultivated as a minor tuber crop, mainly in tropical Asia. Introduced in China, Madagascar, Caribbean Indian subcontinent, southeast Asia, New Guinea. This species is cultivated largely about Calcutta, and a starch is made from its tubers. Firminger (1874) says this is a very distinct kind of yam, the tubers about the size, form and color of large kidney potatoes; when well cooked, it has a greater resemblance in mealiness and flavour to the potato than any other yam he knows of. It is much cultivated in the Philippines by the natives and is much esteemed.

D. fasciculata is called kandka by the Marathi people of Maharashtra, India, and was considered by some as 'probably the tastiest' of all the *Dioscorea* to be found in the Goa/Konkan area.

D. fandra

Dioscorea fandra H. Perrier is found in dry forests and is widespread across southwest Madagascar, especially Toliara province. Andriamparany *et al.* (2015) also reported that *D. fandra* occurred mainly at lower altitudes in open forest habitats near roads, where harvest intensity was very high, and seemed to perform better in sandy soils with low soil nutrient levels and low silt content. Stems are left twining. Tubers are cylindrical in shape, fibrous, with a white peel, and grow in a vertical direction (Botany VM, 2018). Tubers may be consumed raw and they are reported to have an excellent yam flavour, described as slightly sweet and similar in flavour to watermelon (Burkill and Perrier 1950; Jeannoda *et al.*, 2007). The high water content of tubers, up to 96% (Jeannoda *et al.*, 2007), makes *D. fandra* a valuable resource in regions that often experience water shortages (Burkill and Perrier, 1950). Tubers are used as human food, but there was no indication of overharvesting, and expanding agriculture was reported by Kennerley and Wilkin (2017) not to present a significant threat to wild populations. *D. fandra* was identified as potential food resource in the Mahafaly region of Madagascar where 70% of the interviewed households were collecting wild yams (Andriamparany *et al.*, 2014).

D. filiformis

Dioscorea filiformis Blume is synonymous with *D. gibbiflora* Hook. f and *D. myriantha* Kunth. Common names include: aroi huwi curuk, dudung, wauh, kiroi, kiru and phakmaeo-daeng and man-thian. It was reported to be distributed in Thailand, Peninsular Malaysia, Philippines and Indonesia. *D. filiformis* commonly occurs in the lowlands, often on limestone cliffs or granite boulders (Groen *et al.*, 2016). Stems twine to the right, and plants often produce bulbils. Tubers are elongated, up to 50 cm long and 2 cm in diameter, with tender white flesh (Chung, 2001). They produce one tuber per growing season, often with the withering tuber of the previous season on top. Tubers are narrowly cylindrical, deeply buried and grow vertically (Wilkin and Thapyai, 2009). Tubers are boiled before being used as a human food in Peninsular Malaysia, but if they are eaten while still raw, they can burn the throat (Groen *et al.*, 2016).

D. flabellispina

Dioscorea flabellispina R. Couto & J.M.A. Braga is native to Brazil and limited to the southeastern Atlantic rainforests at typically 0 to 200 m, on rocky outcrops, with 'low humidity levels and a fair amount of light'. *D. flabellispina* produces several fibrous roots interspersed with various fusiform, filipendulous tubers (Fig. 8.13) that are white to yellowish in both the skin and flesh, growing about 5 cm below the ground surface (Couto *et al.*, 2015).

Fig. 8.13. Tubers of *D. flabellispina*. Couto et al. (2015) with permission.

D. fordii

Dioscorea fordii Prain & Burkill is indigenous to China (Fujian, Guangdong, Guangxi, Hunan and Zhejiang provinces), in mixed forests, on mountain slopes, in valleys, along rivers and by roadsides from near sea level to 1,200m. *D. fordii* produces large, starchy tubers that grow vertically in the soil (Ding and Gilbert, 2010). *D. fordii* is cultivated in China for both food and in traditional Chinese medicine, but Yan *et al.* (2011) reported that the lack of *D. fordii* healthy planting material has restricted its production and therefore they successfully developed a protocol for its *in vitro* propagation. They used a temporary immersion system based on the principle of temporary contact between plants and liquid medium used for micropropagation.

Wu *et al.* (2014) found that the composition of *D. fordii* tubers on a dry weight basis was approximately 76.5% starch, 9.9% protein and 1.03% fibre. Wu *et al.* (2015) analysed three landraces of *D. fordii*, on a dry matter basis, and reported the following analysis: starch 75.9–77.1%, protein 9.8–10.2%, fibre 0.92–1.14%, Ca 285.6–300.5 mg kg^{-1}, Mg 345.8–356.3 mg kg^{-1} and Zn 17.9–18.5 mg kg^{-1}. Wu *et al.* (2014) showed a lack of diversity by evaluating 21 landraces of *D. opposite*, *D. alata*, *D. persimilis* and *D. fordii* and observed high levels of polymorphism, 95.3% for inter-simple sequence repeat and 93.5% for sequence related amplified polymorphism. Wu *et al.* (2015) also studied a group of core germplasm selections containing 25 *Dioscorea* landraces in order to create an effective classification of usage by characterizing their nutritive and medicinal compositions all on a dry weight basis, which they achieved from analysis of the tubers and principal component analysis scores. All the landraces they tested had high contents of starch (60.7–80.6%), protein (6.3–12.2%), minerals (especially Mg 326.8–544.7 mg kg^{-1}) and essential amino acids. Allantoin and dioscin varied considerably between the landraces from 0.62 to 1.49% and 0.032 to 0.092%, respectively all on a dry weight basis. The quality variability of 25 yam landraces was clearly separated using the unweighted pair group method with arithmetic mean clustering and principal component analysis. Using an eigen value ≥1 as the cut off, the first three principal components accounted for most of the total variability (62%).

D. floribunda

Dioscorea floribunda M. Martens and Galeotti is native to El Salvador, Honduras, Guatemala, Belize and Mexico, where its common name is Galeotti. It occurs in moist or wet thickets or forests. In India, TNAU (2019) reported that *D. floribunda* can be grown in tropical and sub-tropical climatic conditions up to 1,500 m above mean sea level. Ecocrop (1993–2007) reported that it grows best in areas where annual daytime temperatures are within the range 25–30°C, can tolerate 15–38°C, and can be killed by temperatures of 9°C or lower. *D. floribunda* grows best in a mean annual rainfall in the range 900–1,200 mm, but tolerates 600–2,300 mm. Optimum cultivation conditions were reported to be in a sunny position in a well-drained, light to medium soil at a pH in the range 6–6.5, but it can tolerate pH 5.8–7.2. Tubers are harvested from wild plants when they are 3–4 years old and are cooked in various ways for local use.

Stems of *D. floribunda* twine to the left and Martin and Ortiz (1962) described the initiation and development of *D. floribunda* tubers. Tubers arise from a ventral bulge between the first shoot and true root of the seedling. This bulge results from meristematic activity of the ground meristem. This is followed by primary growth of tubers by multiplication and thickening of the primary meristem and by differentiation of the cells so produced. The tissues of the tuber, arising directly from the ground meristem, are the storage parenchyma that enclose bundles of xylem and phloem, with a parenchymatous cortex containing a well-defined cambium-like layer and cork cells. They observed that a true periderm existed transitorily. On the dorsal surface of tubers, new aerial shoots are produced and on the ventral surface new adventitious roots are produced. The starch supply of the tuber is concentrated in the older tissue. Starch levels therefore tend to occur in greatest quantity in the ventral portions.

In India, *D. floribunda* is grown as a medicinal plant where ‘Arka Upkar’ and ‘Pusa 1’ are two popular cultivars (TNAU, 2019) that were found to have a diosgenin content of 3.0 to 3.5% in tubers harvested 2 years after planting. Sengupta *et al.* (1984) also reported that *D. floribunda* is a good source of diosgenin. Nath and Mukherjee (2000) evaluated extracts of their tubers against *Meloidogyne incognita* both *in vitro* and *in vivo* and found that the extract had strong nematocidal properties.

TNAU (2019) reported that *D. floribunda* can be propagated through single node leaf cuttings or tuber pieces and Sengupta *et al.* (1984) developed a successful rapid form of propagation using callus cultures from node and internode segments.

D. futschauensis

Dioscorea futschauensis Uline ex R. Knuth is native to scrub forests, mountain slopes, ravines and roadsides from near sea level to 700 m in Fujian, northern Guangdong, northeastern Guangxi, Hunan and southern Zhejiang provinces of China (Ding and Gilbert, 2010). Tubers are yellowish brown in colour, irregularly cylindric in shape, up to 3.5 cm in diameter and ‘hard’. Diosgenin-

3-O-α-L-rhamnopyranosyl-(1 --> 4)-β-D-glucopyranoside was isolated from *D. futschauensis* tubers by Wang *et al.* (2004). They found that it inhibited the proliferation of specified human cancer cells and suggested that it may be a good candidate as a chemotherapeutic agent to treat human colon carcinoma. Liu *et al.* (2002) isolated a new anti-neoplastic spirostanol saponin and three known compounds (prosapogenin A of dioscin, dioscin and gracillin) from *D. futschauensis* tubers, which showed bioactivity against rice blast fungal disease (*Pyricularia* (*Magnaporthe*) *oryzae*).

D. galeottiana

Dioscorea galeottiana Kunth was reported to be synonymous with *D. convolvulacea* var. *galeottiana* (Kunth) Uline, *D. grandiflora* M. Martens and Galeotti, *D. lobata* Uline and *D. lobata* var. *lasiophylla* Uline ex R. Knuth (Govaerts *et al.*, 2007). It is a Central American species whose native range includes Mexico. Rojas *et al.* (1999) established *D. galeottiana* cell suspension cultures and tested four fungal strains, *Epicoccum nigrum*, *Fusarium* sp., *Mucor* sp. and *Alternaria tenuis* extracted from tubers, seeds and leaves of *D. galeottiana* on yield of diosgenin. They found that diosgenin production increased from 0.10 g^{-1} to 3.98 g^{-1} dry weight when cells were cultivated in the light in a growth medium containing these cultures but with limited in phosphate and sucrose. The addition of 1.3 g L^{-1} of autoclaved *Alternaria tenuis* cell culture growing in the dark induced the production of 0.04 mg diosgenin g^{-1} dry weight.

D. garrettii

Dioscorea garrettii Prain & Burkill occurs in the Yunnan province of China at 1,300–1,400 m and also in Thailand, but Sagwansupyakom and Chantraprasong (1984) reported that it is rarely found there. Ding *et al.* (2000) reported that the stems twine to the left but the 'rootstock was unknown' and 'this species is known from only a single collection within China'.

D. glabra

Dioscorea glabra Roxburgh, is also referred to as *D. glabra* var. *longifolia* Prain & Burkill, *D. hongkongense* Uline ex R. Knuth, *D. oppositifolia* L., *D. nummularia* Lamarck and *D. nummularia* Roxburgh. Common names include: guang ye shu yu and luntak. In Thailand it is called man-dong, man-sai and soronda kanda. Its origin is thought to be Uttar Pradesh of India where it is found in open areas, in mixed deciduous forest, evergreen forest, hill evergreen forest, pine forest, rock outcrop vegetation, beach forest and margins of swampy areas at elevations from near sea level up to 1,500 m. Ding and Gilbert (2010) also reported that it occurs in evergreen broad-leaved forests, scrub forests, on mountain slopes, valley sides and by roadsides at 200–1,500 m in western Guangxi and southern Yunnan provinces of China as well as Bhutan, Cambodia, India, Indonesia, Laos, Malaysia, Myanmar, Thailand and Vietnam. *D. glabra* plants can tolerate significant disturbance, and are found in almost all areas, but, in Thailand, mostly in managed or wet habitats, often growing with *D. bulbifera* (Wilkin and Thapyai, 2009; Botany VM, 2018).

Stems twine to the right and plants do not produce bulbils. Tubers develop from short, thick rhizomes growing deeply in the ground. Tubers are corky, pale yellow or brown in colour, form singly or in clusters and are cylindrical in shape with white edible flesh that turns yellow on drying (Fig. 8.14). The depth of formation of the tubers was found to be 35 cm, the length of tubers 34–50 cm and their diameter around 3–5 cm. Behera *et al.* (2009) reported that the tubers can be formed both below and above ground and they found about five tubers per plant. Tubers taper towards the crown, are annually replaced, narrowly cylindric, globose to clavate in shape and the apex is subtended by a woody crown and parenchyma that is largely fibrous (Wilkin and Thapyai,

Fig. 8.14. *D. glabra*. Photograph: Dr K. Abraham, CTCRI Emeritus, C-63, Gandhipuram, Sreekariam, Trivandrum 695017, India, reproduced with permission.

2009; Botany VM, 2018). Behera *et al.* (2009) reported that from planting to the first leaf senescence was up to 120 days, and the time from planting to tuber initiation was 74 days on average. The growth period of the tuber from planting to maturity was on average 192 days.

Maneenoon *et al.* (2008) reported that *Dioscorea* is the main source of carbohydrate for the Sakai tribe at Banthad Range in Thailand. They found 15 *Dioscorea* species growing in the area, one of them being *D. glabra*. Thirteen of these are consumed by the Sakai the remaining two were inedible. *D. glabra* was also reported to be growing wild, as well as being cultivated in Orissa in India. Tubers are a cheap source of carbohydrates, but can become gluey when cooked (Behera *et al.*, 2009). *D. glabra* was also reported to produce edible fruit that are eaten in parts of India. Maneenoon *et al.* (2008) gave the following analysis of the tubers per 100 g fresh weight: moisture 66.62 g, carbohydrate 29.65 g, protein 1.60 g, fat 0.22 g, ash 0.79 g, fibre 1.12 g 100 g^{-1} and energy 126.98 kcal.

D. gracillima

Dioscorea gracillima Miq. occurs in open forests and on mountain slopes at 200–2,200 m in southern Anhui, northern Fujian, southwestern Hubei, eastern Hunan, Jiangxi and Zhejiang provinces of China as well as in Japan. Its stems twine to the left and tubers are irregular in shape and grow horizontally in the soil (Ding and Gilbert, 2010).

D. gracilipes

Dioscorea gracilipes Prain & Burkill (*D. gracillipes*). Wilkin and Thapyai (2009) reported that *D. gracillipes* was endemic to Thailand, growing on open, rugged limestone slopes and in evergreen forests at 20–200 m.

D. hamiltonii

Dioscorea hamiltonii Hook.f. Common names include: adavi pemdalam, brara, karu kamda, man tien, murmujja amiala, sika kanda tan pawa, tha lang ta, thakan budo and venni kalasu. It occurs, and probably originated, in South East Asia and is found across the Indian sub-continent from foothills up to 1,400 m. It was reported to be rare in the wild but found especially in the interior of moist deciduous forests of the Himalayas, from Nepal to Bhutan, Assam, Myanmar and throughout Indochina (Kaladhar *et al.*, 2008). It is found in the forests of Chittagong and Chittagong hill tracts of India and occasionally on slopes by forest borders, but less so on the plains (Burkill, 1935). A single tuber is produced each growing season, often with the withering tuber of the previous season on top. Bulbils are absent. Tubers have a white surface and flesh and are smooth, cylindric to clavate in shape, subtended by a woody crown and grow vertically in the soil. They are up to 45 cm long and up to 8 cm in diameter (Kaladhar *et al.*, 2008; Wilkin and Thapyai, 2009; Botany VM, 2018).

Tubers are edible when boiled. Prain and Burkill (1936) commented that it is among the most favoured edible yams in several countries including Vietnam and northeast India. Tubers are also used for the treatment of dysentery (Kaladhar *et al.*, 2008) and the Chakma people (a community that inhabits parts of Bangladesh, India and

Myanmar) use the skin and the extracted juice of tubers as a vesicatory and a paste made from the leaves for the treatment of jaundice (Yusuf *et al.*, 2009). Methanolic and ethyl acetate leaf extracts from *D. hamiltonii* have been shown to have good antimicrobial activity against certain organisms that were tested by Kaladhar *et al.* (2008). Mohan and Kalidass (2010) gave the following analysis of tubers on a fresh weight basis: 78.73% moisture, 4.37 ± 0.09 crude protein, 10.24 ± 0.22 crude lipid, 4.15 ± 0.03 crude fibre, 8.7 ± 0.09 ash, 72.53 N free extractives, 1670.44 g 100g^{-1} calorific value (kJ 100g^{-1} dry matter), 12.35 ± 0.03 starch, 11.76 ± 0.12 niacin and 30.16 ± 0.12 ascorbic acid g 100g^{-1}. They also gave *in vitro* protein digestibility as 5.86% and *in vitro* starch digestibility as 66.30%. Anti-nutritional factors identified by Mohan and Kalidass (2010) were: total free phenolics 0.27 ± 0.05 g 100g^{-1}, tannins 0.02 ± 0.01 g 100g^{-1}, hydrogen cyanide 0.04 ± 0.02, total oxalate 0.28 ± 0.09 g 100g^{-1}, amylase inhibitor 1.54 AIU mg^{-1} soluble starch and trypsin inhibitor 12.6 TIU mg^{-1}protein.

D. hastifolia

Dioscorea hastifolia Nees. Coursey (1967) stated that it is a native of Western Australia where the tubers are eaten by indigenous Australians who call it warrine. Hedrick (1919) wrote: ‘Australia. The tubers are largely consumed by the aborigines for food, and this is the only plant on which they bestow any kind of cultivation.’ Maiden (1889) reported that *D. hastifolia* is ‘One of the hardiest of the yams. The tubers are largely consumed by the local Aborigines for food.’ Hallam (1986) found that in Western Australia *D. hastifolia* was cultivated further south than any other yam in environments that were very different from those of their natural habitat of open woodland with granitic and basaltic soils.

It is a perennial, tuberous climber with yellow flowers produced in spikes between May and July in Australia (Marchant *et al.*, 1987). Tubers are boiled and eaten like potatoes.

Denham (2007) wrote that the tubers of *D. hastifolia*

> exhibit phenotypic responses depending upon the type of soil in which they are growing. Where soils are loose and friable, tubers are larger; whereas tubers are smaller in shallow, stony and compacted soils. For most tuberous plants in Australia, the recurrent digging of tubers from a perennial plant will have loosened the surrounding soil and encouraged greater tuber growth.

D. hemsleyi

Dioscorea hemsleyi Prain & Burkill is synonymous with *D. mairei* H. Léveillé. Stems twine to the left, tubers are cylindrical, mucilaginous and occurs in open scrub forests and on mountain slopes at 1,000–3,000 m in Guangxi, Guizhou, Sichuan and Yunnan provinces of China as well as Cambodia, Laos, Myanmar and Vietnam (Ding and Gilbert, 2010).

D. heteropoda

Dioscorea heteropoda Baker. Common names include: bemandry, hofika, igname de la forêt, majôla, ofaka and oviala. It is found at high altitude on the central plateau of Madagascar, distributed in tapia or wooded grassland forests in the central area where it grows in open rocky areas often in or close to forests at elevations from about 900 m, but possibly as low as 600 m (Weber *et al.*, 2005; Botany VM, 2018). It was identified as a species threatened with extinction (Kennerley and Wilkin, 2017). Stems are left twining. Tubers grow vertically in the soil and are cylindrical in shape with a white skin. Tubers are of good quality and are consumed after simple cooking methods without special preparation (Burkill and Perrier, 1950).

D. hirtiflora

Dioscorea hirtiflora Benth is native to tropical Africa from Sierra Leone, through

Democratic Republic of the Congo to Kenya, south to Angola, Zambia, Zimbabwe and Mozambique. In Zambia its common name is Lusala. Tubers are up to 6 m long and about 5 cm in diameter, cylindrical in shape, descending vertically into the soil and produced annually. The stems twine right-handed (Burkil, 1985–2004; Wilkin and Timberlake, 2009; Botany VM, 2018).

It was reported by Burkil (1985–2004) that the tubers of *D. hirtiflora* are toxic and caustic and are not normally eaten, but they can be detoxified and eaten in times of emergency or famine. However, Zulu *et al.* (2019) found that *D. hirtiflora* collected from forests is an important edible wild tuber in the local economy of the southern province of Zambia where it provides a seasonally important food supply and income to rural households and satisfies demand from urban populations. They also found that tubers were sold in markets across southern Zambia, and of 278 households interviewed, 83% collected, 96% consumed and 59% sold *D. hirtiflora*, but populations of this yam were generally perceived to be declining in the wild. They found that they were collected mainly from March to September by women.

D. hispida

Dioscorea hispida Dennst is synonymous with *D. daemona* Roxburgh, *D. lunata* Roth., *D. mollissima* Blume, *D. triphylla*, L. *Helmia daemona* (Roxburgh) Kunth and *H. hirsuta* (Blume) Kunth. Common names include: Asiatic baichandi, Asiatic bitter yam, bhul kand, bitter yam, chanda gadda, củnê, dâm-lô:ng k'duöch, doestestics, dog's ballocks, dukar kand, gadoeng, gadong mabok, gadong mabok, gadongan, gadung, gayos, hastyaluka, hwà koy, intoxicating yam, karot, karukandu, kavalakodi, killoi, kloi, kloi-huanieo, kloi-nok, koi, kywae, kywe, maranpash poll, mo, nami, ondo, palidumpa and pashpoli, pei perendai, periperendai, podava kelengu, puli dumpa, sikapa, tellaagini geddalu, thella chanda gadda, thella gadda, ubi arak and ubi gadong.

D. hispida originated in South East Asia and grows wild in India, Indonesia, the Philippines, southwestern China, Taiwan, Malaysia and Papua New Guinea. There is no appreciable distribution or cultivation outside these areas and even in South East Asia the plant is rarely cultivated. *D. hispida* occurs mostly in rainforest areas and is found mainly at lower elevations in the Philippines, but has been found growing at altitudes of up to 1,200 m in the Himalayas (Onwueme, 2016). Hedrick (1919) wrote:

> In 1894 W. G. Boorsma (*Mededeelingen wuit's Lands Plantentuin*, xiii) separated from it an alkaloid to which he gave the name of dioscorine. This alkaloid has been studied by Plugge and Schutte (A.J.P., iv, 1897) who assigned to it the formula $C_{13}H_{19}NO_2$ and found it to be a convulsant poison, resembling closely in its action picrotoxin, but much more feeble

Onwueme (1978, 2016) reported that the best yields were obtained if they were planted on mounds or ridges. In cultivation, the crop is usually staked. Weeding is done at regular intervals, but the use of fertilizers is rare. No serious diseases and pests were reported. Yields of up to 20 tonnes ha^{-1} are obtainable. It has been suggested that the preparation of flour or starch from the tubers should proceed as follows: washing of the tubers, pulping, treating with lime water containing potassium permanganate and final separation.

D. hispida is a climbing plant twining to the left, usually with a prickly stem, particularly towards the base (Fig. 8.15). Tubers are renewed annually from a superficial corm, from which they grow as globose lobes, sometimes slightly elongated and covered with masses of fibrous roots, pale yellow to light grey in colour with white to lemon-yellow flesh. Three to five tubers are borne on each plant, which may be fused into a cluster weighing up to 35 kg; individual tubers weigh 5–16 kg (Fig. 8.16) (Rao and Beri, 1952; Kay, 1987; Wilkin and Thapyai, 2009; Onwueme, 2016; Botany VM, 2018).

The tubers are mature some 12 months after planting. Harvesting is usually easy since tubers are commonly produced just below the soil surface and are harvested by

Fig. 8.15. *D. hispida*. Photograph taken at Sarinya's yam farm in Chachoengsao province in the eastern region of Thailand, March 2018.

Fig. 8.16. *D. hispida* Photograph: Dr K. Abraham, CTCRI Emeritus, C-63, Gandhipuram, Sreekariam, Trivandrum 695017, India, reproduced with permission.

digging using hand tools. The tubers may be collected throughout the year, therefore storage is not necessary, but Kay (1987) reported that they can also be successfully stored buried in moist sand. Harvesting occurs from wild stands, or from cultivated plants that are at least 12 months old. Manual harvesting with the aid of a fork or digging stick is the usual practice (Onwueme, 1978, 2016).

Composition and uses

Kay (1987) gave an approximate analysis of *D. hispida* tubers as: water 78%, carbohydrate 18%, protein 1.81%, fat 0.16%, fibre 0.93% and ash 0.69% on a fresh weight basis. On a dry weight basis, the tubers contained 0.2 to 0.7% diosgenin and 0.044% dioscorine, which is distributed throughout the plant, and also 81.45–81.89% carbohydrates and 7.20–9.12% albuminoids (Anonymous, 2017). Proximate composition of *D. hispida* tubers in g per 100 g fresh weight was: 5.2 protein, 15.0 carbohydrate, 4.3 fat, 74.0 moisture, 2.0 fibre and 134.0 kcal energy (Rajyalakshmi and Geervani, 1994). The proximate composition of tubers per 100 g edible portion was also given by Burkill (1951) as: water 78 g, protein 1.81 g, fat 1.6 g, carbohydrates 18 g, fibre 0.9 g, and ash 0.7 g. On a dry weight basis, the tuber also contains 0.2–0.7% diosgenin and 0.044% dioscorine. Commercial starch is extracted from the tubers, which were shown to contain 88.34% starch as well as 5.28% protein, 5.33% fibre, 0.23% fat and 0.66% ash.

Some tubers from India have been found to contain ovoid, non-stratified starch grains with a longitudinal diameter of 35–40μm, and a gelatinization temperature of 85°C. Unlike cassava or potato starches, the viscosity of *D. hispida* starch does not decline appreciably after prolonged heating. Its amylose content has been reported as 10.24%. Bhandari *et al.* (2003) gave the following analysis of *D. bulbifera, D. versicolor, D. deltoidea* and *D. triphylla* on a fresh weight basis: crude protein 1.6–3.1, ash 0.5–1.2, crude fat 0.2–0.3, crude fibre 0.6–1.5%, K 250–560 mg 100 g^{-1}, Na 4.15–17.8 mg 100 g^{-1}, P 33.1–61.6 mg 100 g^{-1}, Ca 14.3–46.9 mg 100 g^{-1}, Mg 18.3–27.3 mg 100 g^{-1}, Cu 0.10–0.21 mg 100 g^{-1}, Fe 0.39–2.92 mg 100 g^{-1}, Mn 0.14–0.35 mg 100 g^{-1} and Zn 0.22–0.53 mg 100 g^{-1}.

Lancaster and Coursey (1984) reported that *D. hispida* tubers, as well as being a rich source of starch, were a good source of phosphorus, calcium and iron, but are extremely poisonous due to the presence of alkaloids of the dioscorine group. Consumption of *D. hispida* tubers without careful preparation can be fatal. Traditionally, the toxicity is overcome by soaking the sliced or grated tubers so that the alkaloid leaches out into the water. Dioscorine can also be removed by slicing or grating, boiling them or covering them in wood ash and placing them under running water or by repeated changes of salt water (Kay, 1987). Sulit (1967) describes a technique used in the Philippines where thin slices of peeled tubers are placed in a basket and submerged in the sea or a solution of salt water for 2 to 3 hours. They are then removed and squeezed under weights for a few hours and then replaced in the baskets and left in a running stream for 36 to 48 hours with occasional stirring. The slices are tested for toxicity by squeezing a drop of liquid into the eye and if the eye smarts, soaking is continued for a further period. It is a matter of concern that drops of the extract are placed in the eye.

Kumoro and Hartati (2015) described a successful technique for dioscorin extraction from *D. hispida* using microwaves. As reported above, the tuber contains dioscorine, but can be consumed after the toxin has been removed. Hudzari *et al.* (2011) reported that a traditional detoxification process was peeling, slicing and soaking in flowing river water for up to 7 days. However, they developed dioscorine removal equipment that was more hygienic and reduced the dioscorine removal time to about 6 hours. Kay (1987) reported *D. hispida* to be a major famine food.

Choudhary *et al.* (2009) reported that *D. hispida* is a medicinal plant in Rajasthan in India as well as being the chief famine food of tropical Asia, as is *D. dumetorum* for Africa. The poison in the tubers is often extracted and used as bait for animals or other devious purposes. It is also used for eliminating fish from shrimp pond cultures. The pounded tubers are sometimes used externally as an antiseptic, and a decoction is drunk to alleviate chronic rheumatism (Rao and Beri, 1952). Kay (1987) reported that the tubers may have been used in criminal poisoning and also as an ingredient, together with *Antiaris toxicaria*, in the preparation of arrow poisons. *A. toxicaria* is a tree of the Moraceae family grown in Java and its neighbouring islands.

Starch extracted from the tubers can be used for culinary or industrial purposes, notably the manufacture of glucose. Also, *D. hispida* grown in the East Indies was reported to produce 'very acrid roots' that were eaten by the Karen or Tinguian people in times of scarcity (Cole, 1922). In Kerala, India, the tubers are cooked with salt, chili, tamarind and turmeric powder and used in curry. In Bangladesh, tubers are used to kill worms in wounds and various parts of *D. hispida* are used to control whitlows, sores, boils and bites of rabbit, jackal or dog. In India, tubers are used for birth control. In Malaysia, pounded leaves are applied to sores of yaw (Burkill, 1951). In Thailand they are used in the preparation of snack foods such as being deep fried with peanuts (Fig. 3.11).

D. hombuka

Dioscorea hombuka H. Perrier is native to Madagascar, where it is distributed in the southwestern dry areas and also in degraded spiny forests. Stems are left twining, and tubers grow vertically in the soil and are cylindrical in shape with white skin and flesh (Botany VM, 2018).

D. hurteri

Dioscorea hurteri R. Hills and Wilkin. Hills *et al.* (2018) described *D. hurteri* 'from KwaZulu-Natal in South Africa ... an undescribed species, related to the *D. buchananii* Benth. species complex but differing in its inflorescence and floral morphology from all other taxa... is known from four localities... within a heavily developed region of South Africa'.

D. hypoglauca

Dioscorea hypoglauca Palibin is synonymous with *D. collettii* var. *hypoglauca* (Palibin) S.J. Pei & C.T. Ting, and occurs in mixed forests and on mountain slopes at 200-1,300 m in southern Anhui, Fujian, northern Guangdong, northeastern Guangxi, southern Henan, Hubei, Hunan, Jiangxi and Zhejiang provinces of China, as well as northern Taiwan (Ding and Gilbert, 2010).

D. inopinata

Dioscorea inopinata Prain & Burkill is called prachuap khiri khan in Thailand, where it was reported as being grown on the Royal Project at Hua Hin, where it is also called ชั่งหัวมัน, which means 'to weigh the tuber'. However, this is perhaps a play on words since ช่างหัวมัน means 'what the hell' or 'it doesn't matter'. Specimens of *D. inopinata* were collected in 2002, for the first time in over 70 years, by Thapyai *et al.* (2004) in the Khao Sam Roi Yot National Park in the Prachuap Khiri Khan province of Thailand and it appears that it was restricted to that area. Its habitat is open vegetation on and around rocky limestone hills and outcrops, from near sea level to about 150 m. *D. inopinata* flowers in July and August and sets fruit from October to December. The flowering period is early in the year compared to most Thai yams. The peak flowering period for yams from north of the Isthmus of Kra is in September and October. Thapyai *et al.* (2004), concluded that this is probably a response to water availability at Khao Sam Roi Yot, which declines rapidly once the rains stop, stimulating early fruit development. Yams, which the stallholder called 'man nok', were bought in a Thai market but the term 'man nok' is used in Thailand for several species of yams and they could be *D. alata* (mostly on its bulbil), *D. brevipetiolata*, *D. filiformis*, *D. garettii*, *D. esculenta* as well as *D. inopinata* (Ass. Prof. Dr. Chirdsak Thapyai personal communication, 2017). There are no bulbils, and the tubers of *D. inopinata* were reported by Thapyai *et al.* (2004) to remain unknown.

D. intermedia

Dioscorea intermedia Thwaites is native to the evergreen forests in India (Kalaburagi) and Sri Lanka (Sankara Rao, 2019) where its common names include eyyar and malayalam. In India, Arackal and Pandurangan (2015) reported that *D. intermedia* grow at 100–1,500 m and 'there are 32 species of *Dioscorea* reported of which 21 are distributed in Western Ghats and about 5 million people are directly or indirectly depend on this crop for their food, feed, medicine etc. ... including *D. intermedia*'. Stems twine to the left, don't produce bulbils and produce two or three elongated deeply buried sessile edible tubers that have a creamy-coloured skin and white or purple coloured flesh.

D. irodensis

Dioscorea irodensis Wilkin, Rajaonah and Randriamb is commonly called bemandry. Wilkin *et al.* (2017) referred to its morphology from samples taken from three sites in the Irodo Valley in Madagascar as 'an undescribed species'. Tubers are harvested in the wild and used for food at a level that appears to be unsustainable and *D. irodensis* has provisionally been considered a critically endangered species.

D. japonica

Dioscorea japonica Thunb. is a native of Japan and is widely cultivated as a food crop in Japan, Korea, China and neighbouring islands. There are many varieties including: *D. japonica* Thunb var. *pseudojaponica* Yamamoto, *D. japonica* Thunb. var. *pseudojaponica* (Hayata) Yamam, *D. japonica* var. *japonica*, *D. japonica* var. *oldhamii* and *D. japonica* var. *pilifera* (Ding and Gilbert, 2010). Coursey (1967) stated that it is considered by some authorities to be a form of *D. opposita*, but Araki *et al.* (1983) reported that *D. japonica* tubers are not so thick as those of *D. opposita* and its yield is lower. They also commented that *D. japonica* tubers have 'excellent qualities'. In addition, a kind of virus infecting *D. opposita* does not infect *D. japonica*. Araki *et al.* tested hybrid plants from a natural cross between *D. japonica* (2n=80) and *D. opposita* (2n=140). The tubers of the hybrids were as long as those of *D. opposita*, intermediate in thickness between those of the parents but less mucilaginous than those of *D. japonica*.

Common names include: cham ma, Chinese yam, dang ma, East Asian mountain yam, glutinous yam, Japanese mountain yam, Japanese yam, jinenjo, pinyin, rìběnshǔyù, shan yao, Taiwanese yam, wild yam and yamaimo. Jinenjo is a variety of *D. japonica* that is used as an ingredient in soba noodles. Ding and Gilbert (2010) reported that *D. japonica* occurs in mixed forests and margins, scrub forests, 'herb communities', on mountain slopes, in valleys, along rivers and streams and by roadsides at 100–1,200 m in Anhui, Fujian, Guangdong, Guangxi, eastern and southern Guizhou, Hubei, Hunan, Jiangsu, Jiangxi, Sichuan, Taiwan and Zhejiang provinces of China as well as Japan and Korea. Coursey (1967) also reported that in addition to the habitats listed above it is common in wooded foothills and that it grows in light, medium or heavy soils from acid to basic but prefers well-drained, moist soil and does not grow well in shaded areas. *D. japonica* stems twine to the right. The tubers are yellowish brown with white or yellowish white flesh, long, cylindrical in shape, up to 3 cm in diameter and grow vertically in the soil (Ding and Gilbert, 2010). However, tubers vary considerably in both size and shape.

Composition and uses

Tubers are used as a food and contain about 1.9% protein, 20% carbohydrate, 0.1% fat and 1% ash (Miyazawa *et al.*, 1996). Hedrick (1919) wrote: 'The roots, cut into slices and boiled, have a very pleasant taste'. Birkill (1935) stated that dried tubers are cut into shavings and used in traditional Chinese medicine and as a tonic. As well as diosgenin, *D. japonica* tubers contain the antimutagenic compounds eudesmol (an oxygenated sesquiterpene that can control appetite and digestion) and paeonol (a phenolic compound used in some traditional Chinese medicine that can have anti-inflammatory and analgesic effects) (Miyazawa *et al.*, 1996). Kim *et al.* (2011b) reported that in Korea tubers are used as a medicine known as 'san yak', but PFAF (2018) assessed the medical usefulness of *D. japonica* as only 2 out of 5. Kim *et al.* (2011b) isolated dioscorolide A and dioscorolide B from the tubers and showed cytotoxicity against tumour cell lines – they suggested that there may be selective toxicity among tumour and normal cells. Chiu *et al.* (2013) investigated the ethanol extract of a selected variety of *D. japonica* and found evidence that it might contain a natural antioxidant and anti-inflammatory agent. Chen *et al.* (2003a) tested uncooked *D. japonica* on mice and came to the conclusion that a 25% uncooked yam diet may benefit upper gut function, which could possibly be used to prevent hypercholesterolemia in humans, but the 50% yam diet negatively affected protein absorption. The tubers have been used in the treatment of diarrhoea, enteritis, enuresis and spermatorrhoea. Fu *et al.* (2006) found that tubers contained high levels of allantoin and allantoic acid that can also have medical uses. Chen *et al.* (2017) tested water and ethanol extracts for several *Dioscorea* species including *D. japonica* and found significant cardioprotective properties via its multiple effects on antioxidant, anti-inflammatory or antiapoptotic activities.

D. kamoonensis

Dioscorea kamoonensis Kunth is synonymous with *D. subfusca* R. Knuth. Stems twine to the left and produce bulbils and edible tubers that grow near the soil surface. Tubers are usually subovoid in shape, up to 3 x 2 cm with fibrous roots (Wilkin and Thapyai 2009; Botany VM, 2018). Ding and Gilbert (2010) reported that stems occur in secondary scrub forests, forest margins and valleys at 500–2,900 m in Fujian, Guangdong, Guangxi, Guizhou, Hubei, Hunan, Jiangxi, Sichuan, Xizang, Yunnan and southern Zhejiang provinces of China. Wilkin and Thapyai (2009) and Botany VM (2018) also reported that *D. kamoonensis* is found growing in the Himalayas, from Kumaun to Sikkim, and in Bhutan, Vietnam, Myanmar and northern Thailand, mainly at altitudes of 1,800-2,200 m. *D. kamoonensis* can be grown in light (sandy), medium (loamy) and heavy (clay) soils, but grows best in well-drained but moist soil, with a suitable pH, but 'cannot grow in the shade' (PFAF, 2019).

D. kerrii

Dioscorea kerrii Prain & Burkill is endemic to northern Thailand and Wilkin and Thapyai (2009) reported that the tubers are 'incompletely known, probably annually replaced, slender, possibly spreading, cylindrical, with clavate thickened apices (c. 3.5 by 0.25 cm piece present on type), crown small, ovoid'. They are found in mixed forests at 600–700 m.

D. kratica

Dioscorea kratica Prain & Burkill is a native of Indochina (Wilkin and Thapyai, 2009). Sagwansupyakom and Chantraprasong (1984) reported that the tubers are edible.

D. laurifolia

Dioscorea laurifolia Wall. ex Hook. f. is native to Peninsular Malaysia and common in the mountains of Malaysia, up to 1,200 m altitude (Chung, 2001). Common names include: akar kemahang, akar kemenyan hantu, kemenyan batu and ghost's benzoin climber. One or two pink- to red-fleshed tubers are produced on each plant annually, which 'do not descend very deeply into the soil'. The flowers have a strong scent of benzoin (Burkill 1951; Chung, 2001; Wilkin and Thapyai 2009; Botany VM, 2018). In Peninsular Malaysia raw tubers are used in traditional medicine for making poultices to treat sores, swellings and bites. Tubers are used for food, but in Thailand Maneenoon *et al.* (2008) reported that *D. laurifolia* is so toxic that it is impossible to detoxify.

D. laxiflora

Dioscorea laxiflora Mart. ex Griseb. *microbotrya* Griseb, is synonymous with *D.* calystegioides Kunth, *D. laxiflora* var. *calystegioides* (Kunth) Uline ex R. Knuth, and *D. laxiflora* var. *cincinnata* Uline ex R. Knuth. It originated in Brazil and is found in the Amazon in Brazil and Venezuela. *D. laxiflora* can occur in forested areas along with *D. multiflora* and *D. marginate.* It is commonly found in patches among trees in the tropical savanna ecoregion, for example in Cerrado in Brazil, and not in open vegetation (Costa, 2003). Karnick (1969) reported that *laxiflora* tubers contain sapogenin steroids.

D. lehmannii

Dioscorea lehmannii Uline. Its native range is Colombia and Venezuela at an altitude of

2,000–3,000 m. In Colombia, Raz *et al.* (2013) reported that *D. lehmanni* tubers contained sapogenins. Pinzón-Rico and Raz (2017) characterized yams in markets in Bogotá during 2010, 2012 and 2013, by semi-structured interviews with vendors. They identified wild-harvested tubers of four native species: *D. coriacea*, *D. lehmannii*, *D. meridensis* and *D. polygonoides*. *D. coriacea* was the most prevalent of these, but tubers of all four were sold under the name 'zarzaparrilla'. All vendors interviewed recommended *Dioscorea* species for treating 'ailments of the blood', with 'blood cleanser' as the most often-cited specific use as well as lowering triglycerides and cholesterol. Total estimates of wild yam sales in Bogotá markets over the study period averaged only 488 kg per week.

D. linearicordata

Dioscorea linearicordata Prain & Burkill. Stems twine to the right. Tubers are yellowish brown with white flesh, cylindrical in shape and grow vertically. They are found in open forests, scrub forests and on mountain slopes at 400–800 m in Guangdong, eastern Guangxi and Hunan provinces of China (Ding and Gilbert, 2010).

D. maciba

Dioscorea maciba Jum. and H. Perrier. Its native range is western Madagascar. Tubers are edible and the favoured yam of people in the Morondava region of Madagascar who harvest them from the wild (Wilkin *et al.*, 2008).

D. madecassa

Dioscorea madecassa H. Perrier is a tertiary wild relative of cultivated yams and is found in humid inland and littoral forests of Madagascar, often on forest margins or in areas of disturbance (Wilkin *et al.*, 2008). It is under threat due to deforestation, expanding agriculture and over-harvesting. Tubers grow vertically in the soil and are cylindrical in shape with white peel and flesh (Botany VM, 2018). Tubers are harvested in the wild and are edible after cooking (Wilkin *et al.*, 2008).

D. mangenotiana

Dioscorea mangenotiana J. Miège. Common names include elephant's yam, ẹgẹu, ẹsụsu and suiduo. It is native to the forests of Liberia, Côte D'Ivoire, Togo, Benin, Central African Republic, Democratic Republic of the Congo and southern Nigeria where it is harvested in the wild and the young tubers are used as food. The main part of the tuber is too woody to eat, but the current year's lobes of the tubers are fleshy and edible, but can only be obtained by laborious digging. Coursey (1967) commented that *D. mangenotiana* is 'of little, if any, economic value', nor is the tuber amenable to multiplication and replanting for the plant will grow only from the whole or most of the tuber (Coursey and Haynes, 1970). Coursey commented that the annual stems may be up to 40 m long and up to 4 cm in diameter at the base with stems twining right-handed from a single massive perennial vertically growing tuber. Tubers are long-lived, increasing year by year with extra lobes, giving one of their common names, because of their likeness to an elephant's foot. In time the tuber may attain as much as 60 kg in weight (Haynes and Coursey, 1969). Miège (1958) noted that *D. mangenotiana* harbours nitrifying bacteria in its lower scale leaves.

D. marginata

Dioscorea marginata Griseb. is native to Brazil, Paraguay and the Valle del Magdalena in Colombia where it can be found growing wild in forests at an altitude from sea level to 770 m (Costa, 2003; Araújo and Alves, 2010). Kirizawa *et al.* (2016) identified *D. marginata* growing in the Reserva Biológica do Alto da Serra de Paranapiacaba, Santo André in Brazil and considered it an endangered species.

D. melanophyma

Dioscorea melanophyma Prain & Burkill is synonymous with *Dioscorea tenii* R. Knuth. Stems twine to the left, tubers are ovoid or pear-shaped, slender and fibrous. It occurs in open scrub forests and forest margins at 1,300–2,500 m in Guizhou, Sichuan, eastern Xizang and Yunnan provinces of China as well as in Nepal (Ding and Gilbert, 2010).

D. membranacea

Diocorea membranacea Pierre ex Prain & Burkill. Its native range is Indochina to Peninsula Malaysia and it is called hua-khao-yen in Thailand. *D. membranacea* tubers are used as an ingredient in many traditional Thai medicines as an alternative or complementary treatment of cancer and AIDS. Pakakrong Thongdeeying *et al.* (2016) showed that the ethanolic extract of tubers has been used as an ingredient in medicines for liver cancer and cholangiocarcinoma treatment and that dioscorealide B was the active ingredient. Also, the crude ethanolic extracts of tubers stimulate natural killer cells that can bind to K562 tumour target cells at concentrations of 10 and 100 ng mL^{-1}, but not at higher concentrations (Sumalee Panthong *et al.*, 2014). K562 cells are part of the NCI-60 cancer cell line panel used by the National Cancer Institute in Thailand. Natural killer cells play a major role in host rejection of both tumours and virally infected cells. Supinya Tewtrakul and Arunporn Itharat (2006) also extracted several anti-allergic substances from the tubers of *D. membranacea*.

D. menglaensis

Dioscorea menglaensis H. Li. is native to evergreen broad-leaved forests and scrub forests at 900-1,500 m in China where its common name is shi shan shu yu. Stems twine to the left and tubers usually have about 3–10 branches that are up to 20 cm long and 5 cm in diameter. They are globose or ellipsoid in shape with a corky yellowish white skin and white flesh (Ding and Gilbert, 2010).

D. meridensis

Dioscorea meridensis Kunth. Its native range is western South America to northwestern Venezuela. In Colombia it grows in the rainforests at altitudes of 500–2,500 m in the Andes. In Ecuador it is called the 'wild jam', and produces a very large, edible tuber. Pinzón-Rico and Raz (2017) characterized yams at a Bogotá market during 2010, 2012 and 2013, by semi-structured interviews with vendors. They identified wild-harvested tubers of four native species, *D. coriacea*, *D. lehmannii*, *D. meridensis* and *D. polygonoides*. All vendors interviewed recommended *Dioscorea* species for treating 'ailments of the blood', with 'blood cleanser' and lowering triglycerides and cholesterol as the most often-cited specific uses. Total estimates of sales of wild yams over the study period averaged 488 kg per week.

D. mexicana

Dioscorea mexicana Scheidw. Common names include: cabeza de negro, Mexican yam and tortoise plant. It is native to southern Mexico and is distributed from northeastern Mexico (Chiapas, Colima, Guerrero, Jalisco, Mexico, Michoacan, Oaxaca and Veracruz) south to Belize, Costa Rica, Guatemala, Honduras, Nicaragua and Panama. It produces a caudex, which is a 'stem rootstock particularly a basal stem structure' from which new growth arises. Its caudices take several years to develop, are covered with corky scales and are produced largely above ground. Tubers are used as a commercial source of diosgenin, which was reported to be 0.3–0.8% (Botany VM, 2018). An American chemist, Russell Marker, developed the extraction and manufacture of hormones from *D. mexicana*, but later trade was focused on *D. composita* instead, as *D. composita* has a higher diosgenin content (Dach, 2013).

D. microbotrya

Dioscorea microbotrya Griseb., whose common name is yaco, is synonymous with *D. microbotrya* var. *grandifolia* Hauman. Its native range is from Brazil to northern Argentina and it has been recorded at 30 locations in Brazil (GBIF, 2019). It has also been recorded in Uruguay (Govaerts *et al.*, 2007) and Argentina where *D. microbotrya* var. *grandifolia* is also described as being a synonym. It was reported to grow from sea level to 1,000 m in Argentina (Flora Argentina, 2018), where it produces a single tuber or seven to nine small 'papas' tubers, up to 6 cm long by up to 3 cm in diameter with 'atropurpúrea' flesh (Xifreda and Seo, 2009). Karnick (1969) reported that *D. microbotrya* tubers contain sapogenin steroids. Pedralli (2002) and Pedralli *et al.* (2002) listed *Dioscorea* species cultivated for their medical properties in order of importance as *D. bulbifera, D. cayenensis, D. dumentorum, D. alata, D. trifida, D laxiflora* and *D. microbotrya*. Schols *et al.* (2003) studied the pollen dimensions of various *Dioscorea* species and found considerable variation and in *D. microbotrya* the largest equatorial axis was 28-(28)-30 μm, the smallest equatorial axis 18-(21)-28 μm and the polar axis was 25-(28)-38 μm.

D. minutiflora

Dioscorea minutiflora Engler is synonymous with *D. armata*. De Wild and Coursey (1967) reported that *D. minutiflora* is closely related to *D. smilacifolia* of which it may even be a sub-species. Common names include: bush yam and wild yam. Muthamia *et al.* (2014) list 123 local names in Kenya for *D. minutiflora* (e.g. mundu-wakinyoni). *D. minutiflora* is native to West Africa where it grows abundantly, principally on the forest edges, but it extends throughout Africa to the Gabon and Democratic Republic of the Congo, Sierra Leone, north Angola, Kenya, Uganda and Madagascar. Species other than *D. minutiflora* are found in Kenya, with low distribution, were *D. alata, D. bulbifera* and *D. odoratissima* (Muthamia *et al.*, 2014). *D. minutiflora* grows in the central highlands of Kenya where it is the major species of *Dioscorea* and is grown for food, mainly by older farmers in the eastern, central, western and coastal regions (Maundu *et al.*, 1999) . It is common in cultivated land near villages (Botany VM, 2018). Stems are left twining and its fibrous tubers normally grow vertically in the soil, rarely horizontally, and have a brown skin (Fig. 8.17). *D. minutiflora* is cultivated at an altitude of 1,500–2,400 m in areas with more than 700 mm rainfall (Maundu *et al.*, 1999). It prefers a sunny situation on dry gritty-sandy soil and grows wild, principally on the edges of forests.

Composition and uses

In Côte d'Ivoire, Alexis and Amani (2013) extracted starch from *Dioscorea* tubers and fractionated it according to its sensitivity to acid hydrolysis. They found that the portion of the starch that was easily hydrolysable was 5% for *D. burkilliana*, 10% for *D. praehensilis*, 12% for *D. minutiflora*, 22% for *D. bulbifera* (tubers), 22% for *D. bulbifera* (bulbils), 30% for *D. hirtiflora* and 36% for *D. dumetorum*. Composition of raw *D. minutiflora* (*D. armata*) tubers per 100 g^{-1} edible portion was: energy 118 kcal, 68.4 g 100 g^{-1} moisture, 3 g 100 g^{-1} protein, 0.1 g 100 g^{-1} fat, 27 g 100 g^{-1} carbohydrate, 10 g 100 g^{-1} fibre, 1.5 g 100 g^{-1} ash, 69 mg 100 g^{-1} calcium and 41 mg 100 g^{-1} phosphorus (FAO, 1965).

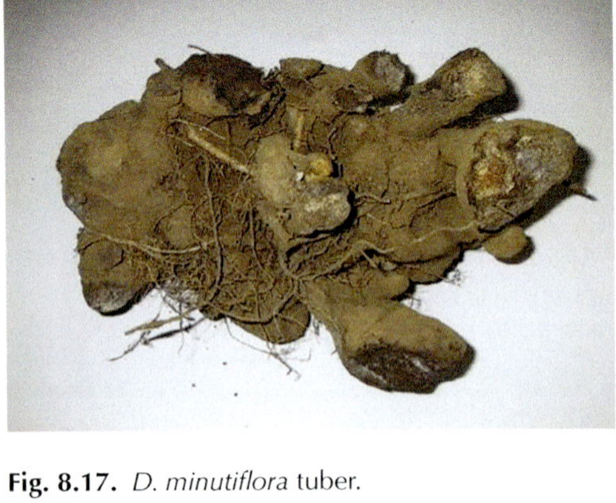

Fig. 8.17. *D. minutiflora* tuber.

D. minutiflora tubers are woody and mainly eaten only in times of food shortage. Only the lower portion of the tuber is eaten when still young and often only after prolonged soaking to make it more palatable (Maundu *et al.*, 1999).

D. monadelpha

Dioscorea monadelpha (Kunth) Griseb. is a native of Bahia, Minas Gerais, Paraná, Rio Grande do Sul, Santa Catarina and São Paulo in Brazil, and is also found in Bolivia and Paraguay. Xifreda and Kirizawa (2003) commented that *D. monadelpha* and *D. subhastata* belong to a nomenclaturally and taxonomically misinterpreted group and that the names *D. similis* and *D. longirachis* can be reduced to synonyms of *D. monadelpha*.

D. multiflora

Dioscorea multiflora Mart. ex Griseb. is native to southern Brazil. Diosgenin is a major secondary metabolite and *D. multiflora* is used as a commercial source of diosgenin in Brazil (Valéria de Souza *et al.*, 2011). Martin (1969) gave the diosgenin content of *D. multiflora* tubers as 0.4–1.4%. Valéria de Souza *et al.* (2011) developed a micropropagation protocol for production of *D. multiflora* that was effective and could be used for commercial production.

D. nako

Dioscrea nako H. Perrier is native to the forests of southwestern Madagascar. One common name is fandra, and its stems are left twining. Tubers grow vertically into the soil and are cylindric in shape, with a white skin. Andriamparany *et al.* (2014) reported that *D. nako* was identified as a potential food resource in the Mahafaly region of Madagascar where 70% of the interviewed households were collecting wild yams. Wild yam species were relatively rare on the adjacent side of the national park where only *D. nako* occurred. Many species of wild yam, including *D. nako*, serve as a food supplement and are therefore over-exploited during food shortage periods, therefore Andriamparany and Buerkert (2019) commented that the domestication of wild species and promotion of their cultivation could reduce this pressure and they showed that several *Dioscorea* species, including *D. nako*, could be propagated by true seed, which could possibly be used to promote their cultivation.

D. natalia

Dioscorea natalia is native to Central America from Costa Rica to Panama, and was reported in the wet regions of both Pacific and Caribbean slopes of Costa Rica. It has been confused with *D. lepida* whose native range is southeastern Mexico to Central America. *D. natalia* is closely related to two other Costa Rican species, *D. racemosa* and *D. standleyi* (Hammel, 2000).

D. nipponica

Dioscorea nipponica Makino is mainly distributed in the northeastern, northern, eastern and central regions of China. Ding and Gilbert (2010) reported that it occurs in mixed forests, on mountain slopes, in ravines and by roadsides, at 1,400–3,200 m in Guizhou, Sichuan, eastern Xizang and Yunnan provinces and they described the tubers as 'Rhizome horizontal, many branched, cylindric, more than 1.5 cm thick; cork layer persistent or readily detached'. Ou-yang *et al.* (2018) reported that, traditionally, tubers were used by Miao and Meng ethnic groups of China to treat rheumatoid arthritis, pain in the legs and lumbar area, Kashin Beck disease, bruises, sprains, chronic bronchitis, cough and asthma. Modern pharmacological studies have discovered that extracts from *D. nipponica* tubers possesses anti-tumour, anti-inflammatory, anti-diuretic, analgesic, anti-tussive, panting-calming and phlegm-dispelling activities, along with enhancing immune function and improving cardiovascular health. They reported that saponin and sapogenins are mainly responsible for most of the pharmacological effects. Kwon *et al.* (2003) found that *D. nipponica* tuber extract contained pancreatic lipase inhibitors that could be used as an anti-obesity measure.

Cho *et al.* (2013) found that dioscine, extracted from *D. nipponica*, exerted a considerable antifungal activity by disrupting the structure in cell membranes of infecting microorganisms, resulting in cell death. Chen *et al.* (2007) tested various growth regulators on micro-tuber induction and showed plantlet regeneration in *D. nipponica* was strongly increased with a combination of N-benzyladenine and α-naphthalene acetic acid. Murashige and Skoog salt media supplemented with sucrose N-benzyladenine and α-naphthalene acetic acid gave the highest frequencies of shoot induction with sucrose being the single most significant medium constituent for micro-tuber growth.

D. nitens

Dioscorea nitens Prain & Burkill. Stems twine to the left, tubers are cylindrical in shape with white flesh. They occur in forests at 1,100–2,600 m in Yunnan province, China (Ding and Gilbert, 2010).

D. nummularia

Dioscorea nummularia Lamarck is synonymous with *D. spicata* Roth. and is native to South East Asia. It is cultivated for its edible tubers in South East Asia, the Pacific islands, Indonesia and Papua New Guinea. *D. nummularia* was reported to be the most important *Dioscorea* species cultivated in New Guinea and some Pacific islands. It is cultivated in many islands but in some it was reported to have been collected from the wild. Hedrick (1919) referring to *D. nummularia*, commented: ‘Tivolo Yam. Moluccas. This yam has cylindrical roots as thick as an arm and of excellent quality’. Malapa *et al.* (2005, 2006) reported that *D. nummularia* was closely related to *D. alata* and confusion over the morphology of *D. alata*, *D. nummularia* and *D. transversa* has been reported in many countries including the Philippines, Indonesia and New Caledonia (Bourret, 1973; Sastrapradja, 1982; Cruz and Ramirez, 1999; Cruz *et al.*, 1999). Malapa *et al.* (2005, 2006) also reported that *D. nummularia* was closely related to *D. alata* and that it was resistant to anthracnose disease (*Colletotrichum gloeosporioides*). Lebot *et al.* (2016) studied *D. nummularia* on nine islands of Vanuatu and collected 110 accessions corresponding to 84 varieties and grouped them into: wild forms under domestication; annual types; perennial-type cultivars; and natural interspecific hybrids between *D. nummularia* and *D. alata*. There was remarkable morphological variation in their stems, spininess, leaf shapes, emerging leaf colour, tuber shapes and tuber flesh colour (Fig. 8.18). Selected accessions had chromosome numbers ranging from 2n = 3x = 60 to 2n = 6x = 120. *D. nummularia* and interspecific hybrids in Vanuatu had a mean dry matter of 33.1% and starch levels of 82.8% (Lebot *et al.*, 2016). Bradbury and Holloway (1988) also analysed the composition of several *Dioscorea* tubers including *D. nummularia* (Tables 8.14, 8.15).

Tubers of some varieties are difficult to harvest because they are large and very long, branched and irregular in shape (Wilson and Hamilton, 1988). Tubers are formed

Fig. 8.18. Tubers of wild *D. nummularia* plants. Lebot *et al.* (2017) with permission.

Table 8.14. Fresh weight composition of *Dioscorea* species (values are means of different samples). Bradbury and Holloway (1988).

Dioscorea species	*alata*	*esculenta*	*bulbifera*	*nummularia*	*pentaphylla*	*rotundata*	*trifida*
Sample (*n*)	16	99	25	12	9	3	3
Moisture %	77.3	74.2	71.9	81.7	82.5	65.7	80.7
Energy kj/100g	406	443	258	266	550	284	580
Protein %	2.06	2.04	1.94	1.65	1.42	1.52	0.53
Starch %	16.7	19.3	23.2	11.7	13.9	30.2	14.2
Sugars %	1.03	0.55	0.22	0.20	0.12	0.32	0.23
Dietary fibre %	1.88	1.15	1.84	1.42	0.66	0.63	1.02
Fat %	0.08	0.06	0.06	0.05	0.03	0.09	0.04
Ash %	0.81	0.82	0.95	0.69	0.76	0.73	0.70

Table 8.15. Fresh weight composition of *Dioscorea* species (values are means of different samples in mg $100g^{-1}$). Bradbury and Holloway (1988).

Dioscorea	*alata*	*esculenta*	*bulbifera*	*nummularia*	*pentaphylla*	*rotundata*	*trifida*
Ca	8.2	7.5	6.5	8.4	13	4.6	8
P	38	39	40	27	26	28	38
Mg	17	26	30	19	23	17	15
Na	3.3	3.1	8.6	2.7	6.1	4.7	2.9
K	318	303	448	346	374	361	350
S	12	16	15	9.0	13	12	8.2
Fe	0.60	0.75	0.38	0.56	0.44	0.60	0.54
Cu	0.15	0.17	0.34	0.21	0.25	0.12	0.13
Zn	0.39	0.46	0.50	0.31	0.36	0.30	0.35
Mn	0.04	0.24	0.04	0.13	0.05	0.03	0.03
Al	0.64	0.51	0.29	0.49	0.62	0.63	0.41
B	0.09	0.07	0.05	0.10	0.17	0.08	0.11

deeply in the soil and may be allowed to grow for 2 or 3 years in order to allow them to reach a size large enough to make the effort of digging worthwhile (Coursey, 1967, Purseglove, 1972). Other cultivars and varieties produce clusters of approximately 5 to 15 small tubers that are easy to harvest by digging. Tubers of *D. nummularis* have a shorter dormancy period than *D. rotundata* and were reported by Wilson and Hamilton (1988) to 'not store as well as *D. rotundata*'.

Curing

Wilson (1980) described a simple method of curing *D. nummularia*. The tubers were first stacked on the ground in a lightly shaded area and covered with grass or mats. Then a canvas tarpaulin was placed over the whole stack to cover the grass or mats ensuring the tarpaulin did not touch the yams. As an alternative, a simple wooden frame could be built and the canvas draped like a tent. She indicated that plastic sheets should not be used for curing since it can make the yams too hot. If there is no canvas tarpaulin, then several layers of sacks or mats could be used.

D. opposita

Dioscorea opposita Thunb. is synonymous with *D. batatas* Decne., *D. decaisneana* Carr., *D. doryphora* Hance, *D. oppositifolia* L., *D. polystachya* Turcz., *D. potaninii* Prain & Burkill, *D. pseudobatatas* (Ham.) Hert., *D. rosthornii* Diels and *D. swinhoei* Rolfe.

Common names include: Chinese yam, Chinese potato, cinnamon vine, cinnamon yam, common yam, Korean yam, nagaimo, wild yam and yam wurzel. It is a native of China and is also widely grown in Japan, Korea, Taiwan and other countries with colder climates in the Far East, since it can tolerate colder temperatures than any other commercially produced yam (Purseglove, 1972). It occurs naturally in forests, scrubland, sides of rivers, roadsides, hillsides and disturbed areas at elevations of 100 to 2,500 m. In the mid-19th century, it was cultivated experimentally in Europe as a substitute for potatoes.

It is a twining herbaceous vine with cinnamon-scented flowers and bulbils produced in the leaf axils. It can produce different forms of underground erect tubers that are up to 1 m long with white flesh (Fig. 8.19). In Japan, three cultivars of *D. opposita* are recognized: 'Naga-Imo' (long and cylindrical), 'Icho-Imo' (palmate) and 'Tsukune-Imo' (globular). Araki *et al.* (1983) reported that the 'Nagaimo' is cultivated widely in Japan and its yield per unit area is high, although the textural qualities of its tubers are inferior to those of other *D. opposita* cultivars such as 'Yamatoimo' and 'Icho-Imo'. Flesh browning that can occur in tubers when cut or damaged seems to be associated with phenolic amines (Kay, 1987). Satoh and Tanabe (1971) also reported that stored tubers, when cut or grated rapidly, discoloured due to the presence of polyphenolic compounds. They suggested that field spraying with maleic hydrazide would suppress browning.

Fig. 8.19. Nagaimo tubers on sale in a California supermarket, 2007.

Composition and uses

Yang *et al.* (2014) reported that in China, Japan and Korea *D. opposita* is used both as a food and as a treatment for digestive problems. They isolated 23 phenolic compounds from *D. opposita* and identified those associated with pancreatic lipase inhibitory activity. *D. opposita* tubers may be eaten raw, which is an exception since most yams must be cooked before consumption due to toxins in the raw state. This is most likely to be because, unlike most other *Dioscorea* species, the tubers of *D. opposita* grown in Japan do not contain any sapogenins, and the whole tubers are briefly soaked in a mixture of vinegar and water to neutralize oxalate crystals found in their skin before being eaten raw. The tuber is starchy, bland and mucilaginous when grated, and is called 'tororo'. The tororo may be mixed with other ingredients including wasabi (*Wasabia japonica*) and spring onions (*Allium cepa*).

Kay (1987) gave a typical analysis of the edible portion of the tubers as: 70 to 80% water, 16 to 29% carbohydrate (mainly starch), 1.11 to 3.1% protein, 0.06 to 1.1% fat, 0.33 to 1% fibre and 0.69 to 1.1% ash. A polysaccharide contained glucose, mannose and galactose was extracted from *D. opposita*, which had health benefits in terms of enhancing immunological activity (Zhao *et al.*, 2005). Yuan (2008) gave the composition of *D. opposita* as including proteins 3.59–8.93%, amino acids 2.31–7.26%, starch 43.7% and sugar 3.39%. Yi *et al.* (2014) sampled yams from different parts of China and measured their diosgenin content. In *D. opposita* no diosgenin was detected in samples from four regions. Ma *et al.* (2017) tested the properties of mucilage obtained from *D. opposita* generated during industrial manufacturing and found that the mucilage could be a sustainable resource of a natural emulsifier obtained from industrial waste.

Harvesting and storage

Harvest maturity of *D. opposita* was reported to be some 6 months after planting. The yams are dug from the soil, but 'Naga-Imo' tubers are difficult to harvest since they are long and cylindrical and may reach a depth of 1 m.

Tubers are consumed mainly directly after harvest, but they are also stored over the winter in clamps or cold stores in Japan (Kay, 1987). Ventilated storage at about 25°C for up to 60 days can also be used (Tindall, 1983; Opara, 1999). Imakawa (1967) reported that fully mature *D. opposita* tubers could be stored for long periods of time at 5°C under low levels of O_2, which was also shown to reduce browning. In China, Zhang *et al.* (2014) stored *D. opposita* tubers in ambient conditions of 10–18°C and 60-80% RH for 45 days and their weight loss was 56.51–67.96% and total sugars, reducing sugars and protein increased by 6.49–9.81%, 1.7–2.27% and 13.02–14.55%, respectively. However, when packaged and stored at 4°C and 60–65% RH there was 'low-temperature sweetening that affected the nutritional potential of tubers'.

D. oppositifolia

Dioscorea oppositifolia L. Gibson and Thomas (2005) reported that *oppositifolia* was synonymous with *D. batatas* Dcne., *D. opposita* Thunb, *D. cayenensis* Lam. var. *pseudobatatas* Hauman, *D. decaisneana* Carr., *D. doryphora* Hance, *D. oppositifolia* L. var. *meeboldtii* Prain & Burkill, *D. oppositifolia* L. var. *linnaei* Prain & Burkill, *D. oppositifolia* L. var. *thwaitesii* Prain & Burkill, *D. polystachya* Turcz., *D. potaninii* Prain & Burkill, *D. rosthornii* Diels, *D. swinhoei*, Rolfe and *D. trinervia*, Roxb. ex Prain & Burkill, with *D. villosa* Smilax as being a similar species. Kew Science (2019) reported that *D. oppositifolia* L. is the accepted name while *D. oppositifolia* var. *dukhunensis*, *D. oppositifolia* var. *linnaei* Prain & Burkill and *D. oppositifolia* var. *thwaitesii* Prain & Burkill are non-accepted names. Other synonyms include: *D. cayenensis* Lamarck var. *pseudobatatas* Hauman, *D. decaisneana* Carrière, *D. divaricata* Blanco, *D. doryphora* Hance, *D. opposita* Thunb., *D. oppositifolia* L. var. *linnaei* Prain & Burkill, *D. opposita* Thunb., *D. oppositifolia* L. var. *meeboldtii* Prain & Burkill, *D. oppositifolia* L. var. *thwaitesii* Prain & Burkill, *D. polystachya* Turcz., *D. potaninii* Prain & Burkill, *D. rosthornii* Diels, *D. swinhoei* Rolfe and *D. trinervia* Roxburgh ex Prain & Burkill. According to Tu (2002) there is confusion regarding the correct taxonomy of *D. oppositifolia* and *D. batatas* Decne is the most common synonym. Purseglove (1972) referred to *D. opposita* Thunb. as being synonymous with *D. batatas* and used the common name of Chinese yam but doesn't mention *D. oppositifolia*. Kay (1987) reported that *D. bulbifera* occurs in a wide variety of forms and many synonyms have appeared in the literature including *D. oppositifolia* Campbell. This is the only reference to *D. oppositifolia* in her book, but she deals in detail with *D. opposite*. Neither Kay nor Coursey (1967) give any alternative names for *D. opposita* and Coursey does not mention *D. oppositifolia* in his book. On this basis *D. opposita*, *D. oppositifolia* and *D. bulbifera* are dealt with separately in this current book. Common names for *D. oppositifolia* include Chinese yam, Chinese potato and cinnamon vine. In India it is also called cinnamon vine as well as pit kanda, since kanda is Hindi for yam. The species name *oppositifolia* refers to the opposite arrangement of its leaves while the species synonym sometimes used is 'batatas', meaning potato (Bailey, 1949).

Tu (2002) commented that *D. oppositifolia* originated in China, India and Sri Lanka and was introduced into the USA where it has become an invasive species.

Both the tuber and bulbils of *D. oppositifolia* are edible, although the bulbils are generally not collected and are not used as food. The tuber, which can weigh up to 2 kg or more, if grown in deep loam soils, has a good flavour and is nutritious. The tuber is also used in Chinese medicine and is sometimes used as an herbal tonic. It stimulates the stomach and spleen and has an effect on the lungs and kidneys. The tuber has been eaten for the treatment of poor appetite, chronic diarrhoea, asthma, dry coughs, frequent or uncontrollable urination, diabetes and emotional instability. Extracts from the tuber have also been used to treat ulcers, boils and abscesses, since tubers contain allantoin, a cell-proliferant that speeds up the healing process. Tu (2002) stated that the 'leaf juice' from *D. oppositifolia* can be used to treat snake bites and scorpion stings. Its roots and tubers contain diosgenin, and extracts have also been used traditionally as a contraceptive and in the treatment of various disorders of the genital organs as well as for asthma and arthritis. *D. oppositifolia* is also grown for its ornamental value, especially its flowers that smell like cinnamon.

D. oppositifolia is a perennial twining vine that has round slender stems. The tuber is cylindrical in shape, up to about 1 m long and 7 to 8 cm in diameter and has a brown smooth surface and white flesh, growing vertically in the soil (Figs 8.20, 8.21). The vine produces bulbils in the leaf axils and each plant can produce some 20 bulbils a year. The bulbils have been observed sprouting new shoots within 2 weeks of formation. It can be propagated both sexually from true seeds and asexually from the bulbils (Tu 2002). *D. oppositifolia* is listed in the South East Exotic Pest Plant Council's Invasive Exotic Pest Plant List for Tennessee, USA as a 'Rank 1 Severe Threat' species, indicating that it is an exotic species that possesses characteristics of an invasive species and could spread easily into native plant communities and displace native vegetation (Tu, 2002).

Composition and uses

The tuber contains about 20% starch, 75% water, 0.1% vitamin B1 and 10 to 15 mg vitamin C. It also contains mucilage, amylase, amino acids and glutamine (Tu, 2002). Proximate composition of *D. oppositifolia* tubers in g per 100 g fresh weight equate to: 1.80 protein, 21.00 carbohydrate, 1.10 fat, 73.00 moisture, 0.90 fibre and 100.00 kcal energy (Rajyalakshmi and Geervani, 1994). Mohan and Kalidass (2010) gave the following

Fig. 8.20. *D. oppositifolia*. Photograph: Dr K. Abraham, CTCRI Emeritus, C-63, Gandhipuram, Sreekariam, Trivandrum 695017, India, reproduced with permission.

Fig. 8.21. *D. oppositifolia* Photograph: Dr K. Abraham, CTCRI Emeritus, C-63, Gandhipuram, Sreekariam, Trivandrum 695017, India, reproduced with permission.

analysis of the tuber per 100 g fresh weight: 78.49% moisture, 9.41 ± 0.04 starch, 7.00 ± 0.07 crude protein, 6.92 ± 0.11 crude lipid, 5.53 ± 0.13 crude fibre, 6.38 ± 0.12 ash, 74.17 nitrogen free extractives, 18.82 ± 0.03 niacin, 26.23 ± 0.12g ascorbic acid and 1616.42 g 100g^{-1} calorific value (kJ 100g^{-1} dry matter). They also gave *in vitro* protein digestibility as 5.29% and *in vitro* starch digestibility as 39.21%. Shajeela *et al.* (2011) gave the following analysis for *D. oppositifolia* var. *dukhumensis* as starch 52.24 ± 0.31 g 100 g^{-1}, niacin 37.14 ± 0.32 g 100 g^{-1}, ascorbic acid 96.42 ± 0.37 g 100 g^{-1}. For

D. oppositifolia var. *oppositifolia* Shajeela *et al.* (2011) gave: starch 46.04 ± 0.1 g 100 g^{-1}, niacin 44.30 ± 0.51 g 100 g^{-1}, ascorbic acid 90.51 ± 0.54 g 100 g^{-1}. *D. oppositifolia* also contains anti-nutritional factors identified by Mohan and Kalidass (2010) per 100g fresh weight as: total free phenolics 0.28 ± 0.03 g, tannins 0.10 ± 0.04 g, hydrogen cyanide 0.14 ± 0.02 mg, total oxalate 0.38 ± 0.05 g, amylase inhibitor 4.13 AIU mg^{-1} soluble starch, trypsin inhibitor 2.1 TIU mg $protein^{-1}$. Shajeela *et al.* (2011) gave the following analysis for antinutritional factors: *D. oppositifolia* var. *dukhumensis* total free phenolics 0.36 ± 0.01 g 100 g^{-1}, tannins 0.24 ± 0.07 g 100 g^{-1}, hydrogen cyanide 0.24 ± 0.02 mg 100 g^{-1}, total oxalate 0.36 ± 0.01 g 100 g^{-1}, amylase inhibitor 2.46 AIU mg^{-1}, soluble starch trypsin inhibitor 13.30 ± 0.09 TIU mg^{-1} protein. For *D. oppositifolia* var. *oppositifolia*. Shajeela *et al.* gave the following total free phenolics 0.56 ± 0.01 g 100 g^{-1}, tannins 0.36 ± 0.11 g 100 g^{-1}, hydrogen cyanide 0.33 ± 0.04 mg 100 g^{-1}, total oxalate 0.46 ± 0.07 g 100 g^{-1}, amylase inhibitor 2.10 AIU mg^{-1}, soluble starch trypsin inhibitor 11.26 ± 0.12 TIU mg^{-1} protein. Arinathan *et al.* (2009) showed that the composition varied considerably between varieties and gave the proximate composition of tubers of *D. oppositifolia* var. *dukhumensis* and *D. oppositifolia* var. *oppositifolia* (Tables 8.16, 8.17). Shajeela *et al.* (2011) showed levels of antinutritional factors of tubers of *D. oppositifolia*; (Table 8.18) and Arinathan *et al.* (2009) showed mineral content (Table 8.19).

Table 8.16. Proximate composition of *D. oppositifolia* var. *dukhumensis* tubers and *D. oppositifolia* var. *oppositifolia* (g 100g^{-1}). Modified from Arinathan *et al.* (2009).

	Level	
Component	var. *dukhumensis*	var. *oppositifolia*
Moisture	81.90 ± 0.18	69.03 ± 0.51
Crude protein	13.80 ± 0.28	6.31 ± 0.35
Crude lipid	6.33 ± 0.34	2.51 ± 0.21
Crude fibre	3.92 ± 0.02	8.97 ± 0.04
Ash	1.60 ± 0.17	6.39 ± 0.26
Nitrogen free extract	74.35	75.82
Calorific value (kJ 100g^{-1})	1,710.75	1,466.20

Table 8.17. Starch, niacin and ascorbic acid content of *D. oppositifolia* var. *oppositifolia* and *D. oppositifolia* var. *dukhumensis* tubers. Modified from Arinathan *et al.* (2009).

	Level	
Component	var. *dukhumensis*	var. *oppositifolia*
Starch g 100g^{-1}	40.37 ± 0.46	48.13 ± 0.03
Niacin mg 100g^{-1}	64.65 ± 0.12	17.64 ± 0.21
Ascorbic acid mg 100g^{-1}	80.57 ± 0.12	104.79 ± 0.31

D. orbiculata

Dioscorea orbiculata Hook.f. PROSEA orbiculata (2019) reported that *D. orbiculata* is the accepted name. Its common names are janggut kelonak, man-tayong, takop, ubi garam and ubi garam takob and its native range is Thailand, Peninsular Malaysia and Indonesia (Sumatra) in moist evergreen forests and possibly mixed deciduous forest at an altitude of 50–200 m (Wilkin and Thapyai, 2009). It is the only Asian yam species with right-twining stems (Burkill, 1951; PROSEA orbiculata, 2019). Burkill reported that underground parts are 'tuberous, with several tubers produced each year at the end of long spreading or descending stalks', with a brown woody crown and white flesh. Tubers are edible and are prepared by boiling in water before they are used as food in Peninsular Malaysia. Raw tubers can burn the throat (Burkill ,1951; Wilkin and Thapyai, 2009; Botany VM, 2018).

D. oryzetorum

Dioscorea oryzetorum Prain & Burkill, synonymous with *D. oryzetorum* var. *angustifolia* and *D. oryzetorum* var. *mediifolia*. It is native to South East Asia including Thailand (Amphur Thongchai, Sikhiu district Chum Phuang Nongson and Bua Yai districts), where

Table 8.18. Antinutritional factors of tubers of *D. oppositifolia*. Modified from Shajeela *et al.* (2011).

Component	Level	
	var. *dukhumensis*	var. *oppositifolia*
Total free phenolics ($g100g^{-1}$)	0.36 ± 0.01	0.56 ± 0.01
Tannins ($g100g^{-1}$)	0.24 ± 0.07	0.36 ± 0.11
Hydrogen cyanide ($mg100g^{-1}$)	0.24 ± 0.02	0.33 ± 0.04
Total oxalate ($g100g^{-1}$)	0.36 ± 0.01	0.46 ± 0.07
Amylase inhibitor AIU mg soluble starch^{-1}	2.46	2.10
Trypsin inhibitor TIU mg protein^{-1}	13.30 ± 0.09	11.26 ± 0.12

its common name is man nok, and Vietnam where its common name is khoai mice. *D. oryzetorum* is classified within the group Enantiophyllum. Prain and Burkill (1927) reported that it produces a single tuber, but Wilkin and Thapyai (2009) and Botany VM (2018) reported that there were often two to three tubers per plant, but there can be more. Botany VM (2018) also reported that tubers are about 50 cm long and up to 7 mm in diameter, spreading, cylindrical to globose in shape and ending in a distinct compact head subtended by a woody crown with short side branches, annually replaced with white clear skin (Fig. 8.22). The stems twine anticlockwise.

D. orangeana

Dioscorea orangeana Wilkin is a newly described species of yam that produces edible tubers. It is found in the dry deciduous forest of Madagascar where it has the common name of angona, but this name is also used for other yam species of northern Madagascar. Its tuber morphology is uncharacteristic of Malagasy yams since it has several digitate lobes instead of just one (Wilkin *et al.*, 2009). It was reported to be endemic in the Forêt d'Orangea near Antsiranana in Madagascar, from which it derives its name, and three specimens have been reported occurring at altitudes between 8 and 100 m (Red List, 2019). It was first discovered in the herbarium of the Museum National d'Histoire Naturelle in Paris, as a pressed specimen collected in 1960 from northern Madagascar, from where it has since been repeatedly collected. Locals heavily harvest several related species of *Dioscorea* with edible tubers from that area. *D. orangeana* is both new to science and threatened by extinction due to deforestation and over-extraction as people collect tubers for personal consumption and to sell at markets (Wilkin *et al.*, 2009). The estimated extent of occurrence and occupancy of *D. orangeana* is an area of only 24 km^2, with just three locations so far identified. This species is therefore assessed as endangered (*Guardian*, 2010).

D. ovinala

Dioscorea ovinala Baker is endemic to west and central Madagascar and occurs from 10 to 1,300 m. Tubers are cylindrical and grow vertically in the soil. They are fibrous with white peel and are cultivated in Thailand (Fig. 8.23). Stems are left twining and leaves obovate, leathery and shallowly lobed. Male and female inflorescences are in spikes and the fruit is a round or ovate capsule containing seeds with long wings at the base (Botany VM, 2018). *D. ovinala* was identified as a potential food resource in the Mahafaly region of Madagascar where 70% of the interviewed households were collecting wild yams (Andriamparany *et al.*, 2014).

D. panthaica

Dioscorea panthaica Prain & Burkill is synonymous with *D. biserialis* Prain & Burkill. Ding and Gilbert (2010) reported that in China *D. panthaica* occurs in scrub forests

Table 8.19. Mineral composition of *D. oppositifolia* tubers (mg 100 g^{-1}). Modified from Arinathan *et al.* (2009).

	Na	K	Ca	Mg	P	Zn	Mn	Fe	Cu
Dioscorea oppositifolia var. *dukhumensis*	123.00 ± 0.38	1,648.00 ± 0.84	230.00 ± 0.33	648.33 ± 0.16	54.08 ± 0.12	1.40 ± 0.01	6.80 ± 0.22	49.10 ± 0.13	11.50 ± 0.28
Dioscorea oppositifolia var. *oppositifolia*	110.18 ± 0.14	1561.00 ± 0.98	880.60 ± 0.44	530.48 ± 0.12	88.46 ± 0.22	5.24 ± 0.13	8.44 ± 0.04	32.00 ± 0.	512.78 ± 0.01

Figure 8.22. Leaf blade of *D. oryzetorum* and young tubers. Photograph taken at Sarinya's yam farm in Chachoengsao province in the eastern region of Thailand, 2018.

Figure 8.23. *D. ovinala* tuber. Photograph taken at Sarinya's yam farm in Chachoengsao province in the eastern region of Thailand, 2018.

and on mountain slopes at 1,000–3,500 m in western Guizhou, southwestern Hubei, northwestern Hunan and western Sichuan provinces, and South East Asia including Thailand. Sagwansupyakom and Chantraprasong (1984) reported that *D. panthaica* was rare. Tubers are cylindrical in shape, fibrous and irregularly branched (Ding and Gilbert, 2010). The diosgenin content of *D. panthaica* tubers was reported to be up to 5.29 mg g^{-1} (Yi *et al.*, 2014). Extracts from its tubers are used in traditional Chinese medicine in the treatment of various physiological conditions, including cardiovascular disease, gastropathy and hypertension (Dong *et al.* 2001; Wang *et al.*, 2015). Dong *et al.* isolated three steroidal saponins from *D. panthaica* tubers. They were identified as 26-O-β-D-glucopyranosyl-3 β, 26-diol-23(S)-methoxy-25(R)-δ 5,20(22)-diene-furosta-3-O-[α-L-rhamnopyranosyl-(1-->2)-O-α-L-rhamnopyranosyl-(1-->4)]-beta-D-glucopyranoside (I), pseudoprotodioscin (II), 26-O-β-D-glucopyranosyl-3 β, 26-diol-25 (R)-δ 5,20(22)-diene-furosta-3-O-α-L-rhamnopyranosyl-(1-->2)-O-β-D-glucopyranoside (III).

D. paradoxa

Dioscorea paradoxa Prain & Burkill is distributed in eastern Thailand and Vietnam, where it is commonly found climbing over bushes on rocky limestone hills (JSTOR, 2018). Sagwansupyakom and Chantraprasong (1984) reported that the tubers were edible. Schols *et al.* (2001, 2003), in their study of pollen morphology of *Dioscorea*, reported that the pollen of *D. paradoxa* had the second largest orbicules of the species included in their survey (0.70 mm on average) with *D. daunea* having the largest (0.77 mm on average).

D. pentaphylla

Dioscorea pentaphylla L. At least two varieties have been identified: *D. pentaphylla* var. *palmata* and *D. pentaphylla* var. *papuana* (Barrau, 1956); or three varieties: *D. pentaphylla* var. *javanica*, *D. pentaphylla* var. *malaica* and *D. pentaphylla* var. *sacerdotalis* (Burkill, 1951). Shajeela *et al.* (2011) also referred to *D. pentaphylla* var. *pentaphylla* and commented that *D. pentaphylla* may sometimes be confused with *D. cumingii*, which is used, to a limited extent, as a food in the Philippines (Burkill, 1951). *D. pentaphylla* is commonly called five-leaf yam and is native to tropical Asia or eastern Polynesia. It was first brought into cultivation somewhere in the Indochina peninsula and is now cultivated throughout the Pacific islands and Indonesia (Coursey, 1967). *D. pentaphylla* produces horseshoe-shaped bulbils in leaf axes, which can be used for propagation. Bulbils are between 3 and 30 mm in diameter and are pitted or tuberculate and can be cooked and eaten. One to three tubers are produced by each plant annually, but there may be up to five – however, this is uncommon. Each tuber may weigh over 1 kg and be up to 60 cm long and 8.5 cm in diameter, but may grow to over a metre in length and be formed over a metre underground. Tubers grow vertically in the soil, are subtended by woody crowns and are variable in shape but usually cylindrical to clavate, sometimes lobed (Prain & Burkill, 1936) (Figs 8.24, 8.25). In Indonesia and Oceania, the tubers are often harvested from wild plants. These plants are also used as hedges on farms, which can form a reserve of food for times of scarcity (Coursey, 1967, Purseglove, 1972). *D. pentaphylla* is a prickly vine, with stems twining to the left (Wilkin and Thapyai, 2009; Botany VM, 2018). The tubers are dark brown in colour with a smooth surface and white flesh and grow laterally in the soil.

Fig. 8.24. *D. pentaphylla* tubers. Photograph: Dr K. Abraham, CTCRI Emeritus, C-63, Gandhipuram, Sreekariam, Trivandrum 695017, India, reproduced with permission.

Composition and uses

Choudhary *et al.* (2009) reported that *D. pentaphylla* is used as a medicinal plant in Rajasthan, India. The powder of dried tubers is administered orally to mothers who have abdominal pain after childbirth. A decoction of tubers is given for curing asthma or coughs and also to animals for early recovery from fractured bones (Swarnkar and Katewa, 2008). Hedrick (1919) stated:

> Tropical Asia. In India, this yam is common in jungles and is found in the South Sea Islands. Wight (1840 to 1853) has never seen it cultivated in India, although the natives dig the tubers to eat. It is cultivated in Amboina and sometimes in Viti. In India, the male flowers are sold in the bazaars and eaten as greens. The tubers are eaten in Viti and Hawai'i. It is a good yam. Graham says the tubers are dreadfully nauseous and intensely bitter even after being boiled. They are put into toddy to render it more potent, as they have intoxicating properties, and a few slices are sufficient. In China, the nauseous tubers are sometimes cooked and eaten.

Coursey (1967) reported that tubers are hardly, if at all, toxic. Mohan and Kalidass

(2010) reported that the chemical content of the tubers was: 93.05% moisture, 15.88 ± 0.07 starch, 9.18 ± 0.18 crude protein, 4.8 ± 0.02 crude lipid, 5.14 ± 0.11 crude fibre, 4.64 ± 0.02 ash, 76.23 nitrogen-free extractives, 1,607.47 g $100g^{-1}$ calorific value (kJ $100g^{-1}$ dry weight), 18.82 ± 0.11 niacin and ascorbic acid 4.59 ± 0.24 g $100g^{-1}$. They also gave *in vitro* protein digestibility as 4.87% and *in vitro* starch digestibility as 86.14%. Proximate composition of *D. pentaphylla* tubers in g per 100 g fresh weight was: 2.80 protein, 15.00 carbohydrate, 0.67 fat, 79.00 moisture, 2.00 fibre and 72 kcal energy (Rajyalakshmi and Geervani, 1994). Shajeela *et al.* (2011) gave the following analysis for *D. pentaphylla* var. *pentaphylla*: starch 55.98 ± 0.51 g 100 g^{-1}, niacin 62.14 ± 0.14 g 100 g^{-1}, ascorbic acid 96.56 ± 0.34 g 100 g^{-1}. Mohan and Kalidass (2010) also identified anti-nutritional factors as: total free phenolics 0.75 ± 0.06 g $100g^{-1}$, tannins 0.44 ± 0.06 g $100g^{-1}$, hydrogen cyanide 0.09 ± 0.03 mg $100g^{-1}$, total oxalate 0.31 ± 0.01 g $100g^{-1}$, amylase inhibitor 1.95 AIU mg soluble $starch^{-1}$ and Trypsin inhibitor 2.86 TIU mg $protein^{-1}$. Shajeela *et al.* (2011) gave the following analysis for antinutritional factors for *D. pentaphylla* var. *pentaphylla*: total free phenolics 0.48 ± 0.05 g 100 g^{-1}, tannins 0.09 ± 0.06 g 100 g^{-1}, hydrogen cyanide 0.18 ± 0.01 mg 100 g^{-1}, total oxalate 0.58 ± 0.05 g 100 g^{-1}, amylase inhibitor 2.46 AIU mg^{-1} soluble starch and trypsin inhibitor 3.66 ± 0.09 TIU mg^{-1} protein (Table 8.20). Proximate composition of *Dioscorea pentaphylla* var. *pentaphylla* tubers was given as calorific value 1652.26 kJ $100g^{-1}$, moisture, 73.46 ± 0.27 g $100g^{-1}$, crude protein 5.38 ± 0.13 g $100g^{-1}$, crude lipid 6.01 ± 0.45 g $100g^{-1}$, crude fibre 7.04 ± 0.07 g $100g^{-1}$, ash 1.58 ± 0.11 g $100g^{-1}$, starch 42.58 ± 0.31 g $100g^{-1}$, niacin, 53.51 ± 0.27 mg $100g^{-1}$, ascorbic acid mg $100g^{-1}$, 91.65 ± 0.38 (Arinathan *et al.*, 2009). Mineral composition: Na 85.24 ± 0.11, K 1341.60 ± 1.41, Ca 640.10 ± 0.54, Mg 440.00 ± 0.32, P 126.10 ± 1.01, Zn 3.22 ± 0.11, Mn 2.32 ± 0.03, Fe 113.48 ± 0.12, Cu 16.60 ± 0.13 mg 100 g^{-1} The chemical content given by USDA is shown in Table 8.21.

Fig. 8.25. *D. pentaphylla*, mature tuber with a withered tuber of the previous year, in a natural forest in Nan Province, northern Thailand. Reproduced with permission of Dr Chirdsak Thapyai.

D. persimilis

Dioscorea persimilis Prain & Burkill produces bulbils as well as tubers, and the stems twine to the right. Tubers are yellowish brown with white flesh. *D. persimilis* occurs in mixed forests, broad-leaved forests, bamboo forests, on mountain slopes, in valleys and by roadsides at 1,000–2,400 m

Table 8.20. Antinutritional factors of tubers of *D. pentaphylla* var. *pentaphylla*; modified from Shajeela *et al.* (2011).

Component	Level
Total free phenolics (g $100g^{-1}$)	0.48 ± 0.05
Tannins (g $100g^{-1}$)	0.09 ± 0.06
Hydrogen cyanide (mg $100g^{-1}$)	0.18 ± 0.01
Total oxalate (g $100g^{-1}$)	0.58 ± 0.05
Amylase inhibitor AIU mg soluble $starch^{-1}$	2.46
Trypsin inhibitor TIU mg $protein^{-1}$	1.26 ± 0.12

Table 8.21. The composition of mountain yam (*Dioscorea pentaphylla*) from Hawaii for $100g^{-1}$ fresh weight (USDA-NRCS 2012).

Component	Level	Component	Level
Water	81.44 g	Thiamin	0.102 mg
Energy	67 kcal	Riboflavin	0.019 mg
Protein	1.34 g	Niacin	0.481 mg
Total lipid (fat)	0.10 g	Vitamin B-6	0.179 mg
Carbohydrate, by difference	16.31 g	Folate	14 µg_DFE
Ca	26 mg	Vitamin B-12	0.00 µg
Fe	0.44 mg	Vitamin A	0 µg_RAE
Mg	12 mg	Vitamin A	0 IU
P	34 mg	Vitamin D (D2 + D3)	0.0 µg
K	418 mg	Vitamin D	0 IU
Na	13 mg	Fatty acids, total saturated	0.022 g
Zn	0.27 mg	Fatty acids, total monounsaturated	0.004 g
Total ascorbic acid	2.6 mg	Fatty acids, total polyunsaturated	0.045 g

in Guangdong, Guangxi, Guizhou, Sichuan and Yunnan provinces of China as well as Thailand and Vietnam (Ding and Gilbert, 2010).

D. petelotii

Dioscorea petelotii Prain & Burkill. Its native range is northern Thailand to northern Vietnam, where it is found growing in hill evergreen forest at 1,100–1,500 m. Its common name in Thailand is man doi. Wilkin and Thapyai (2009) described the tubers of *D. petelotii* as:

> up to 40 cm in diam. and at least 5 kg in mass, shape variable, usually globose to subglobose with several shallow lobes, perennially replaced, shallowly vertically buried but sometimes approaching horizontal when growing in shallow soil on a steep slope; crown of tuber bearing slender rigid roots, very hard and woody, not clearly differentiated from tuber, tuber periderm hard to corky, often dark brown or grey-brown in colour, parenchyma pale yellow to pinkish with a little mucilage.

D. pierrei

Dioscorea pierrei Pr. and Burk. Its native range is Indochina including eastern and southern Thailand, Bau-Chau and Ton-Man. Common names in Vietnam include: từ nước and từ pierre. Tubers are reported to be edible, and in India they are also used to make a soap and a shampoo and to kill lice (Sagwansupyakom and Chantraprasong, 1984).

D. piperifolia

Dioscorea piperifolia Humb. and Bonpl. ex Willd. is synonymous with *D. glandulosa* (Griseb.) Klotzsch ex Kunth and *D. piperifolia* Humb. and Bonpl. ex Wild. var. *glandulosa*. No common name could be found. It is a Brazilian species native to the lowlands and also to the Andes, and was reported to grow in dry forests in parts of Argentina and Bolivia. Hedrick (1919) notes that in South America the tubers are used as a food and Coursey (1967) comments that it was used as a food by the Amerindians but was little cultivated. However, he also notes that in spite of this it was cultivated in Argentina, Colombia, Ecuador, Guyana, Peru, Bolivia and Venezuela. Hornung *et al.* (2017) also reported that tubers are edible and the characteristics of starches extracted from *D. piperifolia* could be useful as a raw material for different commercial applications.

D. pilosiuscula

Dioscorea pilosiuscula Bertero ex Spreng, which is the accepted name, is synonymous with *D. pilosiuscula* var. *panamensis* R. Knuth and *D. sapindoides* C. Presl. Its common name is bulbous yam. It is native to primary forests in various Central American countries including Panama (Veraguas and Santa Fe) at an altitude of 75–730 m, Costa Rica, Mexico, and in several Caribbean Islands including the Dominican Republic at 250–400 m (Tropicos, 2021). IPGRI/IITA (1997) lists *D. pilosiuscula* as 'producing edible tubers'.

D. piscatorum

Dioscorea piscatorum Prain & Burkill. Its native range is Malaysia, Indonesia, Thailand and the Philippines (Govaerts *et al.*, 2007). Prain and Burkill (1936) commented that the tubers were said to be poorly known but were mucilaginous and contained saponins. They were used by native people of Malaysia and the Sakai tribe in Thailand as a fish poison, but Maneenoon *et al.* (2008) reported that tubers can be eaten after prolonged boiling.

D. poilanei

Dioscorea poilanei Prain & Burkill occurs in forests and scrub from near sea level to 200 m in Hainan province of China and Cambodia, Laos, Malaysia, Thailand and Vietnam (Ding and Gilbert, 2010). Stems twine to the left. Tubers grow up to only about 1.5 cm in diameter, are cylindrical in shape and irregularly branched and grow horizontally in the soil. Of the species reported by Tang and Eisenbrand (1992), *D. poilanei* had the lowest sapogenin content at 0.07% compared to 5.9% for *D. zingiberensis*, which was the highest.

D. polygonoides

Dioscorea polygonoides Humb. and Bonpl. ex Willd is commonly called Jamaican bitter yam or bitter jessie. It grows wild in northern Caribbean islands (at elevations of 150–900 m in Jamaica), Central America (including Mexico), Argentina and Brazil, mainly in rainforests and at the margins of clearings on limestone soils (Adams, 1972). *D. polygonoides* produces large edible tubers (Fig. 8.26) that are sometimes gathered from the wild. Tubers are consumed locally and also dried and made into a coffee-type drink. In Jamaica the tubers are consumed in rural areas, mostly during times of drought or food shortage. Tubers contain various bioactive components, especially high levels of sapogenin, which may increase oxidative stress in various tissues (Stennett *et al.*, 2014). In Colombia, Raz *et al.* (2013) reported that *D. polygonoides* contained sapogenins. McKoy *et al.* (2014) investigated their potent cholesterol-lowering capability and evaluated the effects of a 5% *D. polygonoides* supplementation on the integrity of the liver and kidneys on two mice strains and found that this supplementation induced oxidative stress. A preparation made from *D. polygonoides* in Jamaica was shown to reduce blood cholesterol concentrations in hypercholesterolemic rats and mice. McKoy *et al.* (2014) concluded that supplementation of the diet with *D. polygonoides* may be therapeutically beneficial in the management of hypercholesterolemia. In other work, McAnuff-Harding *et al.* (2006) showed that the consumption of extracts of *D. polygonoides* tubers may be useful in the management of hypercholesterolemia often

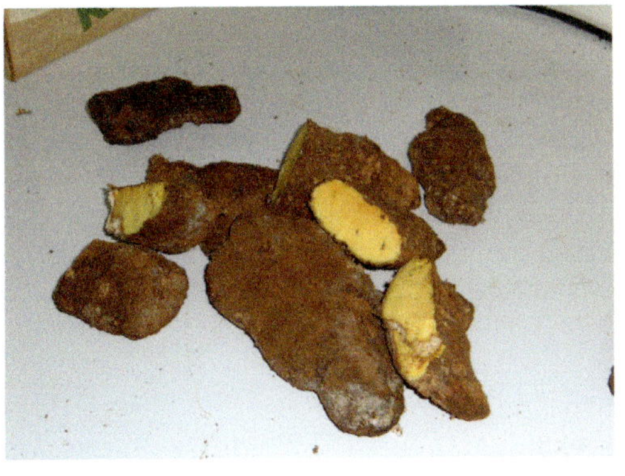

Fig. 8.26. *Dioscorea polygonoides* tubers. Photograph courtesy of Dr Dewayne Stennett, University of the West Indies, Jamaica.

associated with diabetes. They showed that sapogenin extract from *D. polygonoides* tubers resulted in a significant decrease ($p = 0.05$) in lactase and maltase activities in the intestine of diabetic rats compared to the diabetic control group. They also showed that this diet significantly reduced ($p = 0.05$) intestinal sucrase activity. Pinzón-Rico and Raz (2017) reported that on the Bogotá market, vendors recommended *Dioscorea* spp., including *D. polygonoides*, for lowering triglycerides and cholesterol.

D. polystachya

Dioscorea polystachya Turcz. is commonly called Chinese yam and was reported to be synonymous with *D. opposita*. Zheng *et al.* (2007) tested the molecular authentication of *D. polystachya* and other *Dioscorea* species and found that the chloroplast DNA sequencing was a rapid, efficient, credible way to distinguish *D. polystachya* from other *Dioscorea* species. They concluded that their results strongly supported the view that the horticulture classification of *D. polystachya* using leaf and tuber characteristics was related at the molecular level. Peng *et al.* (2016) subsequently also examined the genetic relationships among 14 *D. polystachya* cultivars and found a high level of polymorphism, but ISAP markers were consistent with their morphological characteristics and with the classification of these cultivars by leaf and tuber shape. Using ISAP they were able to discriminate the popular local *D. polystachya* cultivar 'Tiegun' from the other 13 cultivars they tested. Ding and Gilbert (2010) reported that the stems of *D. polystachya* twine to the right and that they produce bulbils. Tubers are cylindrical in shape, up to 1 m long with white flesh and grow vertically in the soil. They occur in forests, scrub forests, 'herb communities', on mountain slopes, along rivers and by roadsides at 100–2,500 m in Anhui, Fujian, eastern Gansu, northern Guangdong, Guangxi, Guizhou, Hebei, Henan, Hubei, Hunan, Jiangsu, Jiangxi, Jilin, Liaoning, southern Shaanxi, Shandong, Sichuan, Taiwan, northern Yunnan and Zhejiang provinces of China as well as Japan and Korea. They are also commonly cultivated and are an important food crop. In the USA, Walck *et al.* (2010) showed that bulbils of *D. polystachya* were mostly dormant in summer or autumn but dormancy could be broken with cold stratification during winter. They found that bulbil germination primarily occurs in spring at the beginning of a favourable period for survival and growth. A low percentage of bulbils of *D. polystachya* collected from July to September produced roots, but no shoots, though these roots died within about a month. Cold stratification markedly increased root and shoot emergence. Bulbils planted outdoors in October produced roots the following March and shoots in April.

D. praehensilis

Dioscorea praehensilis Benth. is considered a forest species (Magwé-Tindo *et al.*, 2018) and is native to the African tropics. It is commonly called safa in Cameroon and anga in Ethiopia. Ngo Ngwe *et al.* (2015) reported that of the species they tested in Cameroon, the wild species *D. praehensilis* and cultivated *D. cayenensis* were those with the highest phylogenetic diversity. Obidiegwu and Akpabio (2017) commented that *D. praehensilis* made a secondary but significant contribution to the domestication process of yams in Nigeria.

Stems are right twining. Tubers are up to 5 cm in diameter and about 60 cm long, cylindrical in shape and grow vertically in the soil. Their skins are white when young but turn brown later (Botany VM, 2018). Often present in the wild are withered tubers of the previous season alongside the new tubers. Hill and Freeman (1903) described the tubers from plants growing at Kew in the UK and stated that the tuber was entirely a subterranean organ, but 'in the natural state the tuberous stem is almost entirely an aerial structure, a fact that was very clearly demonstrated by a photograph of the wild plant'.

Tubers of *D. praehensilis* are cut into small pieces, boiled in water and used as food by the Shinasha people of Ethiopia, who also use dried tuber powder as a treatment for ulcers (Mosissa and Abraha, 2018). In Cameroon, *D. praehensilis* and *D. semperflorens* were reported to be the most important species in the diet of the Baka people. The tubers are used as a starch food reserve, since tuber production peaks during the dry season when food shortages may occur (Dounias, 1993, 2001; Hamon *et al.* 1995). In Côte d'Ivoire, *D. praehensilis* was also reported to be an important food crop (Sahoré *et al.*, 2005). Sato (2006) reported that '*Dioscorea praehensilis* is known as the most reliable staple food in Africa'. In the year 2000 he found six species of yam growing wild in a forest area of southeastern Cameroon with *D. praehensilis* the most densely distributed, followed by *D. minutiflora*, *D. burkilliana*, *D. semperflorens*, an unidentified *Dioscorea* species and *D. smilacifolia*. He estimated the productivity of *D. praehensilis* tubers as 118 kg per hectare, which seems very low. Ngo Ngwe *et al.* (2015) reported that *D. praehensilis* was an important source of diosgenin and Martin (1969) gave its diosgenin content as 0.18%.

D. prazeri

Dioscorea prazeri Prain & Burkill is native to India and Myanmar (Anonymous, 2015) and also occurs in the central areas of Thailand (Sagwansupyakom and Chantraprasong, 1984). It is a climbing plant that twines to the left and produces globose to clavate bulbils that are occasionally present in leaf axils particularly where stems touch the soil. Tubers were reported to be about 0.8–15 cm by 1.5–9 cm, branched, spreading, shallowly buried with hard flesh (Wilkin and Thapyai, 2009; Botany VM, 2018). In Thailand Maneenoon *et al.* (2008) reported that tubers are shallowly buried, produced in clusters and are branched, with pale yellow flesh and peel that is rough and brown or grey. Maneenoon *et al.* found 15 species of *Dioscorea* in the Banthad Range of Thailand of which 13 were edible but only 8 of these (*D. calcicola*, *D. filiformis* Griseb., *D. glabra*, *D. orbiculata*, *D. pentaphylla*, *D. pyrifolia*, *D. stemonoides* and *D. wallichii*) were main sources of carbohydrate for the Sakai, a native Thai tribe. The remaining edible species, *D. daunea*, *D. membranacea*, *D. piscatorum*, *D. prazeri* and an unidentified *Dioscorea* species were found to be bitter and had an unpleasant taste but were often consumed during food shortages and famines. In India, *D. prazeri* is also used in the manufacture of soap and shampoo in order to kill lice, but the Sakai people do not use *D. prazeri* as soap nor shampoo (Maneenoon *et al.*, 2008) and tubers were reported to be poisonous (Sagwansupyakom and Chantraprasong. 1984), with a diosgenin content of 2.1% (Anonymous, 2015).

D. preussii

Dioscorea preussii Pax. is also called *D. preussii* subsp. *hylophila* (Harms) Wilkin. No common name could be found. It is cultivated in tropical Africa from Senegal to the Sudan and south to Angola, Zambia and Mozambique. Obidiegwu and Akpabio (2017) commented that *D. preussii* has made a secondary but significant contribution to the domestication process of yams in Nigeria. It produces one or two tubers on each plant (previous and current year), that are at least 50 cm long and 2 cm wide, narrowly cylindrical in shape, tapering gradually towards the crown, sometimes lobed. They grow vertically into the soil (Wilkin and Timberlake, 2009; Botany VM, 2018). The tubers are very caustic and are only used as a famine food since they need prolonged soaking and washing over a period of 15 days before being suitable to eat. The composition of the edible portion of raw tubers was given as 116 kcal energy, 68.8%, water, 26.4% carbohydrate, 3.0% protein, 0.1% fat, 1.3% fibre, 1.7% ash, 60 mg 100 g^{-1} Ca and 50 mg 100 g^{-1} P (FAO, 1965).

D. proteiformis

Dioscorea proteiformis H. Perrier is endemic to the east coast of Madagascar with a narrow distribution along the coast from Taolagnaro in Toliara province in the south to Analalava in Antsiranana province in the north. They are found growing in dry littoral forests on sandy soils or in lowland humid forest on laterite soils, both of which are among the fastest disappearing habitats in the country. *D. proteiformis* have been reported to be found in only seven locations with its specific habitat requirements and was therefore assessed as 'near threatened', almost meeting the criteria for 'vulnerable' (Kennerley and Wilkin, 2017).

Fig. 8.27. *D. pseudotomentosa* tubers at Sarinya's yam farm in Chachoengsao province, Thailand, 2018.

D. pseudonitens

Dioscorea pseudonitens Prain & Burkill. In Thailand, Sagwansupyakom and Chantraprasong (1984) described it as rare. Wilkin and Thapyai (2009) described tubers as cylindrical in shape, vertically orientated and subtended by a small crown and up to 16 cm long and 1.5 cm in diameter. It is found growing in evergreen hill forests (with a preference for *Pinus kesiya*) at 1,000 to 1,500 m.

D. pseudotomentosa

Dioscorea pseudotomentosa Prain & Burkill is endemic to Thailand where its common name is man saeng hin. Wilkin and Thapyai (2009) described the tubers as at least 20 cm long and up to 6 cm in diameter, cylindrical in shape, spreading, often branched with some of the branches, and possibly apices, swollen and globose (Fig. 8.27). They are found growing in limestone soils in open forest at 100–200 m.

D. pteropoda

Dioscorea pteropoda Boiv. ex H. Perr. occurs in northern Madagascar, where it is known as totongana, exclusively growing in limestone soils. It was assessed to be 'vulnerable' by Kennerley and Wilkin (2017).

D. puber

Dioscorea puber Blume Enum. is synonymous with *D. anguina* Roxburgh and *D. cornifolia* Kunth. Common names include: chekka alu, kasha kanda, kosa and kukai sanga. It is native to northern India but is also found in southern India, Sumatra, Java and Malaysia as well as in the Himalayas from Nepal to Sikkim at altitudes up to 1,500 m, and in Taiwan, Philippines and New Guinea (Coursey, 1967). It produces large bulbils in the leaf axils and one or two tubers each year. Tubers are cylindrical in shape and up to 90 cm long and 30 to 38 cm wide, with brown hairy skin and pale yellow flesh, growing vertically in the soil up to 2 m deep on long stalks. Coursey reported that the tubers have an unpleasant smell and are therefore only used for food in times of famine when they can be eaten as vegetables. Kumar *et al.* (2013) stated that in Odisha, India, both bulbils and tubers are used as vegetables mainly during periods when other staples are in short supply. Bulbils are cooked and taken as a medicine to cure colic pain (Sheikh *et al.*, 2013). *D. puber* tubers are also a source of diosgenin and Asha and Nair (2005) screened several *Dioscorea* species from southwestern Ghats in India and reported the maximum diosgenin yield of those they tested was in *D. pubera* followed by *D. spicata*, *D. hispida* and *D. hamiltonii*. Groen *et al.* (2016) reported that in Java, ripe fruits have never been observed on *D. puber* plants.

D. pubera

Dioscorea pubera Blume is native to the Indochina region (Asiedu *et al.*, 1997) and is also distributed throughout the temperate and tropical Americas and China, in wet regions of the Himalayas, central Nepal, western Malaysia and Bhutan (Coursey, 1967). *D. pubera* produces tubers (Fig. 8.28) and bulbils, both of which are used for food. Bulbils are cooked and used as a medicine to cure colic pain (Sheikh *et al.*, 2013). Coursey (1967) reported that a short-day length generally accelerates formation of bulbils. Thirteen *Dioscorea* species were found by Kumar *et al.* (2017) in Similipal Biosphere Reserve in India with *D. pubera* among the six most common. Asha and Nair (2005) screened several *Dioscorea* species from Southwestern Ghats in India and reported the maximum diosgenin yield was in *D. pubera*. The Ho people, a Kolarian ethnic group from central India, collect different types of wild plants, including tubers from *Dioscorea* species, from the forest and use them for food and medicine (Ota *et al.*, 2013).

D. purpurea

Dioscorea purpurea Roxburgh is synonymous with *D. alata* var. *purpurea* and is also called Pondicherry sweet potato. Coursey (1967) stated that the name *D. purpurea* has been confused with some forms of *D. alata* that have tubers with purplish flesh. Kay (1987) stated 'the flesh of some cultivars of *D. alata* can be pink or even deep reddish-purple and these forms have been classified as *D. purpurea* Roxb. and *D. afropurpurea* Roxb., but this is not generally accepted'. Hedrick (1919) wrote: 'East Indies. The Pondicherry sweet potato is known only in a cultivated state, and was brought to India from Mauritius, where it is much grown ... The tuber is of a dull, crimson-red outside and of a glistening white within'.

Chen *et al.* (2017) tested water and ethanol extracts of *D. purpurea*, as well as extracts of *D. japonica* and *D. alata*, and found significant cardioprotective properties via multiple effects on antioxidant, anti-inflammatory and antiapoptotic activities. In China *D. purpurea* flour was tested as a composite in bread. It was found that 20% substitution of wheat flour with *D. purpurea* flour into bread resulted in similar volume and sensory evaluation to that of bread made with 100% wheat bread (Purwandari *et al.*, 2011). *D. purpurea* flour can also be incorporated into noodles to give hypolipidemic and antioxidative effects, while maintaining the sensory characteristics and consumer acceptability of the noodles (Lin *et al.*, 2006).

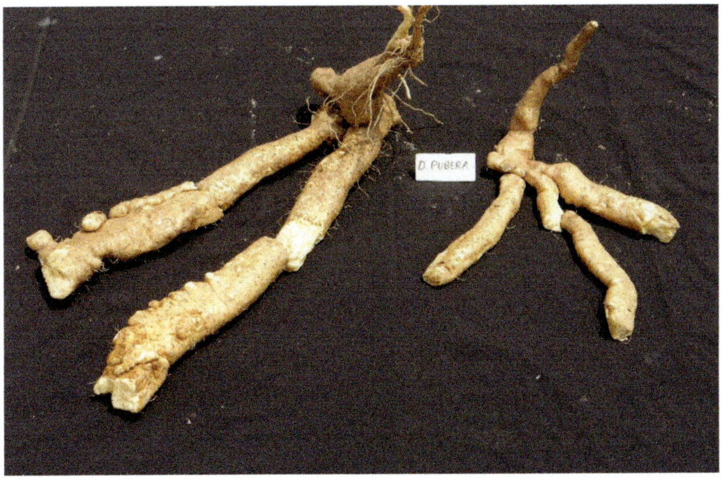

Fig. 8.28. *D. pubera* tubers. Photograph: Dr K. Abraham, CTCRI Emeritus, C-63, Gandhipuram, Sreekariam, Trivandrum 695017, India, reproduced with permission.

D. pyrenaica

Dioscorea pyrenaica Bubani and Bordère ex Gren. is synonymous with *Borderea pyrenaica* (Bubani and Bordère ex Gren.) Miégeville and is called Pyrenean yam. It is a dwarf species native to the central Pyrenees, between Spain and France, in limestone screes and rocks. In parts of the French Pyrenees, it is protected as vulnerable and considered a priority species, classified under the International Union for the Conservation of Nature and included in the National List of Protected Species in France and the Red Book of Endangered Flora of France (García and Antor, 1995). It is critically endangered and Segarra-Moragues *et al.* (2005) stated that it occurs in the southern most Spanish Pyrenean range where it grows in inaccessible crevices of a single limestone cliff with individual plants spatially subdivided into two sub-populations located on the upper and lower parts of the cliff, and vertically separated by about 150 m.

D. pyrenaica produces one to four erect, branching stems up to about 20 cm high. The leaves are heart-shaped and the flowers small and greenish in racemes with bell-shaped male flowers and star-shaped female flowers that produce capsules up to 2 cm long. Plants reach maturity when they are 20 to 30 years old at which time they flower for the first time. Tubers are small, rounded, blackish, with bulbils clustered at the top of each tuber. Tubers remain intact over many years, while leaves and stems die back and are regenerated each year. Tubers can live for a very long time (in some cases as long as 300 years) and their age can be estimated by counting yearly scars left in the tuber (Morales and Bosch, 2015).

D. pyrifolia

Dioscorea pyrifolia Kunth. is synonymous with *D. diepenhorstii* Miquel, *D. oppositifolia* L. and *D. zollingeriana* Kunth. It occurs in the eastern Himalayas, Assam, Thailand, Malaysia, Vietnam and Indonesia (Sumatra, West Java, Kalimantan) and is grown and sold in markets on the east coast of Malaysia (Burkill, 1966). Common names include: akar kemenyan paya, badak, hngo, huwi upas, ilus, marsh benzoin climber, ubi babi and ubi itik. Tubers are pale yellow to pale brown and the flesh is yellow or whitish and mucilaginous (Maneenoon *et al.*, 2008). Bulbils are absent. Tubers are slender and cylindrical (or globose to ovoid) with a spreading habit, up to about 5 cm in diameter and about 50 cm long subtended by a woody crown, annually replaced. More than one tuber may be produced per year, up to about 2.5 m deep in the soil (Wilkin and Thapyai, 2009; Groen, *et al.*, 2016). The edible tubers are gathered from the wild for local use (Burkill, 1966). In Peninsular Malaysia the tubers are eaten after baking or being boiled in water two or three times. Elmi Sharlina *et al.* (2017) subjected *D. pyrifolia* tubers to a fish embryo toxicity test, which showed that the starch was not toxic and that it was suitable for both food and non-food use. Raw tubers are also used in Malaysian traditional medicine as poultices for sores, swellings and bites (Groen *et al.*, 2016). Elmi Sharlina *et al.* (2017) reported that starch from *D. pyrifolia* tubers contained 44.47 ± 1.86% amylose, 4.84 ± 0.29% moisture, 0.88 ± 0.21% ash, 1.34 ± 0.11% proteins and 92.73 ± 0.48% carbohydrates. X-ray diffraction analysis showed a type-C starch with a relative crystallinity of 23.31 ± 2.41% and was polyhedral, with a diameter of 2.8 to 5.6 μm and average size of 3.93 ± 1.47 μg. Initial, peak and finishing gelatinization temperatures for the starch were 71.51 ± 0.07, 75.05 ± 0.15 and 78.25 ± 0.18°C, respectively; the gelatinization enthalpy was 3.86 ± 0.02 J g^{-1}.

D. quartiniana

Dioscorea quartiniana A. Rich. is named after the botanist Richard Quartin-Dillon, and is native to Benin, Botswana, Burundi, Chad, Democratic Republic of the Congo, Côte d'Ivoire, Eritrea, Ethiopia, Gambia, Ghana, Kenya, Madagascar, Malawi, Mozambique, Namibia, Nigeria, Rwanda, Sierra

Leone, South Africa, Sudan, Tanzania, Uganda, Zambia and Zimbabwe, where it occurs in forests, grassland and rocky areas (Contu, 2013). Ruffo *et al.* (2002) described its habitat as 'openings in upland and lowland rainforest, riverine forest and at forest edges, termite mounds, scrub and thickets, often on hillsides and near rock outcrops and grasslands, at elevations from 100 to 2,300 metres'. Tubers grow vertically in the soil and are cylindric or ovoid in shape, with fibres and ramifications and a brown to white skin. Stems are left twining, and bulbils are rare (Botany VM, 2018). *D. quartiniana* is also used as an ornamental. It is very variable in both floral and vegetative morphology, but these variations are continuous in nature and cannot be used to delimit taxa (Wilkin *et al.*, 2009). The tubers are edible and frequently harvested from the wild for local use, but it is most commonly eaten only as a famine food. However, Hanelt *et al.* (2001) reported that it is occasionally cultivated for food in eastern Nigeria and Cameroon but they reported that both the fruits and tubers are poisonous. It is occasionally cultivated as a food crop in parts of East Africa and can sometimes be seen being sold in local markets (Mansfeld's World Database of Agriculture and Horticultural Crops, 2019). In Tanzania, before being used as a food, the tubers are peeled and soaked in water for several days, then washed, sliced and dried in the sun. The dried slices are pounded into flour and used for uji or ugali (Ruffo *et al.*, 2002).

D. quinqueloba

Dioscorea quinqueloba Thunberg is also called kaede-dokoro, which means maple leaved tokoro in Japanese (tokoro = edible roots in Japanese). Hedrick (1919) wrote: 'Japan. This species is an edible yam of Japan' that grows in Honshu, west of the central region, Shikoku, Kyushu, Okinawa as well as in Korea and China where it is found in the mountains and on forest edges (Martin, 1969). *D. quinqueloba* tubers are used widely in traditional medical practice, but in Korea, Kim *et al.* (2011b) reported that they encountered cases of acute renal failure associated with ingestion of *D. quinqueloba*. They emphasized that this tuber should be correctly prepared for traditional medicine, otherwise it may cause life-threatening acute kidney injury. Also, in Korea Kim *et al.* (2014) reported that *D. quinqueloba* tuber extracts are used as an alternative therapy for cardiovascular disease and various other medical conditions. However, a man was admitted to hospital with complaints of skin rash and burning sensation after he had ingested a raw extract of *D. quinqueloba* as a traditional remedy. After examination he was reported to be suffering from acute interstitial nephritis following ingestion of *D. quinqueloba* associated with the skin rash, eosinophilia and increased plasma immunoglobulin E level.

D. racemosa

Dioscorea racemosa (Klotzsch) Uline is native to the Pacific coastal regions of Colombia growing at 0 to 400 m, specifically in Choco in Bahía Solano (Raz and Agudelo Zamora, 2020). No other information could be found on *D. racemose*.

D. retusa

Dioscorea retusa Mast. is synonymous with *D. microcuspis* Bake and *D. tysonii* Baker. It is native to South Africa (Transvaal, Natal, Eastern Cape) in grassland and forests, often in rocky areas of granite and basalt, at altitudes between 750 and 1,700 m (Wilkin, 1999; Botany VM, 2018). *D. retusa* was reported to differ from *D. quartiniana* in its strongly undulate leaflet margins (these are rarely found in *D. quartiniana* and the undulation is less marked even when it is present), and in the male flowers by the more lax inflorescence, which is not catkin-like due to the 1.5–2.7 mm long floral bracts, which are usually shorter than the tepals. Tubers of *D. retusa* are fusiform to cylindrical in shape and fleshy (Wilkin, 1999; Foden and Potter, 2005; Botany VM, 2018).

D. rockii

Dioscorea rockii Prain & Burkill is native to northern Thailand where it is restricted to forests between 1,000 and 1,700 m altitude (Thapyai *et al.*, 2005c). Thapyai also reported that *D. rockii* was known from a single specimen from northern Thailand collected in 1922, but they subsequently discovered male plants and tubers (Fig. 8.29).

D. rotundata

Dioscorea rotundata Poir. Some authorities have claimed that *D. rotundata* and *D. cayenensis* are a single species and *D. rotundata* has been referred to as *D. cayenensis* subsp. *rotundata* (Poir.) J. Miège, and others as *D. rotundata-cayenensis* complex. Okeke (2004) investigated the taxonomic position of *D. rotundata*, *D. cayenensis* and *D. pruinosa* and found diagnostic characteristics that were more than enough to warrant the recognition of each as separate species (Table 8.9). Ramser *et al.* (1997) found that *D. rotundata* and *D. cayenensis* accessions were clearly separated from each other, supporting the concept that they are distinct species. They analysed samples using random amplified microsatellite polymorphism (RAMPO) and identified a series of diagnostic bands that clearly distinguished between *D. rotundata* and *D. cayenensis*. This view is supported by hybridization experiments, which indicated that *D. rotundata* and *D. cayenensis* are almost completely inter-sterile.

Common names include: eboe yam, eight-months yam, Guinea yam, name blanco, negro yam, Portuguese yam, proper yam, white Guinea yam and white yam. *D. rotundata* is believed to be indigenous to both rainforests and savannahs in the area of West Africa stretching from Côte d'Ivoire to Cameroon and is generally considered to be the best edible yam in that region (Coursey, 1968). Tubers are normally cylindrical in shape, but can be palmate (Fig. 8.30). They have brown skin and white flesh and normally weigh around 2 to 5 kg, but can be a much as 25 kg (Martin and Sadik, 1977). An enormous number of varieties and cultivars have been selected over time. For example, Martin (1973) lists the common names of 54 of what he calls 'varieties' mainly from West Africa (Table 8.22), but some of these names are just generic for the particular *Dioscorea* species that is grown and used locally.

Composition and uses

FAO (1965) analysed combined samples of *D. cayenensis* and *D. rotundata* (Table 8.23). Opara (1999) reported that the composition

Fig. 8.29. *Dioscorea rockii* tuber. Reproduced with permission of Dr Chirdsak Thapyai.

Fig. 8.30. *D. rotundata* tubers for sale in Huddersfield, UK, January 2018.

Table 8.22. Selected varieties of the *D. rotundata-D. cayenensis* complex. Modified from Martin (1973).

Source	Name	Tuber notes
Dahomey (Benin)	'Ekepe'	Long, smooth tubers
Dahomey (Benin)	'Soussou'	Uniform tubers
Dahomey (Benin)	'Douba Yeserou'	Large, well-shaped tubers
Dahomey (Benin)	'Douba Yeserou'	Very large, branched
Dahomey (Benin)	'Dodo'	Compact tubers
Dahomey (Benin)	'Beterou'	Long, club shaped tubers
Dahomey (Benin)	'Foucou'	Short, stout tubers
Dahomey (Benin)	'Yassou Sika'	Several tubers per plant
Dahomey (Benin)	'Baniowre Bagarow'	Irregular tubers
Ghana	'Besu'	Compact tubers
Ghana	'Diaje'	Large tubers
Ghana	'Borofoo'	Early, large tubers
Ghana	'Bombetinga'	Long tubers
Ghana	'Punakonna'	Elongated heart-shaped tubers
Ghana	'Abana'	Long tubers
Guadeloupe	'Grosse Gaille Corrosol'	Large tubers
Ivory Coast	–	Stout tubers
Ivory Coast	–	Very smooth short tubers
Ivory Coast	'Akandou'	Smooth tubers
Jamaica	'Negro'	High yields
Martinique	'Saint Prix'	Large tubers
Nigeria	'Pape'	Large, irregular shaped tubers
Nigeria	'Dagi'	–
Nigeria	'Zaria'	Very large, irregular shaped tubers
Nigeria	'Awure'	Elongated globe tubers
Nigeria	'Gbare'	Emooth tapering tubers
Nigeria	'Etentu'	Heavy tapering cylindrical tubers
Nigeria	'Iyawoo1orun'	Heavy tapering cylindrical tubers
Nigeria	'Fe1e'	Large cylindrical tubers
Nigeria	'Efon'	Elongated heart shaped tubers
Nigeria	'Boki'	Smooth tubers
Nigeria	'Lakoko-Ayin'	Smooth, compact tubers
Nigeria	'Iro'	–
Nigeria	'Jocha'	Smooth tubers
Nigeria	'Inowe'	Stout tubers
Nigeria	'A1afulu'	Long, smooth tubers
Nigeria	'Awudo'	Large, conical tubers
Nigeria	'Ejiji'	Large globose tubers
Nigeria	'Unegba'	Large, somewhat irregular tubers
Nigeria	'Aga'	Long, smooth tubers
Nigeria	'Ukom'	Smooth, stout tubers
Nigeria	'Obiatorogu'	Pointed globes tubers
Nigeria	'Obiatorogu'	Early germination and maturity
Nigeria	'Obiatorogu'	Irregular tubers
Nigeria	'Obiatorogu'	Long, conical tubers
Nigeria	'Obiatorogu'	Many tubers
Nigeria	'Obiatorogu'	Many tubers
Nigeria	'Obiatorogu'	Long, smooth tubers
Nigeria	'Igangan'	Large tubers
Puerto Rico	'Guinea Blanco'	Smooth tubers,
Puerto Rico	'Guinea Blanco'	Smooth tubers
Sierra Leone	'Abani'	Top-shaped tubers
Sierra Leone	'Abana'	Large smooth globose tubers
West Africa	'Vedos'	Several tubers

Table 8.23. Composition of *D. cayenensis-D. rotundata* per 100 g edible portion (FAO, 1965).

Energy kcal	Moisture	Protein	Fat	Carbohydrate	Fibre	Ash	Calcium	Phosphorus	Iron
71	80.8	1.5	0.1	16.4	0.6	1.2	36	17	5.2
	70.5-85.9	1.4–1.6			0.3–0.7	0.8–1.4	30–48	6–39	4.9–5.4

of *D. rotundata* tubers per 100g fresh weight was: water 80 g, carbohydrate 16g, protein 1.5 g and fibre 0.5–0.6 g. Knoth (1993) gave the following analysis on a fresh weight basis: 58 to 80% moisture, 15 to 23% carbohydrates, 0.1 to 0.2% fats and 1.1 to 2.0% crude protein. Alinnor and Akalezi (2010) reported the composition of *D. rotundata* tubers as: moisture 54.5%, ash 1.4%, crude fat 2.7%, crude protein 0.09%, crude fibre 0.7%, carbohydrate 40.6% and available energy 731.75 kJ. Polycarp *et al.* (2012) gave the mean chemical content of two cultivars ('Labrekor' and 'Pona') of Ghanaian *D. rotundata* tubers as: moisture 58–63%, crude fat 0.41-0.46%, crude protein 4.02–4.43%, fibre 1.25–1.68%, carbohydrates 85.5–87.3%, energy 1539–1575 kJ $100g^{-1}$, tannins 4.6-6.9%, phytates 2.5–2.6%, oxalates 0.58–0.59%, iron 5.0–6.8%, copper 0.20–0.25%, manganese 1.2–1.8% and zinc 6.3–6.6%. In Ghana, Coursey and Aidoo (1966) measured the ascorbic acid levels of several *D. rotundata* varieties and found that they varied between 6.5–11.6 mg 100^{-1}. Abubakar and Gana (2017) showed tubers of *D. cayenensis* contained significantly higher ($p < 0.05$) magnesium (16.3 ± 1.52 mg/100g), phytate (31.2 ± 0.40 mg) and lower fibre (0.9 ± 0.03 g 100 g^{-1}) than *D. rotundata* or *D. alata*. New white yam (*D. rotundata*) exhibited significantly ($p < 0.05$) higher content in moisture (52.5 ± 8.40 g^{-1}), fibre (2.0 ± 0.02 g^{-1}), calcium (13.7 ± 0.5 mg^{-1}); lower concentration of phytate (11.3 ± 0.08 0.5 mg^{-1}) and moderate concentrations of other parameters studied. Water yam contains significantly ($p < 0.05$) higher protein (5.77 ± 0.2%), potassium (683 ± 3.6 mg $100g^{-1}$) and moderate amounts of other parameters studied. *D. cayenensis* contained significantly ($p < 0.05$) higher magnesium (16.3 ± 1.52 mg $100g^{-1}$), phytate (31.2 ± 0.40 mg^{-1}), lower fibre (0.9 ± 0.03 g $100g^{-1}$) and moderate amounts of other parameters. Old white yam (*D. rotundata*) contained significantly ($p < 0.05$) higher carbohydrate (37.47 ± 0.5 g $100g^{-1}$), lower cyanide (3.2 ± 0.15 mg kg^{-1}) and moderate amounts of other parameters studied.

Osunde and Orhevba (2011) found that during storage of *D. rotundata* tubers in barns in Nigeria for 6 months there was a reduction in crude protein (as high as 30 to 50%) and reductions in both starch and mineral content while fibre and sugar levels increased by as much as 38 to 49%. The effects on mineral content are surprising because the actual levels will remain the same during storage, but can change as a proportion to other chemicals. Osunde (2008) showed that tubers that had been stored for 150 days in traditional barns were preferred, by a sensory evaluation panel, to those that had been freshly harvested. Osunde also reported that *D. rotundata* stored in traditional barns had high losses in moisture, dry matter, crude protein and ascorbic acid after 120 days storage. In Côte d'Ivoire, Kouakou Dje *et al.* (2010) stored tubers of *D. alata* and *D. cayenensis-rotundata* for 6 months in a ventilated store where the conditions were 26.6 ± 3°C and 82 ± 5% RH. The total phenolic content varied in different parts of the tuber and decreased during storage. Protein content was about 7.9% and generally did not vary significantly ($p = 0.05$) during storage although a slight fall was observed in the distal part of one variety after 6 months. Lipids were low, only about 0.2% of the dry matter and didn't vary significantly ($p = 0.05$) during storage. Weight loss was 31% due to sprouting and dehydration after 110 days of storage under ambient conditions in the *D. rotundata* cultivar 'Oshei'. Also, the starch content decreased by approximately 3.5 to 4.5 g 100 g^{-1} while sugar and fibre levels increased slightly (Afoakwa and Sefa-Dedeh, 2001). In the apical and basal regions, the starch content decreased during storage while the sugars and α-amylase activity increased (Muthukumarasamy and Panneerselvam, 2000). The phenolic content of tubers can change during development. Hamadina and Craufurd (2015) found that the highest mean total free phenolics content in *D. rotundata* occurred after about 127 days after planting (2.46 AU units g^{-1} dry weight) and the lowest at vine senescence (1.79 AU units g^{-1} dry weight). After vine senescence, the mean total free phenolics declined, but individual phenolic compounds did not show any discernible decline in concentration over time.

Harvesting and handling

Harvesting of *D. rotundata* is usually about 7 to 9 months after planting, and coincides with when the vines die down (Purseglove, 1972; Martin and Sadik, 1977; Kay, 1987). Harvesting is by hand, but some success has been achieved with mechanical harvesting for *D. rotundata* plants grown on high ridges (Kay, 1987). Harvest damage or postharvest damage can result in rapid deterioration. The respiration rate of tubers that were cut into two or deeply cut rose rapidly during handling and there was an induction or increase in activity of certain enzymes, notably amylase and acid invertase. Invertase activity was highest in the layers of cells adjacent to the wound, but also developed in the region where the periderm formed. Concomitantly, there was a mobilization of food reserves resulting in an increase in reducing and non-reducing sugars in the tissue adjacent to the wound (Passam *et al.*, 1977). Physical injury during postharvest handling of yams is common in Jamaica. Cutting off the 'head' end of the tuber at harvest, either for retention as planting material or because this portion is inedible, is common practice and can lead to higher postharvest losses (Thompson, 1972a).

Curing

See Chapter 5 (p. 64) for detailed discussion of curing. Cured *D. rotundata* tubers had a lower weight loss throughout storage compared to those that had not been cured. However, cured tubers sprouted earlier by about 2 weeks when stored at tropical ambient temperatures, probably due to the increased cambial activity associated with curing. Weight loss of *D. rotundata* was considerably reduced when tubers were cured (Fig. 8.31, Tables 8.24, 8.25) and 14% of the tuber surface of non-cured tubers was covered with mould after 75 days storage compared with zero for those that had been cured. Curing also reduced the necrotic tissue from 8% to only 2%. The reduction in weight loss due to curing was effective whether the tubers were stored in Jamaican ambient conditions or in cold storage (Fig. 5.7). Ikeemobi *et al.* (1986) showed that lipolytic acyl hydrolase activity decreased in tubers after wounding and disappeared by the fourth day. However, there was two to nine times increased activity of lipoxygenase peroxidase and PPO with maximum values on the fourth and fifth days respectively after wounding. Lipoxygenase and PPO activity, measured in the proximal half of tubers, were much higher than the corresponding activity in the distal half.

Exposure of the tubers to 35 to 40°C and 95 to 100% RH for 24 hours was found to initiate the curing process (Thompson, 1972a). Passam *et al.* (1977) found that suberization of yam tubers occurred at 17°C in 4 to 5 days, at 25°C in 3 to 4 days and at 35°C in 2 days and that periderm formation took more than 10 days at 17°C, 5 days at 25°C and 4 days at 35°C. H.J. Passam (unpublished personal communication) reported that *D. rotundata* grown in Greece were cured routinely at about 35°C and 85 to 90% RH and their storage life was 5 to 6 months. Nnodu and Nwankiti (1986) recommended 26°C and 92% RH for 11 to 15 days. Adesuyi (1973) recommended 25 to 30°C and 90% RH for 5 days, both for *D. rotundata* tubers.

When applied commercially in Jamaica there were problems of bacterial soft rots developing during subsequent storage. Loading the tubers into the curing room, heating them until the surface of the tubers reached 40°C and then injecting the steam to increase the humidity overcame these

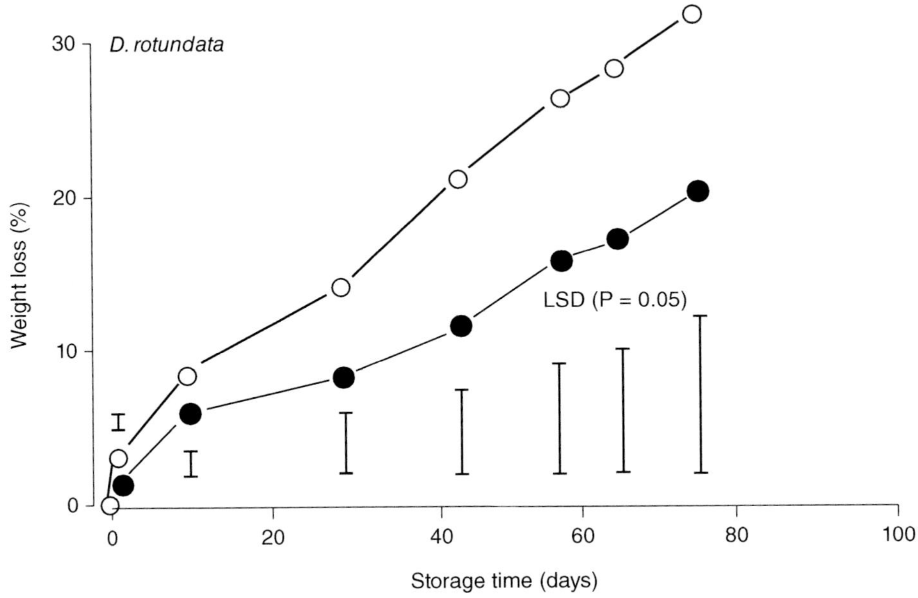

Fig. 8.31. Effects of curing on the percentage fresh weight loss of *D. rotundata* during storage in Jamaican ambient conditions. ○ = not cured, ● = cured. Vertical bars are probability levels (p = 0.05).

Table 8.24. Effects of cutting off the head end of the tuber at harvest on the weight loss of *D. rotundata* tubers during storage in Jamaican ambient conditions (Thompson, 1972b). The mean fungal score over that storage period was 1.9 for cut tubers and 0.9 for those not cut where 0 = none and 5 = maximum.

	Days in storage		
	12	25	34
Cut	9.8	16.8	22.6
Not cut	7.1	11.8	16.5

problems. Tubers cured in this way had reduced weight loss, no fungal infection, no necrotic tissue and no bacterial rots after storage for three weeks in Jamaican ambient room temperature conditions to simulate export marketing conditions (J.E. Cecil, B.O. Been and A.K. Thompson, unpublished).

Storage

In tropical ambient conditions Kay (1987) indicated that sound *D. rotundata* tubers can be stored for up to 4 months, but desiccation and sprouting could be problems. In Nigeria *D. rotundata* tubers are commonly stored in specially constructed barns (Fig. 5.1). In several West African countries, including Nigeria, Ghana and Côte d'Ivoire, yams are stored in pits, where the earth, which was excavated in making the pit, is placed to form a low wall around the edge. In a comparison between barn storage and pit storage for yams in Nigeria, Ezeike (1985) recorded weight losses over a five-month period of 60% for barn storage but only 15–25% for pit storage. Yams stored in slatted wooden trays in the coastal region of Cameroon had losses of 29 to 47% in only 2 months (Lyonga, 1985). Osunde and Orhevba (2011) stored *D. rotundata* tubers in traditional yam barns in Nigeria for 5 months. The barns were erected in the open air, but where sufficient shade and ventilation were available. One barn had intermittent forced-air flow using a fan. Half the tubers in the barn had been soaked in water containing neem bark for 12 hours, the other half were soaked in water, both prior to storage. The temperature in the barn, where there was a fan, was slightly lower than that of the barn without a fan. The temperature in the barn with a fan fluctuated between 20 and 36°C

Table 8.25. Effects of curing at 40°C and 98% RH for 1 day and storage temperature on fresh weight loss, fungal score, (where 0 = none and 5 = maximum) and necrotic tissue of *D. rotundata* during 127 days storage (Thompson, 1972b).

Storage conditions	Weight loss %		Fungal score		Necrotic tissue	
	Not cured	Cured	Not cured	Cured	Not cured	Cured
Ambient	23.3	17.9	0.0	0.0	4	3
13°C	9.8	8.0	1.4	1.2	9	10

with an average of 29°C while that in the barn without a fan fluctuated between 23 and 38°C, with an average of 33°C during the storage period. The humidity in the barn with a fan ranged between 26 and 60% RH with an average of 38% RH while that in the barn without a fan ranged between 23 and 55% RH with an average of 44% RH over the storage period. There was an average of 7% less sprout weight in the barn with a fan than in the barn without a fan. Opara (1999) reported that the optimum conditions for storage of what they referred to as 'white yam' and 'Guinea yam' was 16°C and 80% RH for several months. In West Africa recommended refrigerated storage conditions were also 16°C and 80% RH for several months (Tindall, 1983).

Chilling injury

At 1.1°C total physiological breakdown of *D. rotundata* tubers occurred within 10 days and chilling injury was also observed on tubers stored at 12.5°C (Coursey, 1968). In later work *D. rotundata* tubers stored at 11.7°C kept in better condition than those stored in tropical ambient conditions. However, when the storage period exceeded a week, the tubers lost weight rapidly and the rate was increased as the exposure time to 11.7°C increased (Fig. 8.32). Tubers also became soft, necrotic and infected with fungi shortly after removal to higher temperatures. This confirms Coursey's findings on chilling injury (Thompson 1972b; Thompson *et al.*, 1977). When tubers were stored at 13, 15, 17 or 19°C, those stored at 13°C quickly developed symptoms of fungal infection while those stored at 15°C developed symptoms later, but before those stored at higher temperatures (Fig. 8.33). This probably means that chilling injury can develop even at 15°C after prolonged exposure. However, in a study of *D. rotundata* in Jamaica, Thompson (1972a) found that storage at 12.5°C for 3 months resulted in no chilling injury and the tubers were in excellent condition. Storing them in polyethylene film bags reduced their weight loss, but had no other perceivable effects (Table 8.26). The tubers were then kept for a further 52 days in ambient conditions and still no symptoms of chilling injury were found, although the reduction in the rate of weight loss remained for those in polyethylene bags. After 3 months cold storage there was no sprouting of the tubers, but they all sprouted within 15 days of removal to ambient conditions irrespective of whether or not they were in polyethylene bags. Also, there was a slight indication that storing tubers in polyethylene bags resulted in a perceivable increased level of fungi on the tubers and some necrosis in their flesh.

Respiration rate

Respiratory activity has been shown to be minimal during the period of dormancy of yam tubers. Rates of respiration varied from 5 to 20 mL CO_2 kg^{-1} h^{-1} fresh weight in healthy *D. rotundata* tubers to over 35 mL CO_2 kg^{-1} h^{-1} fresh weight in decaying tubers. Respiratory weight loss was shown to contribute significantly to total storage loss and yams stored for 5 months, which may lose up to 10% of dry matter content through

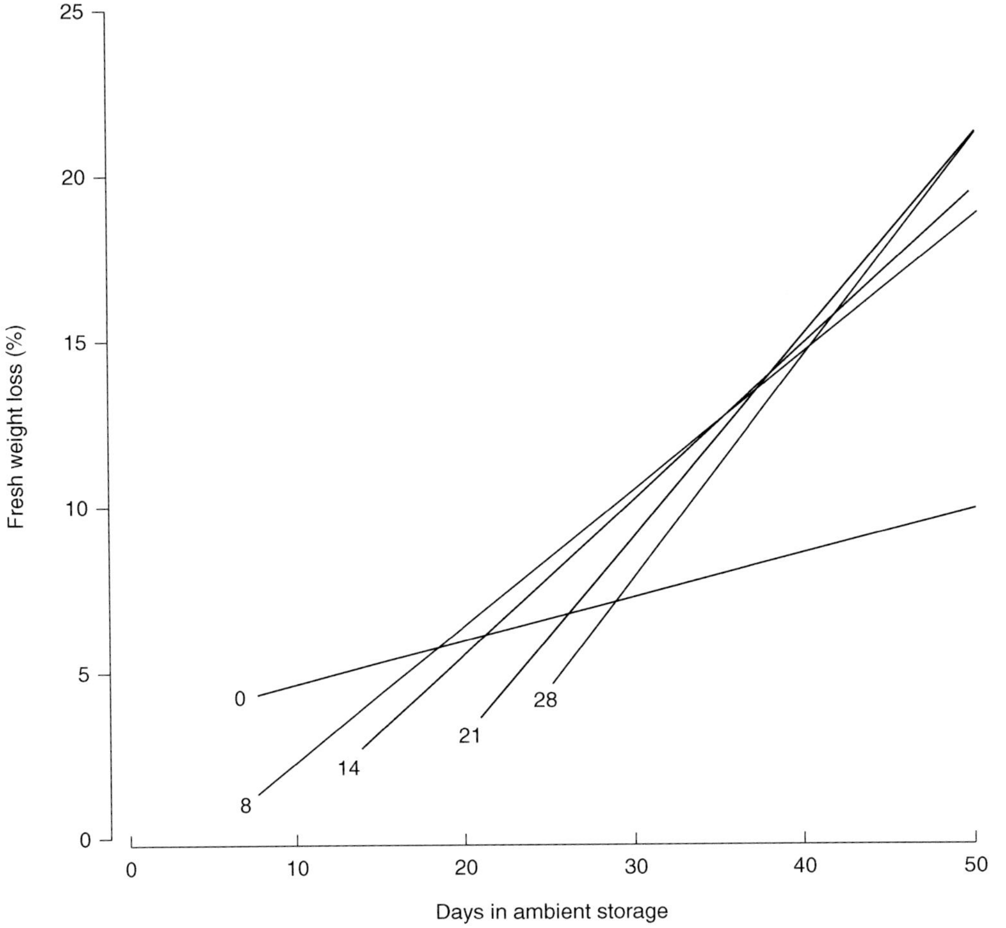

Days at 11.7°C	Regression coefficient	Weight loss % day^{-1}
0	0.984	0.13
8	0.996	0.45
14	0.999	0.46
21	0.984	0.60
28	0.971	0.66

Fig. 8.32. Effects of days in storage at 11.7°C prior to removal to Jamaican ambient conditions on the weight loss of *D. rotundata* (Thompson, *et al.*, 1973c).

respiration (Coursey, 1967). Also, Passam *et al.* (1978) showed that at harvest the rate of respiration was high: 15 mL CO_2 kg^{-1} h^{-1} fresh weight at 25°C and 29 mL kg^{-1} h^{-1} fresh weight at 35°C. However, as the tubers entered dormancy this rate declined to levels as low as 3 mL CO_2 kg^{-1} h^{-1} fresh weight at 25°C and 8 mL kg fresh $weight^{-1}$ h^{-1} at 35°C. The respiration rate then increased again at breakage of dormancy to over 20 mL CO_2 kg^{-1} h^{-1} fresh weight.

Sprouting

Ile *et al.* (2006) described three phases that occurred during the storage of *D. rotundata*.

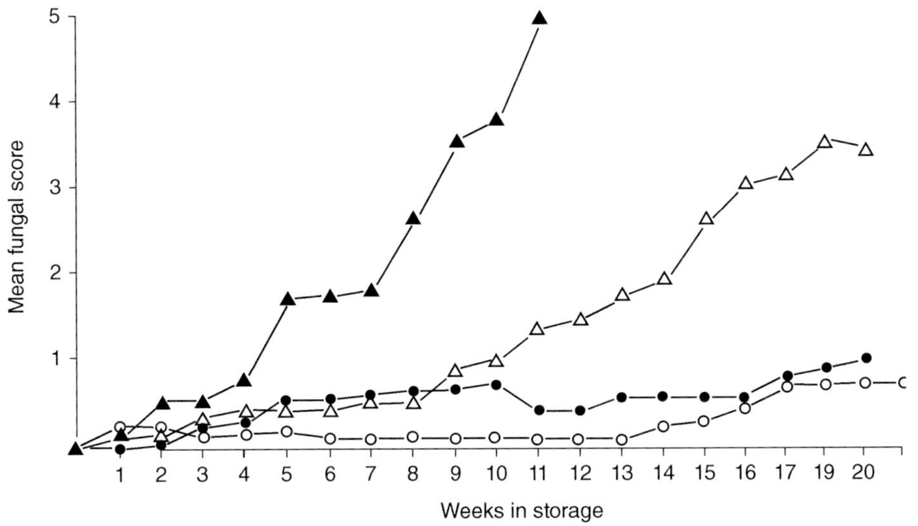

Fig. 8.33. Effects of temperature of the development of mould on the surface of *D. rotundata* tubers during storage: o = 19°C, ● = 17°C, △ = 15°C, ▲ = 13°C (mean score where 0 = none and 5 = surface entirely covered).

Table 8.26. Effects of storing *D. rotundata* tubers in polyethylene bags or not wrapped on their fresh weight loss after 96 days at 12.5°C and 52 days subsequently in Jamaican ambient conditions. Their levels of internal necrosis, fungal score (0–5) and the number of days before sprouting were observed after removal from cold storage, during 52 days subsequent storage in Jamaican ambient conditions.

Wrapping treatment	96 days at 12.5°C	52 subsequent days at ambient		After total of 148 days of storage	
	Weight loss	Weight loss	Necrosis	Fungi	Sprouting
Not wrapped	6.8%	11.5%	0	0	15 days
Polyethylene	1.4%	1.4%	2%	0.7	15 days

The first phase was when the tuber was dormant and did not sprout, whatever the storage conditions. The second phase was when the meristem started to form in the parenchyma tissue just below the tuber's periderm. The third phase was when the tubers sprouted. They postulated that sprout suppressants can only act on *D. rotundata* after the start of phase two. Dormancy and sprouting were shown to be affected by the maturity of the tubers at harvest as well as postharvest conditions. In ambient conditions in Jamaica, sprouting could occur after only 4 weeks (Thompson, 1972a), while in Nigerian ambient conditions sprouting occurred within 8 weeks (Coursey, 1961). Passam (1982) and Opara (1999) reported the dormancy period of *D. rotundata* from Nigeria as 12 to 14 weeks. In India tubers of the *D. rotundata* cultivar 'Sree Latha' stored at room temperature of 30 to 32°C and 80 to 85% RH in the dark sprouted 70 to 80 days after harvest (Muthukumarasamy and Panneerselvam, 2000).

Hamadina and Craufurd (2015) found that the total free phenolics content in *D. rotundata* tubers declined between harvest and sprouting, which might indicate that free phenolics have a role in tuber dormancy. They also measured the concentration of specific phenolic compounds during storage, which were higher in dormant tubers than in sprouting tubers.

Tschannen (2003) reported that gibberellic acid applied to *D. rotundata* tubers directly after harvest prolonged dormancy by 9 to 11 weeks and reduced respiration rate at the time of sprouting. Osunde and Orhevba (2011) reported that tubers treated with either neem slurry or neem bark extract (*Azadirachta indica*) had less sprouting than those not treated. The effect of maleic hydrazide in controlling sprouting in yam showed that soaking the tubers in 1,000 µL L^{-1} solutions for 10 hours before storage reduced the rate of sprouting by 8% (Ramanujam and Nair, 1982). However, Osunde (2008) reported that maleic hydrazide was not effective in inhibiting sprouting in *D. rotundata* tubers. Nonanol ($CH_3(CH_2)_8OH$, which occurs in orange oil) soaked into filter papers and placed inside polyethylene bags alongside tubers, delayed tuber sprouting, but increased the level of flesh necrosis. In contrast to potatoes (*Solanum tuberosum*), where the buds that will sprout are on the tuber surface, the meristematic tissue in yam tubers is below the surface. Sprout suppressants such as CIPC (chloroisopropyl phenylcarbamate) that are effective in potatoes as soon as they are placed in storage have been shown to have little or no effect on yams. A proprietary potato sprout suppressant containing CIPC and IPC (isopropyl-N-phenyl carbamate) in a dust formulation had no effect on sprouting. Orhevba and Osunde (2006) also reported that CIPC did not have any effect on sprouting of *D. rotundata* tubers. Low temperature storage can control sprouting and no sprouting occurred on any tubers at 13°C during 5 months storage, but there was chilling injury at temperatures below 15°C when stored for over a month (Thompson *et al.*, 1973a) so this method was not appropriate. Thompson *et al.* concluded that the only way of preventing sprouting when yams are stored beyond their normal dormancy period was to use cold storage just above the temperature that will cause chilling injury.

Irradiation at doses of between 7.5 and 15 krad inhibited sprouting of *D. rotundata* during 6 months storage without inducing adverse changes in acceptability or 'physiological properties' (Vasudevan and Jos, 1992). They also reported differences in varietal responses to γ irradiation. Osunde (2008) reported that an average dose of 120 Grays (1 Gray = 100 rad) and a dose rate of 114 Grays h^{-1} applied to the *D. rotundata* cultivar 'Asana' stored in a barn or simply kept on the ground showed that irradiation reduced sprouting during 6-month storage in both storage types. However, there was less rotting in the non-irradiated yams stored on the ground than the irradiated ones. Food products made from irradiated yams were judged to be better in quality than those made from the non-irradiated ones (Vasudevan and Jos, 1992).

Modified atmosphere packaging

Packaging yam tubers in plastic film bags has been tested and used commercially in order to prolong their marketable life. Thompson (1972b) dipped cured *D. rotundata* tubers in 500 µL L^{-1} thiabendazole, then dried and individually sealed them in 0.04 mm thick polyethylene film bags. During storage for 5 months at 12–13°C tubers lost only 3% in weight, did not sprout, had no internal necrosis and no surface lenticel proliferation. Lenticel proliferation has previously been shown to occur on plastic wrapped yam tubers in ambient temperatures and can be seen as small white dots on the tuber surface. Clark *et al.* (2013), working with sweet potatoes, reported that lenticels are more or less sealed off with suberized parenchyma cells, but under certain conditions cells beneath the lenticel divide and push out through the opening of the lenticel. These cells are not protected with suberin and have large air spaces between them allowing microorganisms on the surface to grow and enter the tuber.

All the *D. rotundata* tubers removed from cold storage to ambient conditions sprouted within 3 weeks. In storage at 12.5°C the rate of weight loss progressively decreased up to about 60 days then there was some indication of a slight increase in the rate, but when they were removed to ambient conditions after 90 days the rate progressively increased rapidly. This coincided with sprouting so the increased rate would be a combination of sprouting and the changes in temperature and humidity. However, tubers in polyethylene bags had a low constant rate of weight loss throughout storage at both 12.5°C and ambient conditions even though the yams inside the bags still began sprouting when they were removed to ambient conditions (Fig. 8.34). Andrade *et al.* (2012) stored 'chunks' of *D. rotundata* tubers in different plastic bags (linear low-density polyethylene/low density polyethylene, biaxially oriented polypropylene/low density polyethylene and polystyrene films). They injected the following gas mixture into each bag: 5% O_2, 5% CO_2, 90% N_2 or 8% O_2, 8% CO_2, 84% N_2, or left the bags with ambient air. The bags were stored at either 5 ± 1°C or room temperature (27 ± 1°C). The respiration rate at 5°C was 2.9 ± 0.9 mL CO_2 kg h^{-1} (graphically) and 3.2 ± 1.0 mL CO_2 kg h^{-1} (analytically). At room temperature the respiration rate was 20.1 mL CO_2 kg h^{-1} (graphically) and 22.4 ± 0.9 mL CO_2 kg h^{-1} (analytically). They reported that none of the modified atmosphere packaging films or gas mixtures used was a good alternative to refrigeration since they gave a shelf life of between 6 and 10 days.

D. sagittata

Dioscorea sagittata Poir is native to northern South America, particularly French Guiana, and has a taxonomic serial number of TSN 43377 (Govaerts *et al.*, 2007). It produces edible tubers growing up to about 1.5 m deep in the soil.

D. sambiranensis

Dioscorea sambiranensis R. Knuth. The common name is angona, but this is also used for other *Dioscorea* species in Madagascar. *D. sambiranensis* is endemic in littoral forests in Antsiranana province of northern Madagascar and also to northwestern Madagascar (Botany VM, 2018). Wilkin *et al.* (2009) described it as '*D. sambiranensis* species complex *D. buckleyana* Wilkin', which

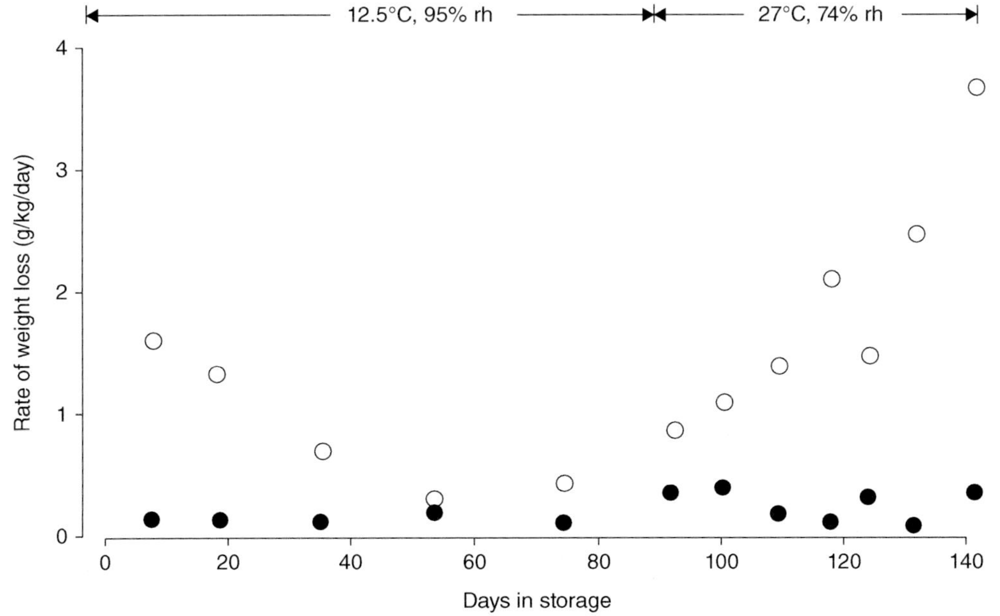

Fig. 8.34. Fresh weight loss rate (g kg^{-1} day^{-1}) of *D. rotundata* tubers that were stored either in 37.5 μm polyethylene bags (●) or not in bags (○). Thompson *et al.* (1973c).

was part of a mixed taxonomic concept under the invalid name of *D. sambiranensis* R. Knuth subsp. *ambrensis* H. Perrier. The species complex comprises of: one common, widespread species from deciduous western vegetation, one widespread in humid eastern forests and four with restricted distributions in evergreen forests of the central highlands and eastern Madagascar (Wilkin *et al.*, 2008). Tubers grow vertically in the soil and are cylindrical in shape with white skin. Stems twine in the left direction (Botany VM, 2018).

D. sanpaulensis

Dioscorea sanpaulensis R. Knut. Kirizawa *et al.* (2016) identified *D. sanpaulensis*, which is listed as endangered, as growing in the Reserva Biológica do Alto da Serra de Paranapiacaba, Santo André in Brazil.

D. sansibarensis

Dioscorea sansibarensis Pax is native to Madagascar and to tropical Africa from Tanzania to Guinea and south to Mozambique. It was reported to be known elsewhere, but as an introduced species. Common names are Zanzibar yam and Yoruba. It is commonly found along forest edges and is distributed in both dry and humid forests and swamps. It is also cultivated near villages (Botany VM, 2018). Obidiegwu and Akpabio (2017) commented that *D. sansibarensis* made a secondary, but significant, contribution to the yam domestication process in Nigeria. It was first introduced for experimental purposes to the Singapore Botanic Gardens, but has since escaped cultivation and is found in the wild. *D. sansibarensis* produces a massive perennial tuber that grows vertically into the soil and is ovoid in shape with brown skin. It also produces brown, hairless bulbils up to 5.5 cm in diameter, sometimes formed in pairs (resembling a pair of testicles) in leaf axils. Stems are usually unarmed and twine to the left (Holttum, 1964; Rao and Tan, 1976; Burkil, 1985–2004). *D. sansibarensis* can occur in many areas of forests, but in particular in disturbed areas such as along paths and roads, as well as in forest clearings. Thereis Choo (2009) reported that when *D. sansibarensis* is left unchecked in forests, it is a fast-growing climber of trees and can completely smother their canopy and shower the ground with thousands of bulbils: 'Once it has established in an area, the forest has little chance of recovery without human intervention'. Although the tuber is primarily considered a famine food, it is spasmodically cultivated in parts of Africa. Tubers are poisonous and are often grown on the edge of yam fields perhaps in the hope that any thieves will steal the toxic tubers instead of other yam species that are considered more valuable within the field. Hammer (1998) speculated that *D. sansibarensis*, considered to have magical properties in parts of Africa, may have been brought to the USA for use in religious rituals by Afro-Cuban groups.

D. sativa

Dioscorea sativa L. is synonymous with *D. anthropophagorum* A. Chev., *D. bulbifera* var. *anthropophagorum* (A. Chev.) Summerh., *D. bulbifera* var. *crispata* (Roxb). Prain, *D. bulbifera* L., *D. bulbifera* var. *elongata* (F.M. Bailey) Prain & Burkill, *D. bulbifera* var. *pulchella* (Roxb). Prain, *D. bulbifera* var. *sativa* Prain, *D. crispata* Roxburgh, D. *heterophylla* Roxburgh, *D. hoffa* Cordem., *D. hofika* Jum. and H. Perrier, *D. korrorensis* R. Knuth, *D. latifolia* Benth., *D. longipetiolata* Baudon, *D. perrieri* R. Knuth, *D. pulchella* Roxburgh, *D. rogersii* Prain andamp; Burkill, *D. sativa* var. *domestica* Makino, *D. sativa* var. *elongata* F.M. Bailey, *D. sativa* var. *rotunda* F.M. Bailey, *D. sylvestris* De Wild, *D. tamifolia* Salisb., *D. tenuiflora* Schltdl., *D. violacea* Baudon, *D. villosa* L. *Helmia bulbifera* (L.) Kunth and *Polynome bulbifera* (L.) Salisb. Coursey (1967) stated that none of the uses of the name *D. sativa* is valid and the use of this taxon in any context should be abandoned. He stated that *D. sativa* is often applied incorrectly to *D. sativa* Thum., which is a synonym of

D. bulbifera L. *D. sativa* Del. is applied to certain Indian varieties of *D. alata* and *D. sativa* Beatson and is synonymous with *D. cayenensis*. Common names for *D. sativa* include: aerial yam, air yam, batata de rama, bitter yam, bonay, bonda, bondanza, bonday, cara de aire, cará de árvore, cara de espinho, cara de sapateiro, cará do ar, cará moela, danda, gunda, hoei oepas, hoi, huang du, igname bulbifere, igname massokor, igname massokor, igname pousse en l'air, kaachil, kaile, kaile manu, kath alu, knollen yam, masoco, massokop, monday, ñame, ñame blanco, ñame bobo, ñame cimarrón, ñame Congo, ñame criollo, ñame de monte, ñame del aire, papa aérea, papa caribe, papa del aire, papa voladora, potato yam and sarau. It is native to Kenya, Zaire, Zambia, Malawi, Zimbabwe and Mozambique. Lankester (1832) reported that *D. sativa* is a native of the East and is supposed to have been transplanted thence to the West Indies, as it has not been found growing in the wild in the Americas but has been found growing extensively in the wild in Sri Lanka. He also reported that it is cultivated in Asia, Africa and the Americas for its edible tubers.

Bulbils are irregularly subglobose in shape and deep purplish in colour and are up to 8 cm in diameter. They are produced at the node of the stem, simultaneously with minute flowers, on a pendulous stem, 25–37 cm long (JSTOR, 2018). Tubers are produced annually and are spherical in shape and have a very bitter taste. They were reported to be only used as a famine food (Purdue University Famine Foods, 2018). In India tubers are boiled then cut into pieces, which are then leached in flowing water to remove the toxic bitter element. An alternate preparation method is to wash the pieces in about ten changes of fresh water. Detoxified tubers may be eaten whole or incorporated into other foods, for example konda or whey, or pounded into flour and used in making bread. Lankester (1832) wrote that the tubers were of good flavour and nutritious and consumed either roasted or boiled or as a substitute for bread. Hedrick (1919) wrote:

> Tropics. Pickering states that this species is found in tropical America and is cultivated by the Waraus of the delta of the Orinoco. The word igname was heard by Vespucius on the coast of Para and was found by Cabral, in 1500, applied in Brazil to a root from which bread was made. This yam was taken by European colonists to the Malayan Archipelago. Its roots, says Seemann, are acrid and require to be soaked before boiling. Browne says it is cultivated in the southern United States for its large, flattened and sometimes palmated roots, which are boiled, roasted and eaten like the potato.

Composition of both tubers and bulbils were given by FAO (1965) (Table 8.27).

D. schimperiana

Dioscorea schimperiana Hochst. ex Kunth is native to tropical Africa (Nigeria to Sudan and Ethiopia and south to Zambia, Malawi, Zimbabwe and Mozambique) in open upland savannah country. In Malawi the plant was reported to be common on termite mounds and sprawls over rocks on river margins (Burkil, 1985–2004). Tubers are solitary and produced annually, can be 60 cm long or longer and about 3 cm in diameter. They are cylindrical in shape, growing vertically into the soil with a brown or red-brown skin and creamy-yellow flesh, often with dense blood-red pigmentation (Wilkin and Timberlake, 2009; Botany VM, 2018). The stems twine right-handed and bear bulbils in the leaf axils. The bulbils are edible and abundant and are harvested from the wild in times of food scarcity. The edible tubers are also harvested for local use, though 'returns are small for the amount of effort required to dig them out of the ground' (Burkil, 1985–2004). Sibanda *et al.* (2000) analysed tubers of both *D. schimperiana* and *D. dumetorum* (Table 8.28) and concluded that '*D. schimperiana* would seem to be a more suitable source of starch for both domestic and industrial purposes than *D. dumetorum*'.

D. scortechinii

Dioscorea scortechinii Prain & Burkill has a wide distribution over the range of China,

Table 8.27. Composition of *D. sativa* mg or g 100 g^{-1} raw edible portion. Modified from FAO (1965).

	Energy kcal	Moisture	Protein	Fat	Total carbohydrate (incl. Fibre)	Fibre	Ash	Calcium	Phosphorus	Iron
Tubers	112	71.0	1.5	0.1	26.5	0.9	0.9	69	29	
Bulbils	78	79.4 ± 77.8–81.0	1.4 ± 1.3–1.4	0.2 ± .1–.3	18.0	1.2	1.0 ± .7–1.2	0.7–1.2	58 ± 45–70	2.0

Table 8.28. Analysis of composition of tubers of *D. schimperiana* and *D. dumetorum*, modified from Sibanda *et al.* (2000).

	D. schimperiana	*D. dumetorum*
Moisture content (%)	14.30	12.35
Amylose content (%)	24.5	14.8
Lipid content (%)	0.44	3.96
Protein content (%)	0.40	1.73
Phosphorus content (%)	19.0	28.8
Ash content (%)	0.90	0.16

Vietnam, Thailand, Assam, eastern Himalayas, Bangladesh, northeast India and Sumatra. Its common name in Vietnam is từ scortechini. *D. scortechinii* has significant infraspecific variation in floral organ size and shape across its distribution in South East Asia (Thapyai, *et al.*, 2005b). *D. scortechinii* produces a single tuber, sometimes alongside the withering tuber from the previous season. Tuber size is similar to *D. pentaphylla* – cylindrical to pyriform in shape and subtended by a woody crown (Wilkin and Thapyai, 2009; Botany VM, 2018). Ding and Gilbert (2010) described *D. scortechinii* var. *parviflora* as having stems twining to the left with brown branched tubers and light brown flesh occurring at 200–1,300 m in the forests of Hainan and possibly Yunnan provinces of China and Vietnam, as well as (also possibly) in Thailand.

D. semperflorens

Dioscorea semperflorens Uline. Its native range is Western Central Tropical Africa (Govaerts *et al.*, 2007). In the Cameroons, *D. semperflorens* was reported to be an important species in the diet of the Baka people since the tubers are used as a starch reserve and tubers mature during the dry season when alternative food sources may be scarce (Dounias, 1993; Hamon *et al.*, 1995; Dounias, 2001). Sato (2006) reported that in the year 2000 he found six species of yams, including *D. semperflorens*, growing wild in a forest area of southeastern Cameroon.

D. sensibarensis

Dioscorea sensibarensis Pax. is synonymous with *D. toxicaria* Bojer, *D. macroura* Harms, *D. welwitschii* Rendle and *D. macabiha* Jum. and H. Perrier. Its common name is Zanzibar yam and it was reported to be widespread in sub-Saharan Africa and Madagascar, occurring at an altitude of 200–650m (Wilkin and Timberlake, 2009). Holttum (1964) commented that *D. sensibarensis* was first introduced for experimental purposes in the Singapore Botanic Gardens but escaped and its aggressive growth is capable of smothering trees. Tubers are up to 50 cm in diameter, depressed globose, flattened below, hollowed towards the centre, developing roundish lobes on the upper surface near ground level. In parts of Africa *D. sensibarensis* is thought to have magical properties, but both the tubers and bulbils are toxic.

D. septemloba

Dioscorea septemloba Thunb. Its native range is China to central and southern Japan (Govaerts *et al.*, 2007). Common names include sevenlobed yam and mianbixie. Tuber extracts are used in traditional Chinese medicine to remove turbid damp and to relieve rheumatic conditions. Tubers may be collected in autumn and winter, washed, cut into slices and dried in the sun (Epilepsy Naturapedia, 2019). Yi *et al.* (2014) sampled yams from different parts of China and measured their diosgenin content. In *D. septemloba* it ranged from 0.78 ± 0.02 to 1.18 ± 0.02 mg g^{-1}. They concluded that 'such variations may presumably be attributed to differences in plant resources'.

D. seriflora

Dioscorea seriflora Jum. and H. Perrier is synonymous with *D. tanalarum* H. Perrier and *D. ovifotsy* H. Perrier. It is endemic to humid tropical forests across much of eastern Madagascar. Tubers grow vertically in the soil and are cylindrical with brown skin and white flesh (Botany VM, 2018).

D. simulans

Dioscorea simulans Prain & Burkill can be found in northern Guangdong, Guangxi and southern Hunan provinces of China in scrub forests and on rocky mountain slopes (Ding and Gilbert, 2010). Stems twine to the left and tubers are black-brown in colour, rough, cylindrical in shape, irregularly branched and grow horizontally in the soil.

D. sinoparviflora

Dioscorea sinoparviflora C.T. Ting M.G. Gilbert & Turland is found in Yunnan province of China at 400–2,000 m, growing in scrub and bamboo forests (Ding and Gilbert, 2010). Ding *et al.* (2000) reported that:

> During preparation of the account of Dioscoreaceae for the Flora of China, it was noticed that one species is illegitimately named, being a later homonym: *Dioscorea parviflora* C.T. Ting (1979), not Philippi (1864). The following new name (nomen novum) is therefore proposed here: *D. sinoparviflora* C.T. Ting, M. G. Gilbert & Turland.

Stems twine to the left and tubers are palmately or irregularly branched, cylindric in shape, with a corky yellow-brown, rough skin, up to 1.5 cm in diameter and grow horizontally in the soil.

D. smilacifolia

Dioscorea smilacifolia De Wild. and T. Durand is native to forest zones from Sierra Leone to western Cameroon and Fernando Po, extending over the Congo basin (JSTOR, 2018). Sato (2006) reported that in the year 2000 he found six species of yams growing wild in a forest area of southeastern Cameroon including *D. smilacifolia*, *D. smilacifolia*, *D. burkilliana* and *D. minutiflora*. These species appear similar, but in Côte d'Ivoire they were recognized as different (Miège, 1958). Coursey (1967) reported that *D. minutiflora* is closely related to *D. smilacifolia* and may even be a sub-species. Leaves have medicinal properties and tubers can be used as food. They are sometimes harvested from the wild for local use, but generally only in times of scarcity. The plant is sometimes cultivated, but mainly only as an insurance against difficult times (Burkil, 1985–2004). Stems twine to the right and sometimes bear bulbils.

D. soso

Dioscorea soso Jum. and H. Perrier. is native to Madagascar and is similar to *D. bemandry* Jum. and H. Perrier. Common names include: bemandry, sosan'ala, sosan-drano, babo gasy, babo menamionga and soso. Diverse forms of *D. soso* have been found in the degraded forest environments of the Réserve Spéciale d'Andranomena in Madagascar that have not been burned and are not in fallow. Several varieties have been identified including 'Babo', 'Sosan'ala' and 'Sosan Drano' (Jeannoda *et al.*, 2007). For the *Dioscorea* species found in both Andranomena and Beroboka, the percentage of users is much higher in Beroboka, even for *D. soso*, which are not very abundant there. Although there are some threats to dry forest habitats in Madagascar, *D. soso* tubers are harvested in the wild and Red List assessments by Rakotondratsimba (2008) and Tostain *et al.* (2007) found *D. soso* to be 'vulnerable' and of 'least concern' respectively, although the criteria used were unclear (Decary, 1946; Debray, *et al.* 1971; Kennerley and Wilkin, 2017). Jeannoda *et al.* (2007) found that the calcium content of one variety of *D. soso* (0.25% of dry weight) was high and generally higher than levels reported for other *Dioscorea* species although levels for *D. ovinala* tubers was higher (0.49% of dry weight), as was *D. bemarivensis*

(0.33% of dry weight). Protein levels of *D. soso* varied considerably between varieties with levels in 'Babo' 2.0 g L^{-1}, 'Sosan Ala' 6.5 g L^{-1} and 'Sosan Drano' 6.2 g L^{-1} from juice extracted from tubers. Coursey (1967) commented that *D. soso* produces sweet sugary tubers, formerly cultivated as a food crop, but was largely displaced by cassava. Jeannoda *et al.* (2007) commented that the watery yams eaten raw (*D. fandra* and *D. soso*) are appreciated only in the African villages where they are grown, probably because locals like their sugary taste. *D. soso* was identified as a potential food resource in the Mahafaly region of Madagascar where 70% of the householders interviewed were collecting wild yams (Andriamparany *et al.*, 2014).

D. spicata

Dioscorea spicata Roth is native to southern India and Sri Lanka. It is synonymous with *D. nummularia* Lamarck, *D. nummularia* var. *glauca* Prain & Burkill, *D. spicata* var. *anamallayana* Prain & Burkill and *D. spicata* var. *parvifolia* Prain & Burkill. Common names include: hard yam, Pacific yam, strong yam and tivolo yam. Local names meaning hard yam or strong yam are often given to cultivars of *D. spicata* because the flesh is drier and has a hard texture after cooking. The term 'strong' is used because the growing tubers can penetrate hard, untilled soils better than other species (Coursey, 1967). The tubers are spindle shaped; Coursey commented that a few poorly differentiated aerial tubers have been observed, but that these are very rare.

D. spicata tubers were reported to be a potent source of natural antioxidants. Mohan and Kalidass (2010) gave the following analysis for *D. spicata*: moisture 89.26%, crude protein 6.38 ± 0.08, crude lipid 4.78 ± 0.12, crude fibre 4.67 ± 0.03, ash 5.18 ± 0.01, N free extractives 78.99, calorific value (kJ $100g^{-1}$ dry matter) 1605.89 g $100g^{-1}$, starch 49.36 ± 0.34, niacin 12.31 ± 0.07 and 21.23 ± 0.12 ascorbic acid g $100g^{-1}$. They also gave *in vitro* protein digestibility as 5.27% and *in vitro* starch digestibility as 76.34%. Shajeela *et al.* (2011) reported the following analysis: starch 41.33 ± 0.33 g 100 g^{-1}, niacin 54.36 ± 0.09 g 100 g^{-1}, ascorbic acid 76.03 ± 0.36 g 100 g^{-1}. Their analysis of anti-nutritional factors is given in Table 8.29. Anti-nutritional factors identified by Mohan and Kalidass (2010) were: total free phenolics 0.38 ± 0.06 g $100g^{-1}$, tannins 0.34 ± 0.06 g $100g^{-1}$, hydrogen cyanide 0.09 ± 0.03 mg $100g^{-1}$, total oxalate 0.33 ± 0.03 g $100g^{-1}$, amylase inhibitor 3.34 AIU mg^{-1} soluble starch, trypsin inhibitor 1.39 TIU mg^{-1} protein.

D. spiculiflora

Dioscorea spiculiflora Hemsl. This species name is accepted and its native range is Central America from Mexico to Panama (Govaerts *et al.*, 2007). Tubers have a dorsiventral orientation, are up to 1 m long and have brown skins.

Martin and Ortiz (1962) reported that tubers arise from a ventral bulge between the first shoot and true root of the seedling as a result of meristematic activity of the ground meristem. Primary growth of the tuber is by multiplication and thickening of the primary meristem and by differentiation of those cells. The tissues of the tuber arising directly from the ground meristem are the storage parenchyma, enclosing bundles of xylem and phloem, with a parenchymatous cortex containing a well-defined cambium-like layer and cork cells. A true periderm exists transitorily. The anatomy of the tuber is more closely related to that of the stem than to that of the root and the starch supply of the tuber is concentrated in the older

Table 8.29. Anti-nutritional factors of tubers of *D. spicata*. Modified from Shajeela *et al.* (2011).

Component	Mean level
Total free phenolics	0.26 ± 0.01 g 100 g^{-1}
Tannins	0.10 ± 0.05 g 100 g^{-1}
Hydrogen cyanide	0.18 ± 0.01 g 100 g^{-1}
Total oxalate	0.44 ± 0.07 g 100 g^{-1}
Amylase inhibitor	3.31 AIU mg soluble $starch^{-1}$
Trypsin inhibitor	1.26 ± 0.12 TIU mg $protein^{-1}$

tissue and therefore tends to occur in greatest quantity in the ventral portions. Preston *et al.* (1964) studied the propagation and growth of *D. spiculiflora* and found that the original weights of tuber pieces was directly correlated to subsequent tuber growth and that plants responded to regular applications of fertilizer with greatly increased shoot and tuber growth. They analysed sapogenin content of tubers and found higher content of sapogenins in the original tuber pieces than in the tubers that were 1 or 2 years old and higher sapogenin content in the outer 1–1.5 mm layer of tubers than in the inner tissue.

D. spongiosa

Dioscorea spongiosa J.Q. Xi M. Mizuno & W.L. Zhao occurs in open forests and scrub forests at 400–800 m in Fujian, northern Guangdong, eastern Guangxi, southwestern Hubei, Hunan, Jiangxi and Zhejiang provinces of China. *D. spongiosa* was previously included within *D. septemloba*, from Japan, but it is now thought that *D. septemloba* should be restricted to Japanese material and the Chinese plants be treated as a distinct species. Tubers are light yellow in colour, cylindrical in shape and much branched. They are up to 5 cm in diameter with soft, spongy flesh. Tubers are an important constituent of traditional medicine (Ding and Gilbert, 2010). Yin *et al.* (2003) extracted four new pregnane glycosides (indirectly a parent of progesterone) from *D. spongiosa* tubers that showed potent inhibition against bone resorption induced by parathyroid hormones in a bone organ culture system.

D. stemonoides

Dioscorea stemonoides Prain & Burkill. Thapyai *et al.* (2004) reported that *D. stemonoides* was distributed sparingly in central, eastern and northern Thailand. Its common name among the Thai Saki people is kungkwad. Tuber characteristics were given by Maneenoon *et al.* (2008) as 'produced in clusters, cylindrical, pale purple mucilaginous flesh and yellow-brown skin'. Proximate composition of *D. stemonoides* tubers, in g per 100 g fresh weight was given by Maneenoon *et al.* as: 0.94 protein, 21.98 carbohydrate, 0.066 fat, 75.57 moisture, 0.73 ash, 0.72 fibre and 92.22 kcal energy.

D. steriscus

Dioscorea steriscus is not included on the Kew Plant List and no authority for this name could be located. However, Tapera and Machacha (2017) reported that *D. steriscus* was widely distributed in the Zimbabwean forests. It is commonly called Zanzibar yam, as is *Dioscorea sensibarensis*, and is found in Botswana, Central African Republic, Democratic Republic of Congo, Kenya, Malawi, Mozambique, Namibia, Tanzania, Uganda, Zambia and Zimbabwe. Its habitat is forest and forest margins, riverine forests and on termite mounds, and sometimes in disturbed areas at an altitude of 200–1,650m. It produces fleshy, spherical tubers, up to about 10 cm in diameter. Bulbils are usually present in leaf axils that are ovoid, up to 1 cm in diameter, dark purple or rarely red and with yellow flesh (Milne-Redhead, 1975; Wilkin and Timberlake, 2009).

Tapera and Machacha (2017) found *D. steriscus* was among the common indigenous vegetables in Zimbabwe and its tubers are consumed in most Zimbabwean provinces. Dzomba and Musekiwa (2014) reported that in Zimbabwe *D. steriscus* tubers are consumed as either a relish or a vegetable. Tubers are also used for medicinal purposes and Dzomba and Musekiwa found that *D. steriscus* tuber extracts possess anti-obesity and anti-cancer activity. Washaya *et al.* (2016) reported that *D. steriscus* tubers contained 93.6 calories kg^{-1} dry matter of energy, 16.8% carbohydrate, 0.83% protein, 2.06% ash and 72.5% moisture. In a survey in the town of Bindura in central Zimbabwe they reported that 45% of 'high density residents' eat *D. steriscus* at least once a week and 20% eat them every day. Dzomba and Musekiwa (2014) and Tapera and Machacha (2017) reported that in Zimbabwe *D. steriscus* tubers are used to prepare folk remedies for several diseases including hypertension, diabetes, heart problems, stomach pains and obesity. They found that the tubers contained bioactive compounds that can act as lipase and α-amylase inhibitors and therefore can be used in the development of functional foods against obesity. Tapera and Machacha gave the total phenolic content as 69 to 78 mg GAE g^{-1} fresh weight depending on which extraction method they used. With this high phenolic content, they concluded that *D. steriscus* has potential applications in the food, cosmetics, plastics and pharmaceutical industries.

D. subcalva

Dioscorea subcalva Prain & Burkill is found in scrub forests, forest margins, in mountain valleys and on slopes at 700–2,600 m in Guangxi, Guizhou, Hunan, Sichuan and Yunnan provinces of China (Ding and Gilbert, 2010). Stems twine to the left; tubers are fibrous and cylindrical in shape with white flesh.

D. subhastata

Dioscorea subhastata Vell. is synonymous with *D. guaranitica* Chodat and Hassl., *D. guaranitica* var. *balansae* Pellegr., *D. guaranitica* f. *membranacea* Chodat and Hassl., *D. guaranitica* f. *subcoriacea* Chodat and Hassl., *D. monadelpha* var. *opaca* Hicken and *D. piratinyensis* R. Knuth. Xifreda (1989) and Xifreda and Kirizawa (2003) commented that *D. subhastata* and *D. monadelpha* are species that belong to a nomenclaturally and taxonomically misinterpreted group, but *D. subhastata* is a lectotype (a plant later selected to serve as the single type specimen for species originally described from a set of syntypes). Its native range is Brazil, Argentina and Bolivia (Govaerts *et al.*, 2007).

D. sylvatica

Dioscorea sylvatica Eckl. also called *Testudinaria sylvatica* (Eckl.) Kunth. is found in Mozambique, Zambia, Zimbabwe, Swaziland and South Africa (Eastern Cape, Free State, Gauteng, KwaZulu-Natal, Limpopo, Mpumalanga and Western Cape). It occurs near sea level up to 1,800 m in a variety of wooded places, including moist bushveld areas, margins of forests, bracken and scrub. Common names include: elephant's foot, elephant's foot yam, ingwevu, ufudu-intelezi, ugebeleweni and wild yam. It is very slow growing, with a generation length estimated to be 30 years, eventually becoming the size of an elephant's foot (40–100 cm in diameter with an average tuber weight between 9 and 30 kg). In rich fertile soil, its tuber will remain mostly underground, but tubers can lie flat on the surface, especially in rocky and hilly locations. Tuber surfaces are divided into regular polygonal plates that become protuberant with age and divided by deep furrows. This thick, woody outer layer of the tuber aids moisture retention and can give some protection against predation (*Encyclopaedia of Succulents*, 2018).

Archibald (1967) found that *D. sylvatica* tubers had up to one-third exposed above the soil, which had been formed at base of the first leaf. He presumed that this was hypocotylar in origin, up to 10 mm diameter in juvenile plants broadening at the base and becoming 'bun-shaped', with a short, conical crown. In adult plants this crown is inconspicuous, marked by a few persistent bases of old shoots and by membranous, deltoid, brown bracts about 1 cm long.

Tubers vary in size and shape according to the terrain. In coastal sand dunes about 25 cm or more of the crown was found to be below the soil surface and their skin was about 12 cm thick. In these conditions, tubers were found to be up to 1 m in diameter, with irregular deeply incised margins and bifurcating lobes that may overlap. The skin is dark brown, corky, hard and finely and irregularly reticulated. On shallow soils or rocky slopes tubers can be up to 20 cm high and 40 cm in diameter, weighing up to 9 kg, with the crown and upper third exposed. Tuber surfaces are brownish-grey, hard, corky, with four- to seven-sided shields up to 10 mm thick by up to 40 mm across. The subterranean parts of the tuber are finely reticulate and corky. In deep forest soils they can be hypogeal, lenticular, with the outer skin finely reticulated and corky. On loose scree slopes tubers can be 20 cm or more below the surface, misshapen by pressure against stones with the outer skin grey, about 2 mm thick, with the inner flesh in all cases smooth, white and brittle with thin, fibrous roots growing only from the base of tubers (Archibald, 1967; Botany VM, 2018; *Encyclopaedia of Succulents*, 2018).

D. sylvatica are used as decorative house plants. Tubers contain diosgenin and are poisonous. Tuber bark extracts were shown to have antibacterial activity against *Escherichia coli* and are used for the treatment of wounds, sores, mastitis and abscesses. Extracts may produce mild inflammation and itching when rubbed on the skin. This is caused, in part, by raphides of calcium oxalate (Kelmanson *et al.*, 2000).

D. tauriglossum

Dioscorea tauriglossum R. Knuth is a native of Brazil. Kirizawa *et al.* (2016) identified eight species growing in the Reserva Biológica do Alto da Serra de Paranapiacaba, Santo André, including *D. tauriglossum*. Pinto Da Luz *et al.* (2020) characterized the pollen of nine *Dioscorea* species in an Atlantic forest in the state of São Paulo, including *D. tauriglossum*. Five of these species were endemic to Brazil: *D. laxiflora*, *D. monadelpha*, *D. olfersiana*, *D. tauriglossum* and *D. trilinguis*. The aim was to clarify the intricate classification of *Dioscorea*. Hayashi *et al.* (2018) analysed the tubers of three Brazilian species and found their starch content was: *D. delicata* 49.25%, *D. tauriglossum* 7.81% and *D. ovata* 8.76% on a fresh weight basis.

D. tenuipes

Dioscorea tenuipes Franchet & Savatier occurs in forests in southern Anhui, Fujian, northern Guangdong, southern Hunan, Jiangxi and Zhejiang provinces of China, and also in Japan, at 800–1,100 m. Stems twine to the left and are cylindrical up to 1.5 cm in diameter with obvious nodes and internodes (Ding and Gilbert, 2010). *D. tenuipes* is a species that produces viable seed and Okagami and Kawai (1977) studied the effects of gibberellic acid on their germination. They found that for complete germination seeds required prechilling in moist conditions before incubation at a higher temperature. They also found that GA_3 promoted germination in the dark and in blue, green or far-red light.

D. tentaculigera

Dioscorea tentaculigera Prain & Burkill. Wilkin and Thapyai (2009) described the tubers as growing vertically in the soil, narrowly cylindrical in shape and up to 20 cm long and up to 1.5 cm diameter, often with a withering tuber from the previous season. Ding and Gilbert (2010) also reported that tubers are cylindrical in shape, grow horizontally and occur in forests and mountain ravines at 1,300–1,500 m. Stems twine to the left. *D. tentaculigera* is found in evergreen forests at 550–1,600 m in northern and southwestern Thailand, Myanmar and Yunnan province in China.

D. togoensis

Dioscorea togoensis R. Knuth is native to savanna and dry forests or rocky places in West Africa including Senegal, northern Nigeria, Gambia and Sierra Leone. Its common name is manding-mandinka. Stems are right-handed, bearing axillary bulbils that are small, grey-brown and 4–16 mm in diameter. Tubers are slender and grow deeply in the soil. *D. togoensis* grows in the wild, but also colonizes well cultivated land that is being allowed to revert to scrub. In Ghana the small bulbils are eaten in the Kita district in times of shortage (Burkill, 1985–2004). Diosgenin content of tubers was reported to be 0.12% (Botany VM, 2018).

D. tokoro

Dioscorea tokoro Makino ex Miyabe is native to eastern Asia and occurs in thickets in lowlands and foothills all over Japan as well as in mixed forests and bamboo forests, usually along ravine sides, from near sea level to 1,000 m (Ohwi 1965; PFAF, 2019). Ding and Gilbert (2010) reported that it also occurs in south Anhui, Fujian, Guizhou, south Henan, Hubei, Hunan, Jiangsu, South Jiangxi, south Sichuan and Zhejiang provinces of China. Common names include oriental yam and oni-dokoro. Stems twine to the left and Ding and Gilbert described the tubers as 'Rhizome horizontal, irregularly branched, subcylindric, 0.7–1 cm thick; cork light brown, rough; roots borne on lower side of rhizome, fibrous'. Although it is grown widely in China and Japan, the tubers have a bitter and pungent taste and are not generally regarded as edible. Some reports also indicate that tubers are said to be poisonous if eaten raw, but in northern parts of Japan they are used as a health-promoting food. The first discovery of disogenin from *D. tokoro* was in the 1930s by Japanese researchers (Fujii and Matsukawa, 1936). It was reported by Oyama *et al.* (2017) that *D. tokoro* contains protodioscin, which is an anti-proliferative compound against human leukemia cells, but the protodioscin/dioscin composition varied in tubers from different regions of Japan. For example, in the region around Fukuoka tubers contained dioscin and gracillin, but no report on protodioscin, while in tubers grown around Iwate, protodioscin was the major saponin and dioscin occurred in only small quantities.

Okagami and Kawai (1977) studied the effects of gibberellic acid on the germination of *D. tokoro* true seed. They found that for complete germination, seeds required pre-chilling in moist conditions before incubation at a higher temperature. They also found that GA_3 inhibited germination in the dark and in red light, but promoted germination in blue and far-red light.

D. tomentosa

Dioscorea tomentosa J. Koenig ex Sprengel is also referred to as *Helmia tomentosa* (J. Koenig ex Sprengel) Kunth. Common names include: chavali, chavankizhangu, chavu, doyala yam, inthi kachchil, inthikachil, kyway pin, nalvaelikizhangu, noolamkizhangu, nuli, nurai, ooyala yam, pindi, shaval kilangu and taragai kanda. It is native to India and Sri Lanka. Tubers are up to 30 cm long and 5 to 7 cm in diameter, have a white and hairy skin with white flesh and can branch laterally (Sasidharan and Pal Fellow, 2019) (Fig. 8.35).

Swarnkar and Katewa (2008) reported that when tubers are used for food, they are sliced and kept overnight in water. The tuber is then cooked as a vegetable. Choudhary *et al.* (2009) reported its use as a medicinal plant in Rajasthan, India and extracts of the tubers are used as a contraceptive by some Indian tribes. Mohan and Kalidass (2010) gave the following composition: 93.65% moisture, 8.31 ± 0.12 g $100g^{-1}$ crude protein, 6.84 ± 0.04 g $100g^{-1}$ crude lipid, 4.38 ± 0.18 g $100g^{-1}$ crude fibre, 6.53 ± 0.03 g $100g^{-1}$ ash, 73.93 g $100g^{-1}$ nitrogen free extractives, 1631.44 g $100g^{-1}$ calorific value (kJ $100g^{-1}$ dry matter), 15.29 ± 0.09 g $100g^{-1}$ starch, 14.1 2 ± 0.02 g $100g^{-1}$ niacin and 24.92 ± 0.21 ascorbic acid g $100g^{-1}$. They also gave *in vitro* protein digestibility as 4.98% and *in vitro* starch

Fig. 8.35. *D. tomentosa* tubers Photograph: Dr K. Abraham, CTCRI Emeritus, C-63, Gandhipuram, Sreekariam, Trivandrum 695017, India, reproduced with permission.

digestibility as 56.84%. Arinathan *et al.* (2009) gave the proximate (Table 8.30) and mineral composition (Table 8.31) of *D. tomentosa* tubers. Shajeela *et al.* (2011) gave the following composition: starch 51.36 ± 0.27 g 100 g^{-1}, niacin 74.12 ± 0.21 mg 100 g^{-1}, ascorbic acid 65.20 ± 0.21 mg 100 g^{-1}. Antinutritional factors identified by Mohan and Kalidass (2010) were: total free phenolics 0.23 ± 0.10 g 100 g^{-1}, tannins 0.07 ± 0.48 g 100 g^{-1}, hydrogen cyanide 0.05 ± 0.01 mg 100 g^{-1}, total oxalate 0.03 ± 0.01 g 100g^{-1}, amylase inhibitor 1.54 AIU mg soluble $starch^{-1}$ and trypsin inhibitor 0.87 TIU mg $protein^{-1}$. Shajeela *et al.* (2011) gave the following analysis for antinutritional factors: total free phenolics 0.41 ± 0.01 g 100 g^{-1}, tannins 0.06 ± 0.01 g 100 g^{-1}, hydrogen cyanide 0.34 ± 0.03 mg 100 g^{-1}, total oxalate 0.31 ± 0.11 g 100 g^{-1}, amylase inhibitor 4.64 AIU mg^{-1}, soluble starch trypsin inhibitor 1.41 ± 0.11 TIU mg^{-1} protein (Table 8.32).

D. torticaulis

Dioscorea torticaulis R. Knuth is native to western Africa. It bears small tubers up to about 2 cm long. Kirizawa *et al.* (2016) identified eight *Dioscorea* species as growing in the Reserva Biológica do Alto da Serra de Paranapiacaba, Santo André, among them *D. torticaulis*, which they claimed should be listed as endangered.

Table 8.30. Proximate composition of *D. tomentosa* tubers (g 100g^{-1} fresh weight). Modified from Arinathan *et al.* (2009).

Component	Level
Moisture	71.86 ± 0.47%
Starch	49.86 ± 0.76%
Crude protein	8.51 ± 0.27%
Crude lipid	5.88 ± 0.19%
Crude fibre	2.24 ± 0.14%
Ash	2.54 ± 0.39%
Nitrogen free extract	80.83%
Calorific value	1713.71 kj 100 g^{-1}
Niacin	88.36 ± 0.12 mg 100 g^{-1}
Ascorbic acid mg	55.68 ± 0.44 mg 100 g^{-1}

D. transversa

Dioscorea transversa R. Br. is synonymous with *D. punctata* R. Br. Confusion over the morphology of *D. alata*, *D. nummularia* and *D. transversa* has been reported in the Philippines, Indonesia and New Caledonia (Bourret, 1973; Sastrapradja, 1982; Cruz and Ramirez, 1999; Cruz *et al.*, 1999). Malapa *et al.* (2006) reported that *D. transversa* and *D. nummularia* were closely related to *D. alata* and both were resistant to anthracnose disease. Lebot (2019) also reported that *D. nummularia* and *D. transversa* were sometimes difficult to distinguish from one another. Common names for *D. transversa* include: kowar, pencil yam, long yam and native yam. They grow in warm, temperate rainforests and moist schlerophyl open forests, mostly along the eastern seaboard of Australia at near sea level to 1,100 m (Herbison-Evans and Ashe, 2008). *D. transversa* is cultivated only in Melanesia and Australia and not in continental Asia. *D. transversa* is sometimes cultivated for food in New Caledonia, and it was introduced to the Antilles in the early 1970s because of its special disease resistance and favourable agricultural characteristics (Mansfeld's World Database of Agriculture and Horticultural Crops, 2019). Lebot (2019) commented that *D. transversa* was hardy, drought tolerant, high yielding and resistant to anthracnose disease and had been introduced into the French West Indies without evident success.

Table 8.31. Mineral composition of *D. tomentosa* tubers (mg 100 g^{-1}). Modified from Arinathan *et al.* (2009).

Na	K	Ca	Mg	P	Zn	Mn	Fe	Cu
35.00 ± 0.08	1345.41 ± 2.31	240.30 ± 0.13	192.00 ± 0.04	98.68 ± 0.62	6.20 ± 0.12	1.10 ± 0.01	23.66 ± 0.04	1.44 ± 0.01

Table 8.32. Antinutritional factors of tubers of *D. tomentosa*. Modified from Shajeela *et al.* (2011).

Component	Level
Total free phenolics	0.41 ± 0.01 g 100 g^{-1}
Tannins	0.06 ± 0.01 g 100 g^{-1}
Hydrogen cyanide	0.34 ± 0.03 g 100 g^{-1}
Total oxalate	0.31 ± 0.11 g 100 g^{-1}
Amylase inhibitor	4.64 AIU mg soluble $starch^{-1}$
Trypsin inhibitor	1.41 ± 0.11 TIU mg $protein^{-1}$

Stems twine to the right. Tubers are edible and typically slender and long. There are two forms: an eastern rainforest form which doesn't have bulbils and a northern form which occurs in open forests and has small bulbils and large tubers (Low, 1988). Tubers are produced deep in the ground and their shape varies, with the most common having a long neck. There are only a few cultivars and all their tubers have white flesh. This is probably the least cultivated of all species but its potential remains to be investigated thoroughly. There are also attractive cultivars with 'neckless and compact tubers', which have a smooth skin (Collings, 2015). *D. transversa* bears late maturing tubers (9–12 months) compared to *D. alata*, and the tubers are greatly valued for their high dry matter content and good organoleptic properties, which were reported to be similar to those of *D. nummularia*. In Australia the tubers are eaten by indigenous Australians, either cooked, or in the case of young tubers, raw (Herbison-Evans and Ashe, 2008). They may be cooked in ashes and eaten and are said to taste like 'crisp slimy potatoes'. A decoction was used as a treatment for skin cancers by indigenous Australians of the Tully area of northeast Queensland.

D. trifida

Dioscorea trifida L. is synonymous with *D. brasiliensis* Willd. and *D. punctata* R. Br. Common names of *D. trifida* include: aja, cara doce, cara branco, couche couche, cush cush, elephant yam, Indian yam, indienne, inhame, inhame roxo, kenke, maona, mapuey, ñame, ñame blanco, ñampi, sacha papa, sachapapa, tabena ñame, yampi, yampie and yampy. *D. trifida* is believed to have originated in South America, probably in Guyana or the Amazonas (Degras, 1993). It is now cultivated throughout northern parts of South America and the Caribbean Islands (to which it was taken by the Arawaks), as far north as the Greater Antilles and has been introduced to other parts of the world, such as Cameroon in the 1970s (Lyonga *et al.*, 1973), although Kay (1987) commented that 'it has not been successfully introduced into other parts of the tropics where it is grown only on a small scale'. One reason why it is not widely cultivated is because of its sensitivity to viruses, especially yam mosaic virus (Arnolin and Lombion, 2005). Also, Solano (1996) reported that one of the limiting factors for the production of *D. trifida* in the Atlantic region of Costa Rica is the lack of availability of high-quality planting material. Hedrick (1919) wrote:

> Guiana and Central America. This species is cultivated as an edible yam. This is the smallest and most delicate of the yams grown in Jamaica. It seldom exceeds eight or nine inches in length and two or three in diameter and is generally smaller. The roots have a pleasant, sweetish taste, very agreeable to most palates.

Martin and Degras (1978) commented that there were two 'botanical varieties' of *D. trifida*. These were *tuberosa* that they put in italics and Genuina that they didn't put in italics and used a capital letter. They also reported that there were 50 clones developed at the INRA station in Guadeloupe and list many cultivars including 'Patte a Cheval', 'Mapuey Largo', 'Cousse-Couche Violette', 'Cousse-Couche Rouge', 'INRA 50', INRA 24' and 'INRA 21' as well as several cultivars from other countries including 'Indien St Lauren' and 'Macouria 3' from Guyana. They also enumerated several factors that are typical of specific cultivars and can be used to identify them, including the number of tubers they produce, the size of tuber clusters, tuber flesh colour, yield of tubers and various leaf and stem characteristics. *D. trifida* is by far the most important of the indigenous American yams (Kay,

1987) and has received intensive study in many countries including Guadeloupe (Rouanet, 1967) where it is highly valued for its organoleptic qualities, although CIRAD (2015) commented that *D. trifida* 'is gradually disappearing as a result of the virus'.

Up to 50 small tubers are produced on each plant at the end of short stolons. Tubers have a brown, thin skin with elongated cracks and are covered with adventitious fibrous roots. Tubers can be up to 20 cm long and vary in form from club-shaped to spherical (Fig. 8.36). Tubers have white flesh but may be tinged with yellow and purple pigments and there are also purple varieties. Ramos-Escudero *et al.* (2010) identified 12 anthocyanin pigments in purple *D. trifida*, the majority of which corresponded to diglucosides acylated with ferulic or p-coumaric acids type of compounds, for example pelargonidin-3-*O*-glucoside-5-*O*-glucoside.

Composition and uses

D. trifida has been a staple food for some indigenous people from time immemorial and is cultivated mainly for its delicious tubers. In parts of the Caribbean it is known as 'the best of the yams'. Tubers are cooked in various ways, including baked or boiled and in Venezuela and Colombia they are mashed or used in soups (Stephens, 2009; Pérez *et al.*, 2010). Opara (1999) reported that the composition of *D. trifida* per 100 g edible tuber portion fresh weight was: water 65–76 g, carbohydrate 38 g, protein 1.9–2.3 g and fibre 0.6 g. Kay (1987) reported the following composition for *D. trifida*: protein 2.54%, fats 0.44% and carbohydrate 38% on a fresh weight basis with the carbohydrate content consisting mainly of starch. In *D. trifida*, starch has been reported to have an amylose content in the range 34.7–43.3% for both white and purple varieties (Rached *et al.*, 2006). The ascorbic acid content was approximately 5.5 mg 100 g^{-1} edible portion, but it was reported to be rapidly lost during storage. Pérez *et al.* (2010) reported that *D. trifida* tubers contain about 38% waxy starch (on fresh weight basis), where the amylose content of landraces cultivated in the Amazons of Venezuela was < 8.1% compared to commercial *D. trifida* genotypes that contained > 8.7%. Gutiérrez *et al.* (2014) also reported the waxy nature of starch from *D. trifida* tubers and modified starch, isolated from *D. trifida* tubers grown in the Venezuelan Amazonian region, showed that differences in the amylose content affected their structure and functional properties. The degree of substitution was more significant in the *D. trifida* starch than that the cassava starch, even if both were within the ranges allowed by the US Food and Drug Administration (FDA). Extracts of *D. trifida* also have medicinal properties. Mollica *et al.* (2013) showed that *D. trifida* extract could have an effect on responses to food allergies and demonstrated its potential for inclusion in food to prevent or treat some allergies.

Fig. 8.36. *D. trifida* tubers from Jamaica, on sale in Huddersfield, UK, January 2018.

Cultivation

D. trifida was reported to be best produced as an annual crop and harvested 10 to 11 months after planting (Kay, 1987). Martin and Degras (1978) reported that in Guadeloupe the annual growth cycle varied between 300 and 340 days, with the best planting material being the small whole tuber, since *D. trifida*, which produce a fairly large number of tubers, can be propagated very easily by reserving a few of the smaller tubers and planting these at the beginning of the next season (Kay, 1987). Large tubers cut into 50–60 g pieces are also used as planting material, and pieces taken from the upper portion of the tuber produced more stems. Solano (1996) reported that in Costa Rica's Atlantic zone, the most important limiting factor of yam production was planting material availability and quality, which they said was due to poor storage conditions and pathogen infection. Storage of tubers for planting was best after pre-harvest spraying with a 0.1% solution of 2,4-D, which maintained 75% of tubers suitable for planting after 3 months as compared to 48% of the non-sprayed tubers. Three storage temperatures (8, 15 or 30°C), three humidity levels (< 60, 76 or 98% RH) and three concentrations of gibberellic acid (0, 100 or 200 mg kg^{-1}), treated for either 12 or 24 hours, were tested on tubers stored for planting. The optimum conditions were found to be 15°C < 60% RH and submersion for 12 hours in 100 mg kg^{-1} gibberellic acid solution.

Yields of *D. trifida* are generally lower than many other currently cultivated *Dioscorea* species mainly because of yam mosaic virus infection. Yields could be as low as 1 to 2 tonnes per hectare. Two years of cultivation of new hybrids in Guadeloupe resulted in yields that varied from 4.6 to 37 tonnes per hectare. Weeding, staking and appropriate soil–water relations, in addition to soil fertility, are the factors that most influence yields. On an experimental basis, yields in Guadeloupe have reached 55 tonnes per hectare, a yield comparable to the best yields of other *Dioscorea* species (Martin and Degras, 1978). During cultivation they are commonly supported with stakes 2–3 m in length.

Harvesting and handling

D. trifida tubers are normally harvested when the foliage begins to die back, which is some 10 or 11 months after planting. Harvesting is by digging using hand tools, but Kay (1987) reported that mechanical harvesting was being developed with cultivars specially bred for this purpose. Physical injury during postharvest handling of yams is common. In a study of handling of *D. trifida* in Jamaica by Thompson (1972b) it was shown that subjecting the tubers to impacts commonly experienced during handling resulted in greatly increased losses during subsequent storage (Table 8.33). The height from which tubers were dropped, of course, affected their deterioration: the greater the height the more rapid the tubers would deteriorate (Table 8.34). Even so, the increased level of deterioration from even the lowest drop (30 cm) indicated that this gave an unacceptable level of injury and tubers must be handled very carefully.

Sprouting

As with other the major *Dioscorea* species, the endogenous factor that limits their maximum storage period is usually sprouting. Martin and Degras (1978) found that sprouting in *D. trifida* could vary considerably

Table 8.33. Effects of dropping tubers of *D. trifida* from a height of 2 m onto a concrete floor on their subsequent losses during storage at about 28°C for 64 days. ⁺Fungal score was 0 = no surface fungal growth, 5 = tuber surface entirely covered with fungi. Necrotic tissue was estimated by cutting the tuber into two lengthways and measuring the area of necrosis and expressing it as a percentage of the total cut surface. Modified from Thompson (1972b).

	% loss in weight	Fungal score⁺	% Internal necrosis	% sprouting
Dropped	47.5	1.5	61	64
Not dropped	23.6	0.2	5	97

Table 8.34. Effects of dropping *D. trifida* tubers from various heights on their deterioration during storage for 86 days in Jamaican ambient conditions. Figures followed by the same letter were not significantly different ($p = 0.05$).

	Height from which tubers were dropped (cm)				
	0	30	60	90	120
Weight loss	29%d	43%c	42%c	55%b	64%a
Fungal score[1]	0.00	0.08	0.16	0.31	0.00
Necrotic tissue	24%	55%	53%	65%	84%

[1]Fungal score: 0 = no fungi, 5 = surface completely covered.

between cultivars. For example, sprouting occurred in 'Cousse-Couche Violette' after about one month and in 'Cousse-Couche Belfort' it was about 50 days. The dormancy period of *D. trifida* from the Caribbean was given as 2 to 4 weeks by Opara (1999) or 4 weeks by Passam (1982). In Jamaica tubers began to sprout within 3 weeks in storage in ambient conditions of 20–29°C and 46–62% RH (Thompson *et al.*, 1973a). Treatment of tubers of two local cultivars in Trinidad immediately after harvest by submergence for 22 hours in water that contained 150 µL L^{-1} GA_3 extended their dormancy period to 20 weeks and decreased tuber weight loss to 17% after 13 weeks storage compared with 26% in the untreated controls (Wickham, 1988).

Storage

Solano (1996) reported that severe losses may occur during storage of *D. trifida* tubers by sprouting and rotting. Kay (1987) reported that tubers can be stored in a cool dry well-ventilated environment, possibly for up to a year, but storage life was normally short under tropical conditions and loss in weight during storage was rapid at over 1% per day. She also reported that treatment with an insecticide and a fungicide was recommended. Tindall (1983) recommended storage of *D. trifida* tubers in barns at 16–18°C and 60–65% RH. Opara (1999) reported that the optimum conditions for storage were 25°C and 60–65% RH for several months. Refrigerated storage recommendations of *D. trifida* tubers were 10°C for several months (Tindall, 1983) and 3°C for 1 month (Kay, 1987; Opara, 1999). This latter recommendation is surprising since yams generally suffer from chilling injury, certainly at 3°C and likely at 10°C. In refrigerated storage trials in Jamaica, Thompson (1972a) found that *D. trifida* tubers stored well at 12°C with about 84% RH, but on removal to ambient temperatures they lost weight rapidly, became soft, developed internal necrosis and had copious surface mould growth, which was concluded to be due to chilling injury. Controlled atmosphere storage of *D. trifida* tubers (no temperature, CO_2 and O_2 levels were specified) was reported to have only a slight to no effect (SeaLand, 1991).

Modified atmosphere packaging

Sealing of *D. trifida* tubers in polyethylene film bags reduced their losses, especially weight loss, but invariably the quantity of fungal mycelium on the tuber surface is increased. In one experiment tubers in polyethylene bags had similar levels of flesh necrosis to those not wrapped or packaged (Table 8.35). However, in another experiment tubers in polyethylene bags had half the internal necrosis (42%) compared with 82% for those not in polyethylene bags when the tubers had been dropped from the height of 2 m before they were stored. In both cases the tubers would be unmarketable.

D. trifoliata

Dioscorea trifoliata Kunth is synonymous with *D. galipanensis*, *D. trifoliata* var. *amazonica*, *D. trifoliata* var. *galipanensis*, *D. triloba*, *Helmia galipanensis* and *H. trifoliate*. It is native to South America including Peru, northern Brazil, Colombia, Venezuela, Panama and Trinidad. Marsden (1812) recorded that *D. trifoliata* was being grown in

Table 8.35. Effects of packaging material on the quality of *D. trifida* after 64 days at 20 to 29°C and 46 to 62% RH. Fungal score was 0 = no surface fungal growth, 5 = tuber surface entirely covered with fungi. Necrotic tissue was estimated by cutting the tuber into two lengthways and measuring the area of necrosis and expressing it as a % of the total cut surface. Modified from Thompson (1972a).

Packaging type	Weight loss %	Fungal score (0 to 5)	% necrotic tissue
Paper bags	24%	0.2	5
Polyethylene bags (0.03 mm thick) with 0.15% of the area as holes	16%	0.2	7
Sealed 0.03 mm thick polyethylene bags with no holes	5%	0.4	4

Malaysia at that time. Tubers are up to 7.5 cm in diameter and borne singly or in clusters that may be fused together; or *trifoliata* may produce a main tuber with several smaller subsidiary tubers in a dense cluster near the soil surface. Tubers are deeply lobed, ovoid to spheroidal in shape, with pale brown skin and the flesh is variable in colour from white to yellow (Kay, 1987; Wilkin, 1999; Botany VM, 2018).

Uphof (1959) reported that *D. trifoliata* tubers have a sweet, floury flavour when cooked. Opara (1999) gave the composition of tubers as: 67 g water, 3.2 g protein, 28 g carbohydrate and 0.8 g fibre per 100 g. The carbohydrate consisted mainly of starch (Coursey, 1967) and was 26.97% for yellow varieties and 21.33% for white varieties (Ojo and Ojo, 2009). Kay (1987) gives a typical analysis of the edible portion of *D. trifoliata* tubers as: water 79%, carbohydrate 17%, protein 2.78%, fat 0.28%, fibre 0.3%, ash 0.72%, calcium 92 mg $100g^{-1}$ and ascorbic acid 6.6 mg $100g^{-1}$. Alozie *et al.* (2009) found that the crude protein content of a wild variety was 11.37 g $l00g^{-1}$ and an 'edible' variety was 7.0 g $100g^{-1}$. Ojo and Ojo (2009) found that the crude protein content was 8.24% for yellow and 5.44% for white varieties. However, after 6 weeks' storage in Nigerian ambient conditions this protein content had fallen to 3.13% for yellow and 2.50% for white. The presence of iron, magnesium, phosphorus, calcium and zinc at high levels indicated that the tubers could have a useful application in the food and pharmaceutical industries. Nimenibo-Uadia (2003) reported that alkaloids present in the tuber extracts provided some biochemical basis for their use in the management of patients with diabetes and confirmed their role as a traditional anti-diabetic remedy.

Marsden (1812) reported that *D. trifoliata* produces 'large but poisonous roots'. Coursey (1967) and Purseglove (1972) reported that tubers of many wild forms of *D. trifoliata* are often very poisonous, but some cultivated varieties have been selected for low toxicity. The toxic varieties have been shown to contain dihydrodiscorine isoclines, a heart stimulant and dioscorcoretine, a hypoglycemic agent (Bevan and Hirst, 1958). The yellow varieties are likely to contain a highly toxic alkaloid, which is a mixture of stereoisomers of dihydrodioscorine, but it can also occur in white fleshed tubers. Alozie *et al.* (2009) reported that tubers of both the edible and wild varieties are processed for food consumption by boiling and, in the case of the wild variety, sliced, tied in jute sacks and left in running water for 3 days in order to remove poisonous and/or bitter compounds. Coursey (1967) previously described the process and reported that tubers were sliced and soaked in frequent changes of water, usually with the addition of common salt. In West Africa, soaked yams were prepared for consumption by flavouring with coconut and sugar after removal of toxins. Other techniques reported from India, Solomon Islands and Madagascar involved boiling whole, mashed or sliced yams before immersing in running water. In parts of India, wood ash or tamarind extract were sometimes added to the water, before boiling, as an aid to 'the removal of acridity' (Karnik, 1969).

D. trilinguis

Dioscorea trilinguis Griseb is synonymous with *D. trilinguis* var. *edwallii* Uline. It is native to southeastern Brazil (Bahia, Rio de Janeiro, São Paulo) and is found growing in dry forests and on clay soil (Contu, 2009). Kirizawa *et al.* (2016) identified eight *Dioscorea* species in Brazil and among them *D. trilinguis* was listed as endangered.

D. undatiloba

Dioscorea undatiloba Baker. Govaerts and Wilkin (2015) listed both *D. undatiloba* and *D. buchananii* as accepted species names, but *D. undatiloba* is the earlier name so *D. buchananii* subsp. *undatiloba* (Baker) Wilkin was reported to be applicable. It is native to South Africa (eastern region, Natal) without a precise locality being provided). Von Teichman *et al.* (1975) surveyed all the species of *Dioscorea* in South Africa and listed *D. undatiloba* among them.

D. variifolia

Dioscorea variifolia Bertero ex Colla is synonymous with *D. acuminata* and D. microphylla. It is native to Madagascar and distributed in dry forests. It is classified as endangered (Royal Botanic Garden Edinburgh, 2019).

D. velutipes

D. velutipes Prain & Burkill is found in northern Thailand (where its common name is man liam), Myanmar as well as Guizhou and Yunnan provinces of China, in forests and on mountain slopes at 500–2,400 m. It was also reported to have been found growing in open areas and in areas of limestone soils in evergreen forests at 1,600 to 1,900 m. Ding and Gilbert (2010) reported that stems twine to the left, but tubers are unknown. Wilkin and Thapyai (2009) also reported that the underground parts were unknown and Sagwansupyakom and Chantraprasong (1984) described *D. velutipes* as 'rare'.

D. versicolor

Dioscorea versicolor Ham. Ex Hook.f. Khare (2010) reported *D. versicolor* to be a synonym of *D. sativa* and *D. bulbifera*. No common name could be found. It is a native of the Kumaon region and the western Himalayas of India (Purdue University, 2019). Bhandari *et al.* (2003) gave the following analysis of *D. bulbifera*, *D. versicolor*, *D. deltoidea* and *D. triphylla* on a fresh weight basis: crude protein 1.6–3.1, ash 0.5–1.2, crude fat 0.2–0.3, crude fibre 0.6–1.5%, K 250–560 mg 100 g^{-1}, Na 4.15–17.8 mg 100 g^{-1}, P 33.1–61.6 mg 100 g^{-1}, Ca 14.3–46.9 mg 100 g^{-1}, Mg 18.3–27.3 mg 100 g^{-1}, Cu 0.10–0.21 mg 100 g^{-1}, Fe 0.39–2.92 mg 100 g^{-1}, Mn 0.14–0.35 mg 100 g^{-1} and Zn 0.22-0.53 mg 100 g^{-1}. In a comparison of four *Dioscorea* species Bhandari and Kawabata (2005) showed that *D. versicolor* had the highest content of cyanogens (6.0 ± 1.4 ppm) compared to *D. bulbifera*, *D. deltoidea* and *D. triphylla* that ranged from 3.2 to 3.3 ppm. When used as a food, the axillary tubers are cut into pieces, steeped in water and boiled. They are also used as food for invalids. The Tripura people of India use extracts from *D. versicolor* tubers as a traditional medicine. Such extracts were found to contain alkaloids, sapanins and tannins and may have effects against cancer (Purdue University, 2019).

D. villosa

Dioscorea villosa L. is synonymous with *D. villosa* L. var. *glabrifolia* (Bartlett) Fernald. Common names include: Atlantic yam, barbasco, China root, colic root, devil's bones, Mexican wild yam, rheumatism root, yamwurzel and yuma. It is native to North America and Coursey (1967) reported that it was specifically native to the eastern part of North

America but was most common in the central and southern regions and wet woodlands from Connecticut to Tennessee and Texas to Minnesota. *D. villosa* is a deciduous perennial herbaceous twiner. Tubers are long, cylindrical in shape, seldom branched, knotted, woody and elongated and grow to up to 12 mm in diameter, with numerous tough, slender roots attached beneath. They are externally pale brown with the surface more or less scaly from the partly detached, thin outer layers (Pengelly and Bennett, 2011). Hedrick (1919) wrote 'N. F. IV. An indigenous, perennial, twining plant, with long, knotty, matted, contorted, ligneous root stocks'.

Composition and uses

In traditional medicine, *D. villosa* tuber extracts are used to treat bilious colic and rheumatism and it has been commonly used by Native Americans in childbirth to help relieve labour pains (Moerman, 1998). Aumsuwan *et al.* (2016) suggested that dioscin extracted from *D. villosa* tubers has potential to be used as an anti-invasive agent in breast cancer. In 1941, Russell Marker discovered that the sapogenins in *D. villosa* tubers could be chemically modified to create natural pregnancy progesterone (Dach, 2013). This cheap source of natural progesterone was used by the pharmaceutical industry to make birth control pills more affordable. Marker's progesterone eventually became the preferred precursor in the industrial preparation of the anti-inflammatory drug cortisone. Important research on sex hormones continued in Mexico, leading to the synthesis of the first useful oral contraceptive in 1951 (ACSIHCL, 1938–1945). Two flavan-3-ol glycosides have been isolated from *D. villosa* (Sautour, *et al.*, 2006) and other constituents identified include phytosterols (sitosterol, stigmasterol, taraxerol), the alkaloid dioscorine, tannins, starch, vitamin B, vitamin C, β-carotene and minerals (Braun and Cohen, 2010). Steroidal saponins are considered to be an important source of anti-fungal agents, and numerous glycosides of diosgenin, including three spirostane saponins found in *D. villosa*, can actively inhibit *Candida albicans* and other human pathogenic yeasts *in vitro* (Sautour *et al.*, 2007).

Cultivation

D. villosa grows best in moist open woodland, thickets and roadsides but it tolerates part shade or high light, from forest to sandy or heavy clay soils and a wide range of pH levels and soil environments. It grows best in well-drained soil (Filyaw, 2006). Cech (2002) reported that *D. villosa* is best cultivated under structures that provide light shade and support for the climbing vines. They are propagated by true seed, basal cuttings and root division, but at least 4 weeks cold stratification is required for the seed to germinate. Root cuttings may produce more than one shoot, which can be cut about 5–7 cm below the shoot keeping fibrous roots attached and planted separately (Cech, 2002). Propagation from tuber division can produce harvestable tubers in 2–3 years (Pengelly and Bennett, 2011).

Harvesting and storage

Plants should be at least 4 years old before harvesting tubers. For optimum concentration of the medically active ingredients, tubers should be harvested after the stems die back. Harvested tubers should be cleaned, washed and graded, and cut into smaller pieces for drying (Cech, 2002; Pengelly and Bennett, 2011). PFAF (2019) stated that tubers should not be stored for longer than a year, since they are 'likely to lose [their] medicinal virtues'.

D. wallichii

Dioscorea wallichii Hook. f. Common names include: cherenga kanda, cheranga, ho, kulu, lao name and tungam sanga. It is distributed from India to China (western Yunnan province) and Peninsula Malaysia, Assam, Bangladesh, Myanmar and Thailand from sea level up to 1,300 m (Ding and Gilbert, 2010). *D. wallichii* was named by Joseph D. Hooker (1817–1911) in honour of Nathaniel Wallich (1786–1854), a surgeon and botanist who was involved in the development of the botanic garden at Calcutta in India. It is found growing in mixed deciduous forests, on mountain slopes and in disturbed areas along roadsides and margins of cultivations (Ding and Gilbert, 2010; Kewscience, 2019). *D. wallichii* bears a single cylindrical shaped tuber that can be up to about 55 or even 100 cm long and 3–6 cm in diameter (Fig. 8.37). Tubers grow deeply in the soil so they can be difficult to harvest. Tubers are white when young, becoming fibrous and pink or yellow towards their narrower base and white and softer towards their broader apex. They are subtended by a small, ovoid to irregular woody crown (Prain and Burkill, 1938; Wilkin and Thapyai, 2009; Ding and Gilbert, 2010; Botany VM, 2018).

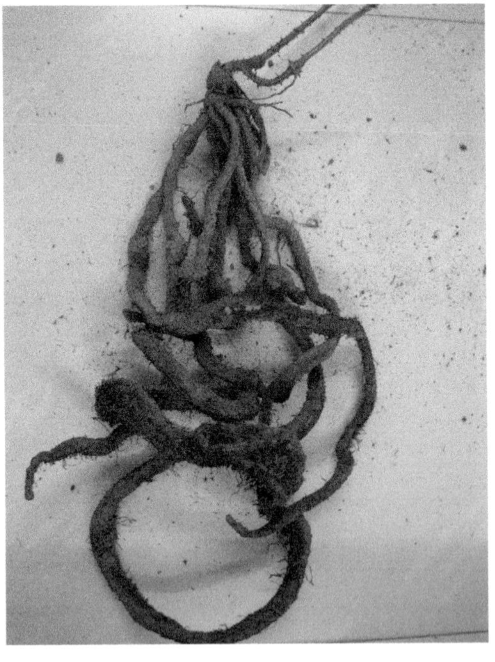

Figure 8.37. *D. wallichii* tubers. Photograph: Dr K. Abraham, CTCRI Emeritus, C-63, Gandhipuram, Sreekariam, Trivandrum 695017, India, reproduced with permission.

Tubers are eaten after thorough washing and cooking in boiling water. Rout and Panda (2010) reported that extracts of *D. wallichii* tubers are used as a medicine against stomach pain in the Mayurbhanj district of Odisha in India. Proximate composition of *D. wallichii* tubers in g per 100 g fresh weight were given as: 2.83 protein, 29.41 carbohydrate, 0.02 fat, 66.52 moisture, 0.79 ash, 0.43 fibre and 129.14kcal energy (Maneenoon *et al.*, 2008). Shajeela *et al.* (2011) gave the composition of *D. wallichi* tubers as: starch 59.30 ± 0.24 g 100 g^{-1}, niacin 52.40 ± 0.37 g 100 g^{-1}, ascorbic acid 88.30 ± 0.29 g 100 g^{-1}, *in vitro* protein digestibility 4.31% and *in vitro* starch digestibility as 49.36% where 1 unit = mg reducing groups1 hr g^{-1} sample. *D. wallichi* also contains antinutritional factors (Table 8.36).

D. xizangensis

Dioscorea xizangensis C.T. Ting is native to China where its common name is zang ci shu yu. Stems twine to the left, its rootstock or tubers are unknown. It is found in broad-leaved forests and on mountain slopes at about 1,200 m in southeastern Xizang in China (Ding and Gilbert, 2010).

D. yunnanensis

Dioscorea yunnanensis Prain & Burkill grows in scrub forests, forest margins and on mountain slopes between 1,000 and 2,800 m in Guizhou and Yunnan provinces of China. Tubers grow vertically in the soil

Table 8.36. Antinutritional factors of tubers of *D. wallichi*. Modified from Shajeela *et al.* (2011).

Component	Level
Total fresh phenolics	0.33 ± 0.02 g 100 g^{-1}
Tannins	0.04 ± 0.02 g 100 g^{-1}
Hydrogen cyanide	0.16 ± 0.05 g 100 g^{-1}
Total oxalate	0.26 ± 0.01 g 100 g^{-1}
Amylase inhibitor	5.27 AIU mg soluble starch^{-1}
Trypsin inhibitor	2.48 ± 0.07 TIU mg protein^{-1}

and are cylindric in shape with white flesh and their stems twine to the left (Ding and Gilbert, 2010). Within this monophyletic clade, four taxa are endemic to southern China (*D. martini*, *D. nitens*, *D. subcalva* and *D. yunnanensis*). These four species were closer to *D. subcalva* var. *submollis* than to *D. subcalva* (Hsu *et al.*, 2013).

D. zingiberensis

Dioscorea zingiberensis C.H. Wright. Its common name is yellow ginger and it grows in eastern Asia, including China and Vietnam, along the banks of rivers and in open forests at elevations around 900–1,000 m. The stems twine to the left and each plant produces one large tuber (Ding and Gilbert, 2010). Zhang *et al.* (2018) reported that more than 70 compounds have been identified in tuber extracts from *D. zingiberensis*, which allowed a wide spectrum of medical biological potential uses, including cardiovascular, anti-thrombosis, hyperlipidemia, neuroprotection, anti-inflammatory and anthelmintic treatments. It is cultivated in China for the production of diosgenin. Yi *et al.* (2014) sampled yams from different parts of China and measured their diosgenin content, which varied from 8.67 ± 0.28 to 19.52 ± 0.51 mg g^{-1}. They concluded that such variations may presumably be attributed to 'differences in plant resources'. Huang *et al.* (2008b) developed an 'eco-friendly process' for utilization of *D. zingiberensis* tubers, which recovered 95% of their diosgenin. Li *et al.* (2010) found the yield of diosgenin obtained from the fermentation of *D. zingibernsis* by *Trichoderma harzianum* was higher than that typically obtained from acid hydrolysis. They tested a three-liquid-phase extraction system, which successfully resulted in the simultaneous separation of diosgenin, untransformed steroidal saponins and glucose. Chen *et al.* (2018) developed an environmental and high-efficiency microbial technology for diosgenin production by fermentation with *Aspergillus awamori* that gave diosgenin purity of $96.9 \pm 2.42\%$.

Chen *et al.* (2003b) developed a protocol for rapid *in vitro* propagation of *D. zingiberensis*, which resulted in more than 85% of the regenerated plantlets surviving and growing vigorously one month after they were transplanted. They found that each plant formed 2 to 4 micro-tubers three months after transplanting. Huang *et al.* (2008a) developed tetraploid lines of *D. zingiberensis* for further selection and breeding into new genotypes with improved production of diosgenin.

References

Abano, E.E. and Amoah, R.S. (2015) Microwave and blanch-assisted drying of white yam (*Dioscorea rotundata*). *Food Science and Nutrition* 3, 586–596.

Abdu, F. (2018) Tanzania: Plant More Yams for Food Security. *Tanzania Daily News (Dar es Salaam)* 28 November 2018, https://allafrica.com/stories/201811280703.html (accessed 16 June 2019).

Abe, M.O. (1973) Adaptability of potato drying to yam processing. *Journal of Milk and Food Technology* 36, 456–462.

Abhyankar, R.K. and Upadhyay, R. (2011) Ethno-medicinal studies of tubers of Hoshangabad, M.P. *Bulletin of Environment, Pharmacology and Life Sciences* 1, 57–59.

Aboagye-Nuamah, F., Offei, S.K., Cornelius, E.W. and Bancroft, R.D. (2005) Severity of spoilage storage rots of white yam (*Dioscorea rotundata* Poir.). *Annals of Applied Biology* 147, 183–190.

Abraham, K. (2002) Genetic potential of sexually propagated seedlings of greater yam, *Dioscorea alata* L. *Journal of Root Crops* 28, 21–25.

Abubakar, S.M. and Gana, F.A. (2017) A comparative assessment of nutrients and anti-nutrients contents of varieties of fried yam (*Dioscorea* specie) commonly consumed in Nigeria. *Bayero Journal of Pure and Applied Sciences* 10, 25–28.

Acevedo-Rodríguez, P. and Strong, M.T. (2005) Monocots and gymnosperms of Puerto Rico and the Virgin Islands. *Contributions from the US National Herbarium* 52, 1–416.

Acevedo-Rodríguez, P. and Strong, M.T. (2012) Catalogue of the seed plants of the West Indies. *Smithsonian Contributions to Botany* 98, 1–1192. Smithsonian Institution, Washington, D.C., USA.

Acha, I.A., Shiwachi, H., Asiedu, R. and Akoroda, M.O. (2004) Effect of auxins on root development in yam (*Dioscorea rotundata*) vine. *Tropical Science* 44, 80–84.

Achebe, C. (1958) *Things Fall Apart*. William Heinemann Ltd.

Achy, J.Y., Koffi, P.K.B., Ekissi, G.S.E., Konan, H.K. and Kouame. L.P. (2016) Assessment of physico-chemical properties and anti-nutritional factors of flour from yam (*Dioscorea bulbifera*) bulbils in Southeast Côte d'Ivoire. *International Journal* of *Advanced Research* 4, 871–887.

Acosta, N. and Ayala, A. (1976) Effects of *Pratylenchus coffeae* and *Scutellonema bradys* alone and in combination on Guinea yam (*Dioscorea rotundata*). *Journal of Nernatology* 8, 315–317.

Acquah, E.T. and Evange, N.W. (1994) The economics of yam (*Dioscorea* species) production in Cameroon: the case of Fako. *Acta Horticulturae* 380, 373–377.

ACSIHCL (1938–1945) American Chemical Society International Historic Chemical Landmarks. The 'Marker Degradation' and Creation of the Mexican Steroid Hormone Industry 1938–1945, http://www.acs.org/content/acs/en/education/whatischemistry/landmarks/progesteronesynthesis.html (accessed 12 October 2018).

Adams, C.D. (1972) *Flowering Plants of Jamaica*. University of the West Indies, Mona, Jamaica.

Adams, Z.S., Wireko Manu, F.D., Agbenorhevi, J. and Oduro, I. (2019) Improved yam-baobab-tamarind flour blends: Its potential use in extrusion cooking. *Scientific African* 6, e00126, https://doi.org/10.1016/j.sciaf.2019.e00126 (accessed 30 March 2021).

Adebowale, A-R., Wahab, A.B., Sobukola, P.O., Adewale, O., Obadina, E.O., Adegunwa, M.O., Sanni, L.O. and Tomlins, K. (2018) The antinutritional and vitamin composition of high-quality yam flour as affected by yam species, pretreatment and drying method. *Food Science and Nutrition* 6, 1985–1990.

Adejo, P.E. (2017) Post-harvest management practices of yam and farmers' information needs in the North-Central of Nigeria. *Journal of Nutraceuticals and Food Science* 2, 9.

Adelberg, J.W. and Simpson, E.P. (2002) Intermittent immersion vessel apparatus and process for plant propagation. In: Aitken-Christie, J., Kozai, T. and Smith M.A.L. (eds) *Automation and environmental control in plant tissue culture*. Kluwer Academic Publishers, The Netherlands, pp. 19–26.

Adeniji, M.O. (1970) Fungi associated with storage decay of yam in Nigeria. *Phytopathology* 60, 590–592.

Adepoju, O.T., Boyejo, O. and Adeniji, P.O. (2018) Effects of processing methods on nutrient and antinutrient composition of yellow yam (*Dioscorea cayenensis*) products. *Food Chemistry* 238, 160–165.

Adesiyan, S.O. and Badra, T. (1982) Granular nematicides for control of the yam nematode *Scutellonema bradys*, and relevant residues in raw tubers. *Journal of Nematology* 12, 213–216.

Adesiyan, S.O. and Odihirin, R.A. (1978) Root knot nematodes as pests of yams (*Dioscorea* species) in southern Nigeria. *Nematologica* 24, 132–134.

Adesiyan, S.O., Odihirin, R.A. and Adeniji, M.O. (1975a) Economic losses caused by the yam nematode, *Scutellonema bradys* in Nigeria. *Plant Disease Reporter* 59, 477–480.

Adesiyan, S.O., Odihirin, R.A. and Adeniji, M.O. (1975b) Histopathology studies of the yam tuber (*Dioscorea rotundata*) infected with *Scutellonema bradys* (Steiner and Le Hew). *International Biodeterioration Bulletin* 11, 48–55.

Adesuyi, N.O. (1973) Advances in yam storage research in Nigeria. *Proceedings of the Third International Symposium on Tropical Root Crops*, Ibadan, Nigeria, 428–433.

Afoakwa, E.O. and Sefa-Dedeh, S. (2001) Chemical composition and quality changes in trifoliate yam *Dioscorea dumetorum* tubers after harvest. *Food Chemistry* 75, 85–91.

Afoakwa, E.O. and Sefa-Dedeh, S. (2002) Changes in cell wall constituents and mechanical properties during post-harvest hardening of *Dioscorea dumetorum* tubers. *Food Research International* 35, 429–434.

Agbaje, G.O., Adegbite, A.A. and Akinlosotu, T.A. (2003) Performance of new hybrid yam (*D. rotundata* Poir) varieties in the Forest Zone of Nigeria. *Tropicultura* 21, 149–152.

Agbede, T.M. and Olasekan, A.A. (2012) Soil properties and yam yield under different tillage systems in a tropical Alfisol. *Archives of Agronomy and Soil Science* 59, 1–15.

Agbo-Egbe, T. and Rickard, J.E. (1991) Study on the factors affecting storage of edible aroids. *Annals of Applied Biology* 119, 121–130.

Agbo-Egbe, T. and Trèche, S. (1995) Evaluation of the chemical composition of Cameroonian yam germplasm. *Journal of Food Composition and Analysis* 8, 274–283.

Agele, S.O., Ayankanmi, T.G. and Kikuno, H. (2010) Effects of synthetic hormone substitutes and genotypes on rooting and mini tuber production of vines cuttings obtained from white yam (*Dioscorea rotundata*, Poir). *African Journal of Biotechnology* 9, 4714–4724.

Ahmad, S.S. and Javed, S. (2007) Exploring the economic value of underutilized plant species in Ayubia National Park. *Pakistan Journal of Botany* 39, 1435–1442.

Aighewi, B.A. and Ekanayake, I.J. (2004) *In-situ* chlorophyll fluorescence and related growth of Guinea yam at different ages. *Tropical Science* 44, 201–206.

Aighewi, B.A. Asiedu, R. Maroya, N. and Balogun, M. (2015) Improved propagation methods to raise the productivity of yam (*Dioscorea rotundata* Poir.). *Food Security* 7, 823–834.

Akinmutimi, A.H. and Onen, G.E. (2008) The response of broiler finisher birds fed graded levels of yam peel meal in place of maize-based diets. *International Journal of Poultry Science* 7, 474–479.

Akissoé, N., Hounhouigan, D.J., Bricas, N., Vernier, P., Nago, C.M. and Olorunda, O.A. (2001) Physical, chemical and sensory evaluation of dried yam (*Dioscorea rotundata*) tubers, flour and amala, a flour-derived product. *Tropical Science* 41, 151–155.

Akom, M., Oti-Boateng, C., Otoo, E. and Dawoe, E. (2015) Effect of Biochar and inorganic fertilizer in yam (*Dioscorea rotundata* Poir.) production in a Forest Agroecological Zone. *Journal of Agricultural Science*, 7, 211, http://dx.doi.org/10.5539/jas.v7n3p211 (accessed 22 May 2018).

Akoroda, M.O. (1983) Variation, heritability and genetic advance of eight characters in white yam. *Theoretical and Applied Genetics* 66, 51–54.

Akoroda, M.O. (1984) Variability, repeatability, character correlation and path coefficient analyses in yellow yam. *Theoretical and Applied Genetics* 69, 217–221, https://doi.org/10.1007/BF00272898 (accessed 1 April 2021)

Akoroda, M.O. and Okonmah, L.U. (1982) Sett production and germplasm maintenance through vine cuttings in yams. *Tropical Agriculture* 59, 311–314.

Alexis, S.D. and Amani, N.G. (2013) Hydrolysis of native starch of some wild yam (*Dioscorea* species) of Côte d'Ivoire *Agriculture Science Developments* 2, 56–58.

Alinnor, I.J. and Akalezi, C.O. (2010) Proximate and mineral compositions of *Dioscorea rotundata* (White Yam) and *Colocasia esculenta* (White Cocoyam). *Pakistan Journal of Nutrition* 8, 998–1001.

Alozie, Y. Akpanabiatu, M.I. Eyong, E.U., Umoh, I.B. and. Alozie, G. (2009) Amino acid composition of *Dioscorea dumetorum* varieties. *Pakistan Journal of Nutrition* 8, 103–105.

Alves, R.M., Grossmann, M.V., Ferrero, C., Zaritzky, N.E., Martino, M.N. and Sierakoski, M. R. (2002) Chemical and functional characterization of products obtained from yam tubers. *Starch-Stärke* 54, 476–481.

Amadioha, A.C., Markson, A.A. (2007) Postharvest control of tuber rot by *Botryodiplodia acerina* using extracts of plant origin. *Archives of Phytopathology and Protection* 40, 359–366.

Amandikwa, C., Iwe, M.O., Uzomah, A. and Olawuni, A.I. (2015) Physico-chemical properties of wheat-yam flour composite bread. *Nigerian Food Journal* 33, 12–17.

Amani, N.G., Buléon, A. and Kamenan, A. (2004) Variability in starch physicochemical and functional properties of yam (*Dioscorea* sp.) cultivated in Côte d'Ivoire. *Journal of the Science of Food and Agriculture*, 84, 2085–2096.

Amponsah, S.K, Rees, D., Abayomi, L., Akowuah, J. O., Bortey, H.M., Bessah, E. and Gyening, K. (2007) Storability of Puna as influenced by sprout control method and stage of harvested yam. *Journal of Root Crops* 43, 10–20.

Amponsah, S.K., Akowuah, J O., Adu-Kwarteng, E. and Bessah, E. (2015) Design and construction of improved yam storage structure using locally-available materials. *International Journal of Research in Agriculture and Forestry* 2, 1–11.

Amusa, N.A. (1997) Fungi associated with anthracnose symptoms of yam (*Dioscorea* species) in south-west Nigeria and their roles in. *Crop Research* 13, 177–183.

Amusa, N.A. and Baiyewu, R.A. (1999) Storage and market disease of yam tubers in southwestern Nigeria. *Ogun Journal of Agricultural Research (Nigeria)* 11, 211–225.

Amusa, N.A., Adegbite, A.A, Muhammed, S. and Baiyewu, R.A. (2003) Yam diseases and its management in Nigeria. *African Journal of Biotechnology* 2, 497–502,

Andrade, R.D., Palacio, J.C., Pacheco, W.A. and Betin, R.A. (2012) Modified atmosphere packaging of chunks yam (*Dioscorea rotundata* Poir.). *Information Technology* 23, 65–72.

Andriamparany, J. and Buerkert, A. (2019) Effect of ultrasonic dormancy breaking on seed germination and seedling growth of three wild yam species (*Dioscorea* species) from SW-Madagascar. *Genetic Resources and Crop Evolution* 66, 1167–1174.

Andriamparany, J.N., Brinkmann, K. Jeannoda, V. and Buerkert, A. (2014) Effects of socio-economic household characteristics on traditional knowledge and usage of wild yams and medicinal plants in the Mahafaly region of south-western Madagascar. *Journal of Ethnobiology and Ethnomedicine* 10, 82–102.

Andriamparany, J.N., Brinkmann, K., Wiehle, M., Jeannoda, V. and Buerkert, A. (2015) Modelling the distribution of four *Dioscorea* species on the Mahafaly Plateau of south-western Madagascar using biotic and abiotic variables *Agriculture, Ecosystems and Environment* 212, 38–48.

Ministry of Agriculture, Animal Industry and Fisheries, Entebbe (1996) Annual Report, 1980–1996, Entebbe, Uganda.

Anonymous (2015) Medicinal Plants. http://medicinalplants.us/dioscorea (accessed 28 December 2019).

Anonymous (2017) The history of the birth control pill, part 2: Barbasco and the roots of hormonal contraception, http://advocatesaz.org/2013/02/12/the-history-of-the-birth-control-pill-part-2-barbasco-and-the-roots-of-hormonal-contraception/dioscorea-composita-2/ (accessed 28 December 2019). *D. kamoonensis*. http://www.flowersofindia.net/catalog/slides/Kumaon%20Yam.html (accessed 26 April 2018).

Arackal, A.J. and Pandurangan, A.G. (2015) Taxonomy and conservation of the genus *Dioscorea* in Western Ghats. *Journal of Root Crops* 41, 3–16.

Araki, H., Harada, T. and Yakuwa, T. (1983) Some characteristics of interspecific hybrids between *Dioscorea japonica* Thunb. and *Dioscorea opposita* Thunb. *Journal of the Japanese Society for Horticultural Science* 52, 153–158.

Araújo, D. and Alves, M. (2010) Climbing plants of a fragmented area of lowland Atlantic forest, Igarassu, Pernambuco (northeastern Brazil). *Phytotaxa* 8, 1–24.

Archibald, E.E.A. (1967) The genus *Dioscorea* in the Cape Province west of East London. *Journal of South African Botany*, 33, 1–46.

Ardener (1954) quoted by: Blench, R and Dendo, M. (2004) Lesser-known African tuber crops and their role in prehistory. Work in progress. http://www.rogerblench.info (accessed 7 May 2021).

Aregahegn, A., Chandravanshi, B.S. and Atlabachew, M. (2013) Levels of major, minor and toxic metals in tubers and flour of *Dioscorea abyssinica* grown in Ethiopia. *African Journal of Food, Agriculture, Nutrition and Development* 13, 7870–7887.

Arinathan, V., Mohan, V.R. and Maruthupandian, A. (2009) Nutritional and anti-nutritional attributes of some under-utilized tubers. *Tropical and Subtropical Agroecosystems* 10, 273–278.

Arnau, G., Bhattacharjee, R., Hana Chair, S.H.N., Malapa, R., Xavier Perrier, A.K., Petro, D., Penet, L. and Pavis, C. (2017) Understanding the genetic diversity and population structure of yam (*Dioscorea alata* L.) using microsatellite markers. *PLoS ONE* 12(3), e0174150. https://doi.org/10.1371/journal.pone.0174150 (accessed 14 July 2019).

Arnold Babwah and Associates (2016) Market study for fresh produce St Lucia, Grenada, Dominica and St Vincent and the Grenadines WUSC Caribbean commissioned report, https://agricarib.org/images/docs/OECS_Market_Study_FINAL_FOR_EXTERNAL_DISTRIBUTION.pdf (accessed 4 November 2019).

Arnolin, R. and Lombion, K. (2005) L'igname cousse-couche *Dioscorea trifida* et les virus: en Guadeloupe, recherche de résistance aux virus sur hybrides d'igname cousse-couche, en génération séminale. *Phytoma* 584, 33–35.

Artschwager, E.F. (1927) Wound periderm formation in the potato as affected by temperature and humidity. *Journal of Agricultural Research* 35, 995–1000.

Artschwager, E.F. and Starrett, R.C. (1931) Suberization and wound periderm formation in sweetpotato and gladiolus, as affected by temperature and humidity. *Journal of Agricultural Research* 43, 353–364.

Arvigo, R. and Balick, M. (1994) *Die Medizin des Regenwaldes: Heilkraft der Maya-Medizin-Die 100 heilenden Kräuter von Belize*. Windpferd, Aitrang, Germany.

Asadu, C.L.A., Egbobe, O.B. and Asiedu, R. (2008) Storage life of eleven cultivars of white yam (*Dioscorea rotundata*) grown with or without NPK fertilizer and the relationship with residual soil NPK. *Agro-Science* 7, 235–241.

Asare-Bediako, E., Showemimo, F.A. and Opoku-Asiama, Y. (2007) Microorganisms associated with rot of minisetts of white yam (*Dioscorea rotundata* Poir.). *Research Journal of Microbiology* 2, 278–283.

Asha, K.I. and Nair, G.M. (2005) Screening of *Dioscorea* species for diosgenin from southern western Ghats of India. *Indian Journal of Plant Genetic Resources* 18, 227–230.

Ashamo, M.O. (2010) Management of the yam moth, *Dasyses rugosella* Stainton, a pest of stored yam tubers (*Dioscorea* spp.) using plant products. 10th International Working Conference on Stored Product Protection Julius Kühn Archiv 425, 793–798.

Asian Plants (2019) http://www.asianplant.net/Dioscoreaceae/Dioscorea_cirrhosa.htm (accessed 19 July 2019).

Asiedu, R. and Sartie, A. (2010) Crops that feed the world 1. Yams: yams for income and food security. *Food Security* 2, 305–315.

Asiedu, R., Wanyera, N.M., Nag, S.Y.C. and Nag, N.Q. (1997) Yams, CGIAR Centres. Cambridge University Press, Cambridge.

ASTI (2019) Agricultural Science and Technology Indicators, https://www.asti.cgiar.org/diiva/cote-divoire/yams (accessed 1 October 2019).

Asumugha, G.N, Njoku, M.E., Okoye, B.C., Aniedu, O.C., Ogbonna, M.C., Anyaegbunam, H.N., Akinpelu, O.A., Ibeagi, O. and Amaefula, A. (2009) An analysis of the supply for seed yams in Nigeria. *African Journal of Business Management* 3, 28–31.

Atu, U., Odurukwe, S. and Ogbuji, R. (1983) Root knot nematode damage to *Dioscorea rotundata*. *Plant Disease* 67, 814–815.

Aumsuwan, P., Khan, S.I., Khan, I.A., Ali, Z., Avula, B., Walker, L.A., Shariat-Madar, Z., Helferich, W.G., Katzenellenbogen, B.S. and Dasmahapatra, A.K. (2016) The anticancer potential of steroidal saponin, dioscin, isolated from wild yam (*Dioscorea villosa*) root extract in invasive human breast cancer cell line MDA-MB-231 *in vitro*. *Archives of Biochemistry and Biophysics* 591, 98–110.

Awuah, R.T and Akrasi, K.O. (2007) Suppression of tuber rot of yam caused by *Aspergillus niger* with a yam *Rhizobacterium*. *African Crop Science Conference Proceedings* 8, 875–879.

Ayala, A. and Acosta, N. (1971) Observations on yam (*Dioscorea alata*) nematodes. *Nematropica* 1, 39–40.

Ayensu, E.S. (1972) Dioscoreales. In: Metcalfe C.R. (ed.) *Anatomy of the monocotyledons*. Oxford University Press, Oxford, pp. 1–182.

Ayensu, E.S. and Coursey, D.G. (1972) Guinea yams. The botany, ethnobotany, use and possible future of yams in West Africa. *Economic Botany* 26, 301–318.

Ayodeji, S.P., Akinnuli, B.O. and Olabanji, O.M. (2014) Development of yam peeling and slicing machine for a yam processing plant. *Journal of Machinery Manufacturing and Automation* 3, 74–83.

Baah, F.D., Maziya-Dixon, B., Asiedu, R., Oduro, I. and Ellis, W.O. (2009) Nutritional and biochemical composition of *D. alata* (*Dioscorea* species) tubers. *Journal of Food, Agriculture and Environment* 7, 373–378.

Bachman, S. (2011) *Dioscorea buchananii* subsp. *undatiloba*, http://www.iucnredlist.org/details/100889960/0 (accessed 8 February 2018).

Badmus, A. A., Raji, A. O. and Akinoso, R. (2013) Effect of process parameters on work index, milling efficiency and some technological properties of yam flour using attrition mill. *Food and Bioprocess Technology* 6, 160–168.

Bahuchet, S. (1988) Food supply uncertainty among the Aka Pygmies (Lobaye, Central African Republic). In: Garine. I. de and Harrison, G.A. (eds) *Coping with Uncertainty in Food Supply*. Clarendon Press, Oxford, pp. 118–149.

Bahuchet, S., McKey, D. and De Garine, I. (1991) Wild yams revisited: Is independence from agriculture possible for rain forest hunter-gatherers? *Human Ecology* 19, 213–243.

Bailey, L.H. (1949) *Manual of Cultivated Plants*. London, Macmillan.

Baimey, H., Coyne, D. and Labuschagne, N. (2006) Effect of fertiliser application on yam nematode (*Scutellonema bradys*) multiplication and consequent damage to yam (*Dioscorea* species) under field and storage conditions in Benin. *International Journal of Pest Management* 52, 63–70.

Baker, E.A., Martin, J.T. and Wilson, A.P. (1966) The distribution of diosgenin in *Dioscorea* species. *Annals of Applied Biology* 58, 203–211.

Balogun, M.O. (2009) Microtubers in yam germplasm conservation and propagation: the status, the prospects and the constraints. *Biotechnology and Molecular Biology Reviews* 4, 1–10.

Balogun, M.O. and Gueye, B. (2013) Status and prospects of biotechnology applications to conservation, propagation and genetic improvement of yam. In: Ramawat, K.G. and Merillon, J-M. (eds) *Bulbous Plants: Biotechnology*. CRC Press, pp. 92–112.

Balogun, M.O., Fawole, I., Ng, S.Y.C., Ng, N.Q., Shiwachi, H. and Kikuno, H. (2006) Interactions among cultural factors in microtuberization of white yam (*Dioscorea rotundata*). *Tropical Science* 46, 55–59.

Balogun, M.O., Maroya, N. and Asiedu, R. (2014) Status and prospects for improving yam seed systems using temporary immersion bioreactors. *African Journal of Biotechnology* 13, 1614–1622.

Bancroft, R., Gray, A., Gallat, S., Crentsil, D. and Westby, A. (1998) The marketing system for fresh yam in Techiman, Ghana, and associated postharvest losses. *Tropical Agriculture* 75, 115–119.

Bancroft, R.D. (2000) Relieving postharvest constraints and identifying opportunities for improving the marketing of fresh yam in Ghana, Crop Postharvest Programme. Final Technical Report, No: R6505.

Bancroft, R.D. Crentsil, D., Panni, J.Y., Aboagye-Nuamah, F. and Krampa L. (2005) The influence of postharvest conditioning and storage protocols on the incidence of rots in white yams (*Dioscorea rotundata* Poir.) in Ghana. *Acta Horticulturae* 682, 2183–2190.

Baquar, S.R. (1980) Chromosome behaviour in Nigerian yams (*Dioscorea*). *Genetica* 54, 1–9.

Barlagne, C., Cornet, D., Blazy, J-M., Diman, J-L. and Ozier-Lafontaine, H. (2017) Consumers' preferences for fresh yam: a focus group study. *Food Science* and *Nutrition* 5, 54–66.

Barrau, J. (1956) Les ignames alimentaires des iles du Pacifique sud. *Journal d'Agriculture Tropicale et de Botanique Appliquée* 3, 385–401.

Bassey, E.E. (2017) Constraints and prospects of yam production in Nigeria. *European Journal of Physical and Agricultural Sciences* 5, 55–64.

Beckford, C., Campbell, D. and Barker, D. (2011) Sustainable food production systems and food security: economic and environmental imperatives in yam cultivation in Trelawny, Jamaica. *Sustainability* 3, 541–561.

Been, B.O., Marriott, J. and Perkins, C. (1975) Wound periderm formation in dasheen and its effects on storage. Proceedings of a Meeting of the Caribbean Food Crop Society held in Trinidad, 13.

Been, B.O., Thompson, A.K. and Perkins, C. (1974) Effects of curing on stored yams. *Proceedings of the Caribbean Food Crops Society* 12, 38–42.

Been, B.O., Perkins, C. and Thompson, A.K. (1976) Yam curing for storage. *Acta Horticulturae* 62, 311–316.

Behera, K.K., Sahoo, S. and Prusti, A. (2009) Relative agronomic performance of different *Dioscorea* species found in different parts of Orissa. *Nature and Science* 7, 23–35.

Belachew Garedew, Bewuketu Haile and Aklilu Ayiza (2017) Distribution, diversity and potential production of yams (*Dioscorea* species) in Sheko District, Southwest Ethiopia. *American Journal of Life Sciences* 5, 86–92.

Ben Yehoshua, S., Kim, J.J. and Shapiro, B. (1989) Elicitation of resistance to the development of decay in sealed citrus fruit by curing. *Acta Horticulturae* 258, 623–630.

Bencini, M.C. (1991) Post-harvest and processing technologies of African staple foods: a technical compendium. FAO Agricultural Service Bulletin 89. FAO, Rome.

Bergh, K., Orozco, P., Gugerty, M.K. and Anderson, C.L. (2012) Yam Value Chain: Nigeria EPAR Brief No. 207, https://evans.uw.edu/sites/default/files/EPAR_UW_Request_207_Yam_Value_Chain_Nigeria_Final.pdf (accessed 5 April 2020).

Berisio, S. (2010) Sierra Leone – from earth to table. FAO, project GTF/RAF/426/ITA.

Bevacqua, R.F. (1994) Origin of horticulture in Southeast Asia and the dispersal of domesticated plants to the Pacific Islands by Polynesian voyagers: The Hawaiian Islands case study. *Hortscience* 29, 1226–1229.

Bevan, C.W.L. and Hirst, J. (1958) A convulsant alkaloid of *Dioscorea dumetorum*. *Chemistry and Industry* 4, 103.

Bevan, C.W.L., Broadbent, J.L. and Hirst, J. (1956) Convulsant alkaloid of *Dioscorea dumetorum*. *Nature* 177, 935.

Bhandari, M.J., Kasai, T. and Kawabata, J. (2003) Nutritional evaluation of wild yam (*Dioscorea* species) tubers of Nepal. *Food Chemistry* 82, 619–623.

Bhandari, M.R. and Kawabata, J. (2005) Bitterness and toxicity in wild yam (*Dioscorea* species) tubers of Nepal. *Plant Foods for Human Nutrition* 60, 129–135.

Bilgili, A.V. (2013) Spatial assessment of soil salinity in the Harran Plain using multiple kriging techniques. *Environmental Monitoring and Assessment* 185, 777–795.

Birkill, I.H. (1935) *A dictionary of economic products of the Malaya peninsula*, 2 volumes. Crown Agents for the Colonies, London.

Blench, R. (2007) New palaeozoogeographical evidence for the settlement of Madagascar. *Azania* 42, 6982.

Blench, R. (2010) Evidence for the Austronesian Voyages in the Indian Ocean. In: Anderson, A., Barrett, J.H. Boyle, K.V. (eds) *The Global Origins and Development of Seafaring*. McDonald Institute for Archaeological Research. pp. 239–248.

Blench, R. and Dendo, M. (2004) Lesser-known African tuber crops and their role in prehistory. www.rogerblench.info/Ethnoscience/Plants/Crops/Tubers/Little-known African tubers.pdf (accessed 30 March 2021).

Blunden, G., Hardman, R. and Trease, G.E. (1963) The anatomy of *Dioscorea belizensis* Lundell. *Journal of Pharmacy and Pharmacology* 15, 394–405.

Boakye, I. and Essuman, E.K. (2016) Effects of storage conditions and time on the physicochemical properties of *Dioscorea rotundata*. *International Journal of Research in Agricultural Sciences* 3, 273–278.

Boateng, M.A., Wireko-Manu, F.D., Agbenorhevi, J.K. and Amoako-Andoh, F.O. (2019) Product development from yam in West Africa: market potential and challenges. In *Roots and tubers in Ghana: overview and selected research papers*. Knust Printing Press.

Bömer, M., Rathnayake, A.I., Visendi, P., Sewe, S.O., Sicat, J.P.A., Silva, G., Kumar, P.L. and Seal, S.E. (2019) Tissue culture and next-generation sequencing: A combined approach for detecting yam (*Dioscorea* spp.) viruses. *Physiological and Molecular Plant Pathology* 105, 54–66.

Boorsma (1894) Meded, P1. Tuin, Batavia, 13. Quoted by Bevan *et al.*, 1956.

Booth, R.H. (1976) Storage of fresh cassava (*Manihot esculenta* Crantz). I. Post-harvest deterioration and its control. *Experimental Agriculture* 12, 103–111.

Bornstein, A.J. (1989) Taxonomic studies in the Piperaceae I. The pedicellate pipers of Mexico and Central America (Piper Subg. Arctottonia). *Journal of the Arnold Arboretum* 70, 1–55.

Bosrotsi, C.K., Addo, A., Dzisi, K.A and Agodzo, S.K. (2017a) Technical performance and economic feasibility assessment of a mechanical yam harvester. *International Journal of Research Studies in Agricultural Sciences* 3, 1–10.

Bosrotsi, C.K., Addo, A., Dzisi, K.A. and Agodzo, S.K. (2017b) Development of a yam harvester using finite element method. *International Journal of Modern Studies in Mechanical Engineering* 3, 35–44.

Botany VM (2018) *D. pierrei*. http://www.botanyvn.com/cnt.asp?param=edirandv=Dioscorea%20pierreiandlist=species (accessed 21 June 2018).

Botondi, R., Vailati, M., Bellincontro, A., Massantini, R., Forniti, R. and Mencarelli. F. (2009) Technological parameters of water curing affect postharvest physiology and storage of marrons (*Castanea sativa* Mill., Marrone Fiorentino). *Postharvest Biology and Technology* 51, 97–103.

Bourke, R.M. and Harwood, T. (2009) *Food and Agriculture in Papua New Guinea*. The Australian National University Press.

Bourke, R.M. and Vlassak, V. (2004) *Estimates of food crop production in Papua New Guinea*. Land Management Group, Research School of Asia and the Pacific. The Australian National University, Canberra.

Bourret, D. (1973) Etude ethnobotanique des Dioscoreacées alimentaires. Ignames de Nouvelle-Calédonie. PhD. dissertation, Paris University, France (quoted by Lebot, 2009).

Bousalem, M. and Loubet, S. (2008) Molecular evidence for a new potyvirus species in yam (*Dioscorea* species) on the island of Guadeloupe. *Plant Pathology* 57, 389–389.

Bradbury, J.H. and Holloway, W.D. (1988) *Chemistry of Tropical Root Crops: Significance for Nutrition and Agriculture in the Pacific*. ACIAR Monograph No. 6, Canberra Australia.

Braun, L. and Cohen, M. (2010) *Herbs and natural supplements*. Sydney: Churchill Livingstone.

Braun, U. (1993) New genera of phytopathogenic Deuteromycetes. *Cryptogamic Botany* 4, 107–114.

Bressan, E., Veasey, E.A., Peroni, N., Felipim, A. and Pacheco dos Santos, K.M. (2007) Collecting yam (*Dioscorea* spp.) and sweet potato (*Ipomoea batatas*) germplasm in traditional agriculture small-holdings in the Vale do Ribeira, São Paulo, Brazil. *Plant Genetic Resources Newsletter* 144, 8–13.

Bridge, J. (1973) Nematodes as pests of yams in Nigeria. *Gent Medelelingen Fakulteit Landbouwwetenschapp* 38, 841–852.

Bridge, J. (1982) Nematodes of yams. In: Miège, J. and Lyonga, S.N., (eds) *Yams*. Oxford University Press, Oxford, pp. 253–264.

Bridge, J., Coyne, D. and Kwoseh, C.K. (2005) Nematode Parasites on Root and Tuber Crops. In: Luc, M., Sikora, R.A. and Bridge, J. (eds) *Plant Parasitic Nematodes in Subtropical and Tropical Agriculture*, 2nd edn. CAB International, Wallingford, UK, pp. 221–258.

Broadbent, J.L. and Schnieden, H. (1958) A comparison of some pharmacological properties of dioscorine and dioscine. *British Journal of Pharmacology* 13, 213–215.

Buckmire, K.U. Benjamin, R.L. and Burris, N. (1986) Root and tuber crops in Grenada. In: Regional Workshop on Root Crops Production and Research in the Caribbean (1985, Guadeloupe). Root crops production and research in the Caribbean: Proceedings. Centro Internacional de Agricultura Tropical (CIAT), Cali, Colombia 51–57.

Burkil, H.M. (1985–2004) *The Useful Plants of West Tropical Africa*. Royal Botanic Gardens, Kew.

Burkill, I.H. (1911) The polarity of the bulbils of *Dioscorea bulbifera* Linn. *Journal of the Asiatic Society of Bengal* (n.s.) 7, 467–469.

Burkill, I.H. (1935) *Dioscorea hispida. A dictionary of the economic products of the Malay peninsula*, Vol. I (A-H). Crown Agents for the Colonies, London, pp. 818–821.

Burkill, I.H. (1951) Dioscoreaceae. *Dioscorea hispida*. In: van Steenis, C.G.G.J. (ed.) *Flora Malesiana*. Series 1. Vol. 4. Noordhoff-Kolff, Djakarta, Indonesia, pp. 318–320.

Burkill, I.H. (1960) The organography and the evolution of Dioscoreaceae, the family of the yams. *Journal of the Linnean Society of London (Botany)* 56, 319–412.

Burkill, I.H. (1966) *A dictionary of the economic products of the Malay Peninsula*. Kuala Lumpur: Government of Malaysia and Singapore.

Burkill, I.H. and Perrier, H. (1950) Dioscoréacées. In: H. Humbert (ed.) Flore de Madagascar et des Comores. Vol. 44, Typographie Firmin-Didot, Paris.

Burton, C.L. (1970) Diseases of tropical vegetables in the Chicago market. *Tropical Agriculture* 47, 303–313.

Burton, W.G. (1989) *The Potato*, 3rd edn. Longman Scientific and Technical, London.

CABI (2017) *Dioscorea alata* (white yam). Invasive species compendium. https://www.cabi.org/isc/datasheet/19293 (accessed 5 December 2017).

CABI (2019) Belize. Invasive species compendium. https://www.cabi.org/isc/datasheet/108387 (accessed 25 February 2020).

Cable, S. (2018) Madagascar yam conservation for livelihoods. https://www.kew.org/science/projects/madagascar-yam-conservation-for-livelihoods (accessed 18 April 2017).

Cable, W.J. and Wilson, J.E. (1984) *Dioscorea nummularia* Lam.: the primary edible yam of Western Samoa. In: Proceedings of the 6th Symposium of the ISTRC, International Potato Centre (CIP), Lima, 599–605.

Cabrera, M., Gómez, R., Espinosa, E., López, J., Medero, V., Basail, M. and Santos, A. (2011) Yam (*Dioscorea alata* L.) microtuber formation in temporary immersion system as planting material. *Biotecnología Aplicada* 28, 268–271.

Caddick, L.R., Wilkin, P., Rudall, P.J., Hedderson, T.A.J. and Chase, M.W. (2002) Yams reclassified: a recircumscription of Dioscoreaceae and Dioscoreales. *Taxon* 51, 03–114.

Cadet, P. and Quénéhervé, P. (1994) Fluctuations naturelles de *Scutellonema bradys* (Nematoda: Hololaimidae) au cours de la croissance et du stockage de l'igname (*Dioscorea alata*) à la Martinique. *Nematologica* 40, 587–600.

Campbell, J.S., Chukwueke, V.O., Teriba, F.A. and Ho-A-Shu, H.V.S. (1962) Some physiological investigations Into the White Lisbon Yam (*Dioscorea alata* L.) II. Growth period and out-of-season production. *Empire Journal of Experimental Agriculture* 30, 232–238.

Campbell-Chin-Sue, H. (1995) Advantages and disadvantages of minisett technology. IICA report, Jamaica.

Campbell-Chin-Sue, H., Bekele, I. and Wickham, L. (1994) Effects of sett size, sett type, mulching and staking on tuber yield produced from yellow yam (*Dioscorea cayenensis* L.) minisetts. Caribbean Food Crops Society, 30th Annual Meeting, 31 July–5 August 1994, St. Thomas, US Virgin Islands.

Carsky, R. J., Asiedu, R. and Cornet D. (2010) Review of soil fertility management for yam-based systems in West Africa. *African Journal of Root and Tuber Crops* 8, 1–17.

Casals, C., Viñas, I., Landl, A., Picouet, P., Torres, R. and Usall, J. (2010) Application of radio frequency heating to control brown rot on peaches and nectarines. *Postharvest Biology and Technology* 58, 218–224.

Casals, C., Elmer, P.A.G., Viñas, I., Teixidó N., Sisquella, M. and Usall, J. (2012) The combination of curing with either chitosan or *Bacillus subtilis* CPA-8 to control brown rot infections caused by *Monilinia fructicola*. *Postharvest Biology and Technology* 64, 126–132.

CBI (2019) Exporting roots and tubers to Europe. https://www.cbi.eu/market-information/fresh-fruit-vegetables/roots-tubers/ (accessed 25 December 2019).

Cech, R. (2002) *Growing at-risk medicinal herbs: cultivation, conservation and ecology*. Williams, Oregan: Horizon Herbs.

Celestine, A.A. and David, O.I. (2015) Comparative nutritional and phytochemical evaluation of the aerial and underground tubers of air potato (*Dioscorea bulbifera*) available in Abakaliki, Ebonyi State, Nigeria. *British Journal of Applied Science and Technology* 11, 1–7.

Chandler, F.L., Haque, S.Q. and Bethell, P. (1983) A systematic approach to improved yam production methods in Barbados. Caribbean Food Crops Society 19th Annual Meeting, Puerto Rico 306–319.

Chen, Y., Fan, J., Yi, F., Luo, Z. and Fu, Y. (2003a) Rapid clonal propagation of *Dioscorea zingiberensis*. *Plant Cell, Tissue and Organ Culture* 73, 75–80.

Chen, C-T., Wang, Z-H., Hsu, C-C., Lin, H-H. and Chen, J-H. (2017) Taiwanese and Japanese yam (*Dioscorea* species) extracts attenuate doxorubicin-induced cardiotoxicity in mice. *Journal of Food and Drug Analysis* 25, 872–880.

Chen, F.Q., Fu, Y., Wang, D.L., Gao, X. and Wang, L. (2007) The Effect of plant growth regulators and sucrose on the micropropagation and microtuberization of *Dioscorea nipponica* Makino. *Journal of Plant Growth Regulation* 26, 38–45.

Chen, Y. and Wu, Y. (1994) Progress in research and manufacturing of steroidal sapogenins in China. *Journal of Herbs Spices Medicinal Plants* 2, 59–70.

Chen, Y., Fan, J., Yi, F., Luo, Z. and Fu, Y. (2003b) Rapid clonal propagation of *Dioscorea zingiberensis*. *Plant Cell, Tissue and Organ Culture* 73, 75–80.

Chen, Y., Tang, Y.M. Yu, S.L. Han, H.W., Kou, J.P., Liu, B.L. and Yu, B.Y. (2015) Advances in the pharmacological activities and mechanisms of diosgenin. *Chinese Journal of Natural Medicines* 13, 578–587.

Chen, Y., Dong, Y., Chi, Y., He, Q., Wu, H. and Ren, Y. (2018) Eco-friendly microbial production of diosgenin from saponins in *Dioscorea zingiberensis* tubers in the presence of *Aspergillus awamori*. *Steroids* 136, 40–46.

Chevalier, A. (1913) *Etudes sur la flore de l'Afrique centrale française*, Vol 1. Challamel, Paris.

Chevalier, A. (1936) Contribution a l'étude de quelques espèces africaines du genre *Dioscorea*. *Bulletin du Muséum National d'Histoire Naturelle* 2eme Serie 8, 520–551.

Chiu, C.S., Deng, J.S., Hsieh, M.T., Fan, M.J., Lee, M.M., Chueh, F.S., Han, C.K., Lin, Y.C and Peng W.H. (2009) Yam (*Dioscorea pseudojaponica* Yamamoto) ameliorates cognition deficit and attenuates oxidative damage in senescent mice induced by d-galactose. *American Journal of Chinese Medicine* 37, 889–902.

Chiu, C.S., Deng, J.S., Chang, H.Y., Chen, Y.C., Lee, M.M., Hou, W.C., Lee, C.Y., Huang, S.S. and Huang, G.J. (2013) Antioxidant and anti-inflammatory properties of Taiwanese yam (*Dioscorea japonica* Thunb. var. *pseudojaponica* (Hayata) Yamam.) and its reference compounds. *Food Chemistry* 141, 1087–1096.

Chiu, H-W., Peng, J-C., Tsai, S-J. and Lui, W-B. (2012) Effect of extrusion processing on antioxidant activities of corn extrudates fortified with various Chinese yams (*Dioscorea* sp.). *Food and Bioprocess Technology* 5, 2462–2473.

Cho, J., Choi, H., Lee, J., Kim, M-S., Sohn, H-Y. and Lee, D.G. (2013) The antifungal activity and membrane-disruptive action of dioscin extracted from *Dioscorea nipponica*. *Biochim. Biochimica et Biophysica Acta* 1828, 1153–1158.

Choi, W-S. (2012) Development of functional beverage using yam (*Dioscorea opposita* Thunb.). *Food Industry and Nutrition* 17, 20–22.

Choudhary, K. Singh, M. and Pillai, U. (2009) *Ethnobotanical survey of Rajasthan – an update*. *American-Eurasian Journal of Botany*, 2, 22–29.

Chung, R.C.K. (2001) *Dioscorea* L. In: van Valkenburg, J.L.C.H. and Bunyapraphatsara, N. (eds) *Plant Resources of South-East Asia* No 12 (2): Medicinal and poisonous plants 2. PROSEA Foundation, Bogor, Indonesia. Database record: prota4u.org/prosea (accessed 28 December 2019).

CIRAD (2015) Guadeloupe: eight new yam varieties are now available to producers, https://www.cirad.fr/en/news/all-news-items/articles/2015/science/guadeloupe-eight-new-yam-varieties-are-now-available-to-producers (accessed 2 November 2019).

CIRAD (2018) https://www.cirad.fr/en/news/all-news-items/articles/2015/science/guadeloupe-eight-new-yam-varieties-are-now-available-to-producers (accessed 22 August 2018).

CJBG (2018) Conservatoire et Jardin botaniques de la Ville de Genève, http://www.ville-ge.ch/musinfo/bd/cjb/africa/index.php?langue=an (accessed 7 February 2018).

Clark, C., Da Silva, W.L. and Villordon, A. (2013) Lenticel proliferation and flooding associated soft rot in sweetpotato. 73rd Annual Meeting Southern Region American Society for Horticultural Science.

Clarke, B.A., Sorhaindo, C.A. and Clarke, A.P. (1986) Preliminary characterization of yam (*Dioscorea* species) germplasm cultivated in Dominica. Caribbean Food Crops Society 22nd Annual Meeting, August 25–29, St Lucia.

CNSA (2013) Evaluation Previsionnelle de La Performance Des Recoltes de La Campagne Agricole de Printemps. Port-au-Prince, Haiti. Ministère de l'Agriculture, des Ressources Naturelles et du Développement Rural.

Cole, F-C. (1922) The Tinguian Social, Religious, and Economic Life of a Philippine Tribe, https://www.gutenberg.org/files/12849/12849-h/12849-h.htm (accessed 24 January 2020).

Collings, R. (2015) Eating Australia – Australian Native Foods. https://brisbanelocalfood.ning.com/group/eating-australia-australian-native-foods/forum/topics/australian-native-long-yam-dioscorea-transversa (accessed 25 September 2019).

Commonwealth of Puerto Rico (2000) Facts and figures on agriculture in Puerto Rico. Department of Agriculture, Office of Agricultural Statistics.

Conservatoire et Jardin Botaniques de la Ville de Genève (2019), http://www.ville-ge.ch/musinfo/bd/cjb/africa/details.php?langue=anandid=11203 (accessed 29 July 2019).

Contu, S. (2009) *Dioscorea trilinguis*. The IUCN Red List of Threatened Species 2009: e.T165733A6104253, http://dx.doi.org/10.2305/IUCN.UK.2009-2.RLTS.T165733A6104253.en (accessed 5 September 2019).

Contu, S. (2013) *Dioscorea quartiniana. The IUCN Red List of Threatened Species* 2013: e.T44392887A44450966. http://dx.doi.org/10.2305/IUCN.UK.2013-2.RLTS.T44392887A44450966.en (accessed 17 September 2019).

Cooke, R.D., Rickard, J.E. and Thompson, A.K. (1988) The storage of tropical root and tuber crops – cassava, yam and edible aroids. *Experimental Agriculture* 24, 457–470.

Corkill, N. L. (1948) The poisonous wild cluster yam *Dioscorea dumetorum* Pax., as a famine crop in the Anglo-Egyptian Sudan. *Annals of Tropical Medicine and Parasitology* 42, 278–287.

Cormier, F., Lawac, F., Maledon, E., Gravillon, M.C., Nudol, E., Mournet, P., Vignes, H., Chaïr, H. and Arnau, G. (2019) A reference high-density genetic map of greater yam (*Dioscorea alata* L.). *Theoretical and Applied Genetics* 132, 1733–1744.

Cornago, D.F., Rumbaoa, R.G.O. and Geronimo, I.M. (2011) Philippine yam (*Dioscorea* species) tubers phenolic content and antioxidant capacity. *Philippine Journal of Science* 140, 145–152.

Cornet, D., Sierra, J. and Bonhomme, R. (2007) Characterization of the photosynthetic pathway of some tropical food yams (*Dioscorea* spp.) using leaf natural 13C abundance. *Photosynthetica* 45, 303–305.

Cornet, D., Sierra, J., Tournebize, R. and Ney, B. (2014) Yams (*Dioscorea* species) plant size hierarchy and yield variability: emergence time is critical. *European Journal of Agronomy* 55, 100–107.

Cornet, D., Sierra, J., Tournebize, R., Gabrielle, B. and Lewis, F. I. (2016) Bayesian network modeling of early growth stages explains yam interplant yield variability and allows for agronomic improvements in West Africa. *European Journal of Agronomy* 75, 80–88.

Correa, A., Galdames, M.D.C. and Stapf, M.N.S. (2004) Catalogo de Plantas Vasculares de Panama. Smithsonian Tropical Research Institute.

Co's Digital Flora of the Philippines (2019) https://www.philippineplants.org/Families/Dioscoreaceae.html (accessed 30 August 2019).

Costa, L.P. (2003) The historical bridge between the Amazon and the Atlantic forest of Brazil. A study of molecular phylogeography with small mammals. *Journal of Biogeography* 30, 71–86.

Costas, R.C., Boneta, E. and Silva, S. (1968) Effect of various cultural practices on yields of yams in Puerto Rico. *The Journal of Agriculture of the University of Puerto Rico* 52, 356–362.

Coursey, D.G. (1961) The magnitude and origins of storage losses in Nigerian yams. *Journal of the Science of Food and Agriculture* 12, 574–580.

Coursey, D.G. (1967) *Yams*. Longmans, London.

Coursey, D.G. (1968) Low temperature injury in yams. *Journal of Food Technology* 3, 143–150.

Coursey, D.G. (1972) The Civilizations of the Yam: Interrelationships of Man and Yams in Africa and the Indo-Pacific Region *Archaeology and Physical Anthropology in Oceania*. 7, 215–233.

Coursey, D.G. (1976) The origins and domestication of yams in Africa. In: Harlan, J.R., DeWet, J.M.J. and Stemler, A.B.L. (eds) *Origin of African Plant Domestication*. Mouton, The Hague, The Netherlands, pp. 383–408.

Coursey, D.G. (1981) Interaction of yam and man. *Journal d'agriculture tropicale botanical applications* 28, 5–21.
Coursey, D.G. and Aidoo, A. (1966) Ascorbic acid levels in Ghanaian yams. *Journal of the Science of Food and Agriculture* 17, 446–449.
Coursey, D.G. and Haynes, P.H. (1970) Root crops and their potential as food in the tropics. *World Crops* 22, 261–265.
Couto, R.S., Lopes, R.C. and Braga, J.M.A. (2015) *Dioscorea flabellispina* (Dioscoreaceae), a new endangered species from the Brazilian Atlantic Rainforest. *Phytotaxa* 231, 89–94.
Couto, R.S., Martins, A.C., Bolson, M., Lopes, R.C., Smidt, E.C. and Braga, J.M.A. (2018) Time calibrated tree of *Dioscorea* (Dioscoreaceae) indicates four origins of yams in the Neotropics since the Eocene. *Botanical Journal of the Linnean Society* 188, 144–160.
Coyne, D., Kajumba, C. and Asiedu, R. (2005a) Expanding guinea yam (*Dioscorea* spp.) Production in Uganda? Results of a short survey in Luweero District. *African Journal of Root and Tuber Crops* 6, 53–58.
Coyne, D.L., Tchabi, A., Baimey, H., Labuschagne, N. and Rotifa, I. (2005b) Distribution and prevalence of nematodes (*Scutellonema bradys* and *Meloidogyne* spp.) on marketed yam (*Dioscorea* spp.) in West Africa. *Field Crops Research* 96, 142–150.
Coyne, D.L., Tchabi, A., Baimey, H., Labuschagne, N. and Rotif, I. (2006) Distribution and prevalence of nematodes (*Scutellonema bradys* and *Meloidogyne* species) on marketed yam (*Dioscorea* species) in West Africa. *Field Crops Research* 96, 142–150.
Coyne, D.L., Kolombia, Y.A., Luambano, N. and Bert, W. (2016) First report of dry rot disease of yam caused by *Scutellonema bradys* in East Africa. *Plant Disease* 100, 1794.
Craufurd, P.Q., Summerfield R.J., Asiedu R. and Vara Prasad P.V. (2001), Dormancy in yams (*Dioscorea* species). *Experimental Agriculture* 37, 147–181.
Cruz, V.M.V. and Ramirez, D.A. (1999) Variation in morphological characters in a collection of yams (*Dioscorea* spp.). *Philippines Journal of Crop Sciences* 24, 27–35.
Cruz, V.M.V., Altoveros, N.C., Mendioro, M.S. and Ramirez, D.A. (1999) Geographical patterns of diversity in the Philippine edible yam collection. *Plant Genetic Resources Newsletter* 119, 7–11.
Cruzado, H.J., Cabanillas, E., Martin, F.W. and Delpin, H. (1964) Seeds rest and seed viability in medicinal *Dioscorea* species. *Proceedings of the American Society for Horticultural Science* 84, 436–440.
CTA (1989) Yam-harvester that pays for itself. *Spore* 21. CTA, Wageningen, the Netherlands, http://hdl.handle.net/10568/45069 (accessed 30 June 2018).
Ćulafić, L., Grubišić, D., Vujičić, R., Volkova, L.A. and Popov, A.S. (1994) Somatic embryo production in vitro in *Dioscorea caucasica* Lipsky and *Dioscorea balcanica* Košanin and cryopreservation of their organogenic callus tissues. *Russian Journal of Plant Physiology* 41, 821–826.
Cumo, C. (2013) *Encyclopedia of Cultivated Plants: from acacia to zinnia*. Santa Barbara, California, ABC-CLIO, LLC.
Czyhriniciw, N. and Jaffe, W. (1951) Modificaciones quemicas durante la conservacion de raices y tuberculos. *Archivos Venexolanos de Nutricion* 2, 49–67.
Dach, J. (2013) Russell Marker and the Origins of Bioidentical Hormones. https://jeffreydachmd.com/2013/06/russell-marker-and-the-origins-of-bioidentical-hormones-by-jeffrey-dach-md/ (accessed 9 July 2019).
Dade, H.A. and Wright, J. (1931) *Year Book*. Department of Agriculture, Gold Coast.
Dallot, S., Guzmán, M, and Bousalem, M. (2001) Occurrence of potyviruses on yam (*Dioscorea* species) in Colombia and first molecular characterization of yam mild mosaic virus. *Plant Disease* 85, 803–804.
Danquah, E.O., Ennin, S.A., Lamptey, J.N.L. and Acheampong, P.P. (2015) Staking options for sustainable yam production in Ghana *Sustainable Agriculture Research* 4, 106–113.
Dansi, A., Zoundjihekpon, J., Mignouna, H.D. and Quin, M. (1997) Collecte d'ignames cultivées du Complexe *Dioscorea cayenensis-rotundata* au Benin. *Plant Genetic Resources Newsletter* 112, 81–85.
Dansi, A., Mignouna, H.D., Zoundjihékpon, J., Sangare, J., Asiedu, R. and Quin, M.F. (1999) Morphological diversity, cultivar groups and possible descent in the cultivated yams (*Dioscorea cayenensis-rotundata*) complex in the Benin Republic. *Genetic Resources and Crop Evolution* 46, 371–388.
Dansi, A., Mignouna, H.D., Pillay, M. and Zok, S. (2001a) Ploidy variation in the cultivated yams (*Dioscorea cayenensis-Dioscorea rotundata* complex) from Cameroon as determined by flow cytometry. *Euphytica* 119, 301–307.
Dansi, A., Pillay, M., Mignouna, H.D., Mondeil, F. and Dainou, O. (2001b) Ploidy level of the cultivated yams (*Dioscorea cayenensis-Dioscorea rotundata* complex) from Benin Republic determined by chromosome counting and flow cytometry. *African Crop Science Journal* 8, 355–364.

Dansi, A., Dantsey-Barry, H., Agré, A.P, Dossou-Aminon, I., Assogba, P., Loko, Y.L, N'Kpenu, E.K, Kombaté, K., Dansi, M. and Vodouhè, R. (2013) Production constraints and farmers' cultivar preference criteria of cultivated yams (*Dioscorea cayenensis-D. rotundata* complex) in Togo. *International Journal of Applied Biology and Pharmaceutical Technology*, 4, 191–199.

Davies, F.S. and Albrigo, L.G. (1994) *Citrus*. CAB International, Wallingford, UK.

Daw, B., Keighery, G.J. and Walley, T. (1997) *Bush Tucker: Plants of the South-West*. Australia, Department of Conservation and Land Management.

De Almeida, A.V.D.L., Muniz, M.F.S., Noronha, M.DeA., De Souza R.C., Filho. G.M. and De Farias, S.P. (2019) *Scutellonema bradys* and *Pratylenchus* spp. associated with weeds in yam fields. *Nematology* 21: 805–811.

de Courcey, J.W. and Headley, W. (1967) The Problem of 'Internal Spotting' of yams *Dioscorea* sp. in Barbados. 5th Annual Meeting, July 24–31, 1967, Paramaribo, Surinam 263589, Caribbean Food Crops Society.

Debray, M., Jacquemin, H. and Razafindrambao, R. (1971) *Contribution à l'inventaire des plantes médicinales de Madagascar*. Travaux et Documents No 8. ORSTOM, Paris, France.

Decary, R. (1946) Plantes et animaux utiles de Madagascar. *Annales du Musée Colonial de Marseille* 54e année, 6e série, 4e volume, 1er et dernier fascicule.

Degras, L., Arnolin, R., Poitout, A. and Suard, C. (1977) Quelques aspects de la biologie des ignames (Dioscorea spp). Les ignames et leur culture. *Annales de l'amélioration des plantes* 27, 1–23.

Degras, L.M. (1993) *The yam, a tropical root crop*. MacMillan Press Ltd, London.

Deliu, C., Tamas, M., Dobrota, C. and Munteanu-Deliu, C. (1992) Kinetics of cellular growth and diosgenin biosynthesis in *Dioscorea caucasica* cell suspensions, cultured in batch system. *Plant Science* 85, 99–105.

Denham, T. (2007) Traditional forms of plant exploitation in Australia and New Guinea: the search for common ground. *Vegetation. History and Archaeobotany* 7, 245–248.

Diby, L.N., Hgaza, V.K., Tie, T.B., Assa, A., Carsky, R., Girardin, O. and Frossard, E. (2009) Productivity of yams (*Dioscorea* species) as affected by soil fertility. *Journal of Animal and Plant Sciences* 5, 494–506.

DIIVA (2020) CGIAR's Diffusion and Impact of Improved Varieties in Africa. https://www.asti.cgiar.org/diiva/uganda/yams (accessed 4 February 2020).

Ding, Z. and Gilbert, M.G. (2010) Flora of China Vol. 24, Page 290 *www.efloras.org › florataxon* (accessed 27 November 2019).

Ding, Z., Gilbert, M.G. and Turland, N.J. (2000) A New Name in Chinese *Dioscorea* (Dioscoreaceae). *Novon* 10, 13.

Djedatin, G., Adeoti, K., Ewedje, E. and Agbangla, C. (2017) Does the domestication of yam wild type *Dioscorea praehensilis* and *Dioscorea abyssinica* have a genetic basis? *International Journal of Sciences* 11, 18–28.

Dong, M., Wu, L.J., Chen, Q. and Wang, B.X. (2001) Isolation and identification of steroidal saponins from *Dioscorea panthaica* Prain et Burkill. *Yao Xue Xue Bao* 36, 42–45.

Dorr, L.J. and Stergios, B. (2003) A new species of *Dioscorea* (dioscoreaceae) from the Andes of Venezuela. *SIDA, Contributions to Botany* 20, 1007–1013.

Dougall, D.K. (1979) Factors affecting the yield of secondary products in plant tissue cultures. In: Sharp, W.R., Larsen, P.O., Paddock, E.F. and Raghavan, V. (eds) *Plant cell and tissue cultures — principles and applications*. Ohio State Univ Press, Columbus, Ohio. 727–743.

Dounias, E. (1993) Perception and use of wild yams by the Baka hunter-gatherers in South Cameroon. In: Hladik, C.M., Hladik, A., Licares, O.F., Pagezy, H., Semple, A. and Hadley, M (eds) *Tropical Forests, People and Food: Biocultural Interactions and Applications to Development*. UNESCO, Paris, pp. 621–632.

Dounias, E. (2001) The Management of Wild yam Tubers by the Baka Pygmies in Southern Cameroon. *African Study Monographs* 26, 135–156.

Downes, K., Chope, G.A. and Terry, L.A. (2009) Effect of curing at different temperatures on biochemical composition of onion (*Allium cepa* L.) skin from three freshly cured and cold stored UK-grown onion cultivars. *Postharvest Biology and Technology* 54, 80–86.

Dumont, R. and Vernier, P. (2000) Domestication of yams (*Dioscorea Cayenensis-Rotundata*) within the Bariba ethnic group in Benin. *Outlook on Agriculture* 29, 137–142.

Dumont, R., Hamon P. and Seignobos C. (1994) Les ignames au Cameroun. CIRAD, CA. Repères, Cultures Annuelles.

Dzomba, P. and Musekiwa, C. (2014) Anti-obesity and antioxidant activity of dietary flavonoids from *Dioscorea steriscus* tubers. *Journal of Coastal Life Medicine* 6, 465–470.

Ecocrop (1993–2007) FAO. http://ecocrop.fao.org/ecocrop/srv/en/home (accessed 30 August 2019).

Ecocrop (2018) http://ecocrop.fao.org/ecocrop/srv/en/cropView?id=5382 (accessed 7 June 2018).

Egesi, C.N., Asiedu, R., Ude, G., Ogunyemi, S. and Egunjobi, J.K. (2006) AFLP marker diversity in water yam (*Dioscorea alata* L.). *Plant Genet Resources* 4, 181–187.

Eitzinger, A., Farrell, A., Rhiney, K., Carmona, S., van Loosen, I. and Taylor, M. (2015) Trinidad and Tobago: Assessing the impact of climate change on cocoa and tomato. CIAT Policy Brief 27. https://cdkn.org/wp-content/uploads/2014/04/CIAT_PB27_TRINIDAD-AND-TOBAGO-ASSESSING-THE-IMPACT-OF-CLIMATE-CHANGE-ON-COCOA-AND-TOMATO.pdf (accessed 6 November 2019).

Eitzinger, A., Läderach, P., Carmona, S., Collet, L., Jean-Simon, L., Dufane, P. and Nowak, A. (2014) Haiti: coffee and mango production under a changing climate. CIAT Policy Brief No.16. Centro Internacional de Agricultura Tropical, Cali, Colombia.

Eke-Ejiofor, J., Deedam, J.N. and Beleya, E.A. (2015) Effect of addition of three leaf yam flour on dough properties and sensory qualities of bread. *Journal of Food Research* 4, 133.

Ekwu, F.C., Ozo, N.O. and Ikegwu, O.J. (2005) Qualities of fufu flour from white yam varieties (*Dioscorea* species). *Nigerian Food Journal* 23, 107–113.

Elmi Sharlina, M.S., Yaacob, W.A., Lazim, A.M., Fazry, S., Lim, S.J., Abdullah, S., Noordin, A. and Kumaran, M. (2017) Physicochemical properties of starch from *Dioscorea pyrifolia* tubers. *Food Chemistry* 220, 225–232.

Emua, S.A. and Fajola, A.O. (1983) Chemical control of two leaf spot diseases of cluster yam (*Dioscorea dumetorum*) caused by *Cercospora contraria* and *Didymosphaeria donacina*. *Plant Disease* 67, 389–391.

Encyclopedia of Succulents (2013–2019) http://www.llifle.com/Encyclopedia/SUCCULENTS/Family/Dioscoreaceae/17507/Dioscorea_elephantipes (accessed 8 August 2019).

Encyclopaedia of Succulents (2018) http://llifle.com/Encyclopedia/SUCCULENTS/Family/Dioscoreaceae/28410/Dioscorea_sylvatica (accessed 17 February 2018).

Encyclopedia of Life (2019) http://eol.org/pages/1120724/overvie (accessed 12 July 2019).

Enesi, R.O., Hauser, S., Lopez-Montez Osonubi, O. (2018) (2 Yam tuber and maize grain yield response to cropping system intensification in south-west Nigeria. *Archives of Agronomy and Soil Science* 64, 953–966.

Eni, A.O., Hughes, J.d'A. and Rey, M.E.C. (2008) Survey of the incidence and distribution of five viruses infecting yams in the major yam producing zones in Benin. *Annals of Applied Biology*, 153, 223–232.

Ennin, S.A., Otoo, E. and Tetteh, F.M. (2009) Ridging, a mechanized alternative to mounding for yam and cassava production. *West African Journal of Applied Ecology* 15 (1).

Enyi, B.A.C. (1970) Yams in Africa. Proc. 2nd International Symposium on Tropical Root and Tuber Crops, College of Tropical Agriculture, University of Hawaii, Honolulu, Vol. 1. 190–193.

Enyi, B.A.C. (1972a) The effects of seed size and spacing on growth and yield of lesser yam (*Dioscorea esculenta*). *The Journal of Agricultural Science* 78, 215–225.

Enyi, B.A.C. (1972b) Effect of staking, nitrogen and potassium on growth and development in lesser yams: *Dioscorea esculenta*. *Annals of Applied Biology* 72, 211–219.

Epilepsy Naturapedia (2019) http://www.epilepsynaturapedia.com/index.php?title=Dioscorea_septemloba (accessed 11 September 2019).

Escalona, M. (2006) Temporary immersion beats traditional techniques on all fronts. *Prophyta Annual* 48–50.

European Food Safety Authority (2019) http://www.cipccompliant.co.uk/regulation/ (accessed 20 November 2019).

Eyitayo, O.A., Anthony, T.O. and Theresas, I. (2010) Economics of seed yam production using minisett technique in Oyo State, Nigeria, *Field Actions Science Reports*, Vol. 4, http://journals.openedition.org/factsreports/659 (accessed 24 May 2018).

Ezeh, N.O. (1998) Economics of production and post-harvest technology. In: Orkwor, G.C., Asiedu, R. and Ekanayake, I.J. (eds) *Food yam*. Advances in research IITA and NRCRI, Nigeria, 187–214.

Ezeike, G.O.I. (1985) Experimental analysis of yam *Dioscorea* species tuber stability in tropical storage. *Transactions of the American Society of Agricultural Engineers* 28, 1641–1645.

Falade, K.O. and Ayetigbo, O.E. (2015) Effects of annealing, acid hydrolysis and citric acid modifications on physical and functional properties of starches from four yam (*Dioscorea* species) cultivars. *Food Hydrocolloid* 43, 529–539.

Falade, K.O., Olurin, T.O., Ike, E.A. and Aworh, O.C. (2007) Effect of pretreatment and temperature on air-drying of *Dioscorea alata* and *Dioscorea rotundata* slices. *Journal of Food Engineering* 80, 1002–1010.

FAO (1965) Food composition table for use in Africa. FAO Rome, http://www.fao.org/docrep/003/X6877E/X6877E04.htm (accessed 15 November 2017).

FAO (1996) Saint Lucia: country report to the FAO international technical conference on plant genetic resources. http://www.fao.org/fileadmin/templates/agphome/documents/PGR/SoW1/americas/SAINTLUC.pdf (accessed 4 November 2019).

FAO (1998) Handling and storage methods for fresh roots and tubers. *Storage and Processing of Roots and Tubers in the Tropics*, Ch. 4. Agro-industries and Post-Harvest Management Service, Agricultural Support Systems Division. http://www.fao.org/3/x5415e/x5415e04.htm (accessed 17 November 2019).

FAO (2008) Second country report on the state of plant genetic resources in St Vincent and the Grenadines, http://www.fao.org/3/i1500e/Saint%20Vincent%20and%20the%20Grenadines.pdf (accessed 3 November 2019).

FAO (2019) Storage of horticultural crops, http://www.fao.org/Wairdocs/X5403E/x5403e09.htm.

FAO Benin (2019) Strengthening linkages between small actors and buyers in the Roots and Tubers sector in Africa, http://www.fao.org/in-action/african-roots-and-tubers/countries/benin/en/ (accessed 3 November 2019).

FAO Saint Lucia (1996) Saint Lucia Agricultural Census 1996 – main results, http://www.fao.org/publications/card/en/c/8beb6d2d-26d0-4f33-97fc-6edb99696cf2/ (accessed 23 December 2019).

FAO Stats (2019) http://www.fao.org/faostat/en/#data/QC (accessed 20 February 2019).

FAO Stats (2020) http://www.fao.org/faostat/en/#data/QC (accessed 6 May 2021).

FAO Sudan (1995) Country report to the FAO international technical conference on plant genetic resources Prepared by: Mahmoud Ahmed Mahmoud, Mohamed Osman Khidir, Mohamed Ahmed Khalifa, Abdel Moneim Bashir El Ahmadi, Hassan Abdel Rahman Musnad and El Tahir Ibraim Mohamed. Khartoum, March 1995, http://www.fao.org/fileadmin/templates/agphome/documents/PGR/SoW1/africa/SUDAN.PDF (accessed 6 November 2019).

FAO Trobriand Islands (2018) Papua New Guinea. http://www.fao.org/papua-new-guinea/news/detail-events/en/c/1176454/ (accessed 24 October 2018).

Fenwick, D.E. and Oakenfull, D. (2006) Saponin content of food plants and some prepared foods. *Journal of Science of Food and Agriculture* 34, 186–191.

Ferede, R., Maziya-Dixon, B., Alamu, O.E., Asiedu, R. (2010) Identification and quantification of major carotenoids of deep yellow-fleshed yam (tropical *Dioscorea dumetorum*). *Journal of Food, Agriculture and Environment* 8, 160–166.

Ferguson, T.U. (1969) Yam agronomy. In Rep. Faculty of Agriculture 1968-69, University of the West Indies, Trinidad, p. 29.

Ferguson, T.U. (1972) The propagation of *Dioscorea* species by vine cuttings: a critical review. *Tropical Root and Tuber Crops Newsletter* 5, 5–7.

Ferguson, T.U. (1973) The effect of sett characteristics and spacing on growth, development and yield of yams (*Dioscorea* spp). PhD thesis, University of the West Indies St Augustine, West Indies.

Ferguson, T.U. (1977) Tuber development in yams; physiological and agronomic implications. Proceedings of the Third Symposium of the International Society for Tropical Root Crops, International Institute for Tropical Agriculture, Nigeria, 72–77.

Ferguson, T.U., Haynes, P.H. and Springer, B.G.F. (1969) A study of variability in yams (*Dioscorea* species). In Caribbean Food Crops Society. University of the West Indies, St Augustine, Trinidad. Vol. 6, 50–58.

Fern, K. (2014) Useful Tropical Plants Database 2014, http://tropical.theferns.info/viewtropical.php?id=Dioscorea+cirrhosa (accessed 8 February 2018).

Filyaw, T. (2006) *Dioscorea villosa*. Plants to watch: Non-timber products from Appalachian forest and field. hhttp://www.appalachianforest.org/ptw_wild_yam.html (accessed 12 October 2017).

Flexport (2020) US import and export data yams, fresh, chilled, frozen or dried HS Code. https://www.flexport.com/data/hs-code/071430-yams-fresh-chilled-frozen-or-dried.

Flora Argentina ((nd)) http://www.floraargentina.edu.ar/espsin.asp?quegenbusco=Dioscorea (accessed 9 June 2018).

Flora de la República de Cuba (2019) http://portal.cybertaxonomy.org/flora-de-la-republica-de-cuba/cdm_dataportal/taxon/43900736-f786-42fc-b5ac-d84f7ea22bb7 (accessed 30 August 2019).

Flora of China (2019) http://www.efloras.org/florataxon.aspx?flora_id=2andtaxon_id=240001191 (accessed 19 July 2019).

Flora of Zimbabwe (2019) https://www.zimbabweflora.co.zw/speciesdata/species.php?species_id=115400 (accessed 7 August 2019).

Florida Exotic Pest Plant Council (2008) *Air potato (D. bulbifera) management plan: recommendations from the Air Potato Task Force*. Florida, USA: Florida Exotic Pest Plant Council.

Flowers of India (2019) https://www.flowersofindia.net/catalog/slides/Spear-Leaved%20Yam.html (accessed 23 August 2019).

Foden, W. and Potter, L. (2005) *Dioscorea retusa* Mast. National Assessment: Red List of South African Plants version 2017.1. (accessed on 18 September 2019).

Food-Info.net (2019) Wageningen University http://www.food-info.net/uk/products/rt/yam.htm (accessed 7 August 2019).

Fresh Plaza (2017) http://www.freshplaza.com/article/170678/China-Domestic-Yam-market-saturated (accessed 11 October 2019).

Fresh Plaza (2018) China: White yams in short supply for many years. https://www.freshplaza.com/article/9052970/china-white-yams-in-short-supply-for-many-years/ (accessed 16 October 2019).

Frossard, E. (2018) Inter- and transdisciplinary research on yam systems for improved food security in West Africa. https://www.openaccessgovernment.org/inter-and-transdisciplinary-research-on-yam-systems-for-improved-food-security-in-west-africa/46755/ (accessed 18 June 2018).

Fu, Y-C.,Ferng, A.L-H. and Huang, P-Y. (2006) Quantitative analysis of allantoin and allantoic acid in yam tuber, mucilage, skin and bulbil of the *Dioscorea* species. *Food Chemistry* 94, 541–549.

Fujii, K. and Matsukawa, T. (1936) Saponins and Sterols. 8. Saponin of *Dioscorea tokoro* Makino. *Journal of Pharmaceutical Society of Japan* 56, 408–414.

Gaillard, C. and Manner, H.I. (2010) Yam cultivation on the east coast of New Caledonia: adaptation of agriculture to social and economic changes. *Australian Geographer* 41, 485–505.

Gamiette, F., Bakry, F. and Ano, G. (1999) Ploidy determination of some yam species (Dioscorea species) by flow cytometry and conventional chromosome counting. *Genetic Resources and Crop Evolution* 46, 19–27.

García, M.B. and Antor, R.J. (1995) Age and size structure in populations of a long-lived dioecious geophyte: *Borderea pyrenaica* (Dioscoreaceae). *International Journal of Plant Science*, 156, 236–243.

García, M.B., Espadaler, X. and Olesen, J.M. (2012) Extreme reproduction and survival of a true cliff hanger: The endangered plant *Borderea chouardii* (Dioscoreaceae). http://www.indefenseofplants.com/blog/2015/2/19/a-real-cliffhanger, https://doi.org/10.1371/journal.pone.0044657 (accessed 26 April 2018).

Garrod, B. Lewis, B.G. Brittain M.J.and Davies W.P. (1982) Studies on the contribution of lignin and suberin to the impedance of wounded carrot root tissue to fungal invasion. *New Phytologist* 90, 99–108.

Gaztambide, S. and Cibes, H.R. (1975) Nutritional deficiencies of yams (*Dioscorea* species) and related effects on yield and leaf composition. *The Journal of Agriculture of the University of Puerto Rico*, 59, 264–273.

GBIF (2019) GBIF Backbone Taxonomy. Checklist dataset. https://www.gbif.org/dataset/d7dddbf4-2cf0-4f39-9b2a-bb099caae36c (accessed 18 December 2019).

Gebre-Mariam, T. and Schmidt P.C. (1998) Some physico-chemical properties of *Dioscorea* starch from Ethiopia. *Starch - Stärke* 50, 241–246.

GEPA (2017) Yams – Potential markets. https://www.gepaghana.org/market-report/yams-potential-markets-2017/). Ghana Export Promotion Authority –www.gepaghana.org (accessed 5 February 2020).

Gerardin, O., Nindjin, C., Farah, Z., Escher, F., Stamp, P. and Otokore, D. (1998) Use of gibberellic acid to prolong dormancy and reduce losses during traditional storage of yams. *Journal of the Science of Food and Agriculture* 77, 172–178.

Ghayur, M.N., Kazim, S.F., Rasheed, H., Khalid, Jumani, M.I. and Choudary, M.I.J. (2011) Ethnopharmacological potential and medicinal uses of miracle herb *Dioscorea* spp. *Chinese Journal of Integrative Medicine* 9, 619–625.

Ghosh, A. (2014) The genus *Dioscorea* L. in Andaman and Nicobar Islands, India. *Indian Journal of Plant Sciences* 4, 7–14.

Gibson, D. and Thomas, J. (2005) *Global Invasive Species Database*, species profile: *Dioscorea oppositifolia*, http://www.iucngisd.org/gisd/species.php?sc=296 (accessed 28 November 2017).

Girardin, O., Farah, Z., Escher, F. and Stamp, P. (1997) Improving village storage of yams in Côte d'Ivoire, https://www.researchgate.net/publication/293153970_Improving_village_storage_of_yams_in_cote_d%27Ivoire (accessed 11 December 2017).

Girardin, O., Nindjin, C., Farah, Z. and Stamp. P. (1998) Use of gibberellic acid to prolong dormancy and reduce losses during traditional storage of yams. *Journal of the Science of Food and Agriculture* 77, 172–178.

Girma, G., Spillane, C. and Gedil, M. (2016) DNA barcoding of the main cultivated yams and selected wild species in the genus *Dioscorea*. *Journal of Systematics and Evolution* 54, 228–237.

Go, H.K., Rahman, M., Kim, G.B., Na, C.S., Song, C.H., Kim, J.S., Kim, S.J. and Kang, H.S. (2015) Antidiabetic effects of yam (*Dioscorea batatas*) and its active constituent, allantoin, in a rat model of streptozotocin-induced diabetes. *Nutrients* 7, 8532–8544.

Gonzales, M.A. and de Rivera, A.C. (1972) Storage of fresh yams *Dioscorea alata* L under controlled conditions. *Journal of Agriculture of the University of Puerto Rico* 56, 46–56.

Gonzalez, V.A. (2006) Yield and natural reaction to anthracnose (*Colletotrichum gloeosporioides*) of four yam cultivars planted without staking. *Journal of Agriculture of the University of Puerto Rico* 90, 75–81.

Gooding, H.J. (1960) West Indian *Dioscorea alata* cultivars. *Tropical Agriculture* 37, 11–30.

Goswami, N., Borthakur, S.K. and Hor, D.K. (2013) The genus *Dioscorea* Linnaeus [Dioscoreaceae] in Assam, India. *Pleione* 7, 73–83.

Govaerts, R. (2012) World Checklist of Dioscoraceae. Royal Botanic Gardens, Kew. London. http://apps.kew.org/wcsp/ (accessed 6 November 2019).

Govaerts, R. (2013) World Checklist of Dioscoraceae. Royal Botanic Gardens, Kew, London http://apps.kew.org/wcsp/ (accessed 6 January 2020).

Govaerts, R. and Wilkin, P. (2015) *World Checklist of Dioscoreaceae*. The Board of Trustees of the Royal Botanic Gardens, Kew: http://www.kew.org/wcsp/monocots/ (accessed 30 January 2015).

Govaerts, R., Wilkin, P. and Saunders, R.M.K (2007) *World Checklist of Dioscoreales: Yams and their Allies*. Kew Publishing, London.

Graham-Acquaah, S., Ayernor, G.S., Bediako-Amoa, B., Saalia, F.S., Afoakwa, E.O. and Abbey, L. (2013) Effect of blanching and frying on textural profile and appearance of yam (*Dioscorea rotundata*) French fries. *Journal of Food Processing and Preservation* 39, 19–29.

Grenada (2017) Yam Varieties in Grenada. *Farming in Grenada*, http://farminggrenada.com/yams/yam-varieties-in-grenada/#comment-110 (accessed 4 November 2019).

Grant, L.G. (1996) Antigua and Barbuda: country report to the FAO international technical conference on plant genetic resources, http://www.fao.org/fileadmin/templates/agphome/documents/PGR/SoW1/americas/ANTIGUA.pdf (accessed 29 May 2020).

Green, K.R. and Simons, S.A. (1994) 'Dead skin' on yams (*Dioscorea alata*) caused by *Colletotrichum gloeosporioides*. *Plant Pathology* 43, 1062–1065.

Groen, L.E. Siemonsma, J.S. and Jansen, P.C.M. (2016) https://uses.plantnet-project.org/en/Dioscorea_puber_ (PROSEA) (accessed 23 June 2018).

Guardian (2010) New to Nature No 8: *Dioscorea orangeana* https://www.theguardian.com/science/2010/aug/27/plants (accessed 18 June 2018).

Gucker, C.L. (2009) Dioscorea spp. In: *Fire Effects Information System* [Online]. U.S. Department of Agriculture, Forest Service, Rocky Mountain Research Station, Fire Sciences Laboratory (Producer), https://www.fs.fed.us /database/feis/plants/vine/diospp/all.html (accessed 29 February 2020).

Gupta, S.K., Kuo, C-L., Chang, H-C., Chan, H-S., Chen, E.C-F., Chueh, F-S. and Tsay, H-S. (2012) *In vitro* propagation and approaches for metabolites production in medicinal plants. *Advances in Botanical Research* 62, 35–54.

Gutiérrez, T.J., Pérez, E., Guzmán, R., Tapia, M.S. and Famá, L. (2014) Physicochemical and functional properties of native and modified by crosslinking, dark-cush-cush yam (*Dioscorea trifida*) and cassava (*Manihot esculenta*) starch. *Journal of Polymer and Biopolymer Physics Chemistry* 2, 1–5.

Guyana (2004) *Yam: Postharvest Care and Market Preparation*. Ministry of Fisheries, Crops and Livestock, New Guyana Marketing Corporation National Agricultural Research Institute Technical Bulletin No. 32.

Guzmán, M. and Buitrago, G. (2000) *Yam: production of seed for biotechnology*. Unibiblos, Bogotá.

Gwa, V.I. and Abdulkadir, K.H. (2017) Biological control using *Trichoderma harzianum* against *Penicillium purpurogenum*, causal agent of white yam tuber (*Dioscorea rotundata* Poir.). *Journal of Bioresearch Communications* 1, 102–108.

Hahn S.K. (1995) Yams, Dioscorea spp. (Dioscoreaceae), in Evolution of Crops Plants, Smartt, J. and Simmonds, N.W. (eds) Longman Group Limited, London, pp. 112–120.

Haiti News (2014) https://www.haitilibre.com/en/news-12006-haiti-agriculture-yams-strategic-resource-generator-of-foreign-currency.html (accessed 27 January 2020).

Hallam, S. (1986) Yams, alluvium and villages on the west coastal plains. In: Ward, G.K. (ed.) *Archaeology at ANZAAS Canberra*. Canberra Archaeological Society, Canberra, pp. 16–132.

Hamadina E.I. (2011) The control of yam tuber dormancy: a framework for manipulation. IITA, Ibadan, Nigeria.

Hamadina, E.I. and Asiedu, R. (2015) Effect of provenance and storage agroecology on duration of yam (*Dioscorea rotundata* Poir.) tuber dormancy. *Agriculture, Forestry and Fisheries* 4, 95–100.

Hamadina, E.I. and Craufurd, P.Q. (2015) Changes in free phenolics contents during tuber development, dormancy and sprouting in White Yam (*Dioscorea rotundata* Poir.). *International Journal of Plant Research* 5, 34–41.

Hammel, B.E. (2000) A New Species of *Dioscorea* from Costa Rica. *Novon* 10, 378–381.

Hammer, R.L. (1998) Diagnosis *Dioscorea*. *Wildland Weeds* 2, 8–10.

Hamon, P. and Touré, B. (1990a) Characterisation of traditional yam varieties belonging to the *Dioscorea cayenensis-rotundata* complex by their isozymic patterns. *Euphytica* 46, 101–107.

Hamon, P. and Touré, B. (1990b) The classification of the cultivated yams (Dioscorea cayenensis-rotundata complex) of West Africa. *Euphytica* 47, 179–187.

Hamon, P., Dumont, R., Zoundjihekpon, J., Tio-Toure, B. and Hamon, S. (1995) Les Ignames Sauvages d'Afrique de l'Ouest: Caracteristiques Morphologiques. Paris: ORSTOM.

Hanelt, P. (2017) Mansfeld's World Database of Agricultural and Horticultural Crops. Gatersleben, Germany: Leibniz Institute of Plant Genetics and Crop Plant Research (IPK). http://mansfeld.ipk-gatersleben.de/apex/f?p=185:1:::NO (accessed 16 October 2019).

Hanelt, P., Büttner, P., Mansfeld, R. and Kilian, R., (2001) *Mansfeld's Encyclopedia of Agricultural and Horticultural Crops*. Institut für Pflanzengenetik und Kulturpflanzenforschung Gatersleben, 3380, Springer, Germany

Harada, Y., Tominaga, T. and Kemuriyama, A. (1982) *Pythium* rot, a new disease of Chinese yam tubers in cold storage. *Bulletin of the Faculty of Agriculture, Hirosaki University* 37, 66–80.

Hariprakash, C.S. and Nambisan, B. (2007) Carbohydrate metabolism during dormancy and sprouting in yam. *Journal of Zhejiang University Science* 8, 283–288.

Harris, T. (2018) Yams in Madagascar https://www.kew.org/blogs/archived-blogs/studying-yams-madagascar (accessed 19 April 2018).

Hawley, W.O. (1956) *Dioscorea* as ornamental foliage plants. *National Horticultural Magazine* 35, 23–29.

Hayashi, A.H., Chu, E.P., Kirizawa, M., de Cássia Leone Figueiredo-Ribeiro, R. and Henrique da Silva, J. (2018) Anatomical and reserve carbohydrates characterization of vegetative organs from Atlantic Rainforest *Dioscorea* species (Dioscoreaceae). 25ª Reunião Anual do Instituto de Botânica, '90 anos do Jardim Botânico de SP:trabalhando em prol da biodiversidade', https://www.infraestruturameioambiente.sp.gov.br/institutodebotanica/wp-content/uploads/sites/235/2019/04/anais_21_22_04_2019.pdf (accessed July 2020).

Haynes, P.H. and Coursey, D.G. (1969) Gigantism in yam. *Tropical Science* 11, 93–96.

Hazarika, E.B., Borah, A. and Mahanta, C.L. (2013) Optimisation of extrusion cooking conditions and characterization of rice (*Oryza sativa*) - sweet potato (*Ipomoea batatas*) and rice-yam (*Dioscorea alata*) based RTE products. *Agricultural Sciences* 4, 12–22.

Headland, T.N. (1987) The wild yam question: How well could independent hunter-gatherers live in a tropical rain forest ecosystem? *Human Ecology* 15, 463–491.

Hedrick, U.P. (1919) (Editor) with updated botanical names by Michael Moore 1919. *Sturtevant's Edible Plants of the World*. Sturtevant E.L. 1892. The Southwest School of Botanical Medicine, http://www.swsbm.comwww.swsbm.com/Ephemera/Sturtevants_Edible_Plants.pdf (accessed 14 May 2012).

Herbal Fire Botanicals (2018) https://www.herbalfire.com/wild-yam-dioscorea-dregeana.html (accessed 25 February 2018)

Herbison-Evans, D and Ashe, C. (2008) *Dioscorea transversa* Australian Native Yam Dioscoreaceae, http://lepidoptera.butterflyhouse.com.au/plants/dios/dioscorea-transversa.html (accessed 4 July 2019).

Heywood, V.H., Tutin, T.G., Heywood, V.H., Burges, N.A., Moore, D.M., Valentine, D.H., Walters, S.M., Webb, D.A. and Tamus, L. (1980) *Flora Europaea*. Cambridge University Press, Cambridge.

Hgaza, V.K., Diby, L.N., Aké, S. and Frossard, E. (2009) Leaf growth and photosynthetic capacity as affected by leaf position, plant nutritional status and growth stage in *Dioscorea alata* L. *Journal of Animal and Plant Sciences* 5, 483–493.

Hgaza, V.K., Diby, L.N., Tié, T.B., Tschannen, A., Aké, S., Assa, A. and Frossard, E. (2011) Growth and distribution of roots of *Dioscorea alata* L. do not respond to mineral fertiliser application. *Open Plant Science Journal* 5, 14–22.

Hgaza, V.K., Diby, L.N., Oberson A., Tschannen, A., Tié, B.T., Sangakkara, U.R., Ake, S. and Frossard, E. (2012) Nitrogen use by yam as affected by mineral fertilizer application. *Agronomy Journal* 104, 1558–1568.

Hill, T.G. and Freeman, W.G. (1903) The Root-Structure of Dioscorea prehensilis: With Plate XIX, and a Figure in the Text. *Annals of Botany* 17, 413–424.

Hills, R. and Wilkin, P. (2017) *Dioscorea cotinifolia*. The IUCN Red List of Threatened Species 2017: e.T100890102A100890104. http://dx.doi.org/10.2305/IUCN.UK.2017-3.RLTS.T100890102A100890104.en (accessed 7 June 2018)

Hills, R., Muthama Muasya, A., Maurin, O. and Wilkin, P. (2018) A threatened new species of *Dioscorea* from KwaZulu-Natal, South Africa, *Dioscorea hurteri* (Dioscoreaceae). *Kew Bulletin* 73, 14. https://doi.org/10.1007/s12225-018-9742-9 (accessed 18 September 2019).

Hladik, A. and Dounias, E. (1993) Wild yams of the African forests as potential food resources. In: Hladik, C.M., Hladik, A., Linares, O.F., Pagezy, H., Semple, A., Hadley, M. (eds) *Tropical forests, people and food: biocultural interactions and applications to development*. Parthenon UNESCO, Paris, France, pp. 163–176.

Hladik, A., Bahuchet, S., Ducatillion, C. and Hladik, C.M. (1984) Les plantes a tubercules de la foret dense d' Afrique centrale. *Revue d'Ecologie* (*Terre et Vie*) 39, 249–290.
Hogan, C.M. (2011) Respiration. *Encyclopedia of Earth.* Mark McGinley and C.J. Cleveland (eds) National Council for Science and the Environment. Washington, D.C.
Hokche. O., Berry, P.E. and Huber, O. (2008) Nuevo Catálogo de la Flora Vascular de Venezuela. Fundaci&# x00F3;n Instituto Botánico de Venezuela.
Holttum, R.E. (1964) *Plant Life in Malaya*. Longmans, Green and Co., London.
Hong, S.K., Kim, W.G., Lee, Y.K., Choi, H.W., Choi, K.J., and Lee, S.Y. (2010) Leaf Spot of yam caused by *Pseudophloeosporella dioscoreae* in Korea. *Mycobiology* 38, 78–80.
Hornung, P.S., do Prado Cordoba, L., da Silveira Lazzarotto, S.R., Schnitzler, E., Lazzarotto, M. and Ribani, R.H. (2017) Brazilian Dioscoreaceas starches. Thermal, structural and rheological properties compared to commercial starches. *Journal of Thermal Analysis and Calorimetry* 127, 1869–1877.
Horrocks, M. and Nunn, P.D. (2007) Evidence for introduced taro (*Colocasia esculenta*) and lesser yam (*Dioscorea esculenta*) in Lapita-era (c. 3050–2500 cal. yr BP) deposits from Bourewa, southwest Viti Levu Island, Fiji. *Journal of Archaeological Science* 34, 739–748.
Horrocks, M., Grant-Mackie, J. and Matisoo-Smith, E. (2008) Introduced taro (*Colocasia esculenta*) and yams (*Dioscorea* spp.) in Podtanean (2700–1800 years BP) deposits from Mé Auré Cave (WMD007), Moindou, New Caledonia. *Journal of Archaeological Science* 35, 169–180.
Hseu, S.H., Lai, W.C., Hung, Y.H. and Deng, T.C. (2010) Yam bacterial leaf spot caused by *Bacillus pumilus. Plant Pathology Bulletin* 19, 213–224.
Hsu, C.L., Hurang, S.L., Chen, W., Weng, Y.M. and Tseng, C.Y. (2004) Qualities and antioxidant properties of bread as affected by the incorporation of yam flour in the formulation. *International Journal of Food Science and Technology* 39, 231–238.
Hsu, F.L., Lin, Y.H., Lee, M.H., Lin, C.L. and Hou, W.C. (2002) Both dioscorin, the tuber storage protein of yam (*Dioscorea alata* cv. Tainong No. 1), and its peptic hydrolysates exhibited angiotensin converting enzyme inhibitory activities. *Journal of Agricultural and Food Chemistry* 50, 6109–6113.
Hsu, K-M., Tsai, J-L., Chen, M-Y., Ku, H-M. and Liu, S-C. (2013) Molecular phylogeny of *Dioscorea* (Dioscoreaceae) in East and Southeast Asia. *Blumea* 58, 21–27.
Huang, H., Gao, S., Chen, L. and Jiao, X. (2008a) *In vitro* induction and identification of autotetraploids of *Dioscorea zingiberensis. In Vitro Cellular and Developmental Biology – Plant* 44, 448–455.
Huang, T.C. (2003) *Flora of Taiwan, Volume Six,* 2nd edn. Editorial Committee of the Flora of Taiwan, Department of Botany, National Taiwan University, Taipei, Taiwan.
Huang, W., Zhang, H., Ni, J., Zuo, H., Qiu, L., Li, H. and Li, H. (2008b) The best utilization of *Dioscorea zingiberensis* C. H. Wright by an eco-friendly process. *Bioresource Technology* 99, 7407–7411.
Hudzari, R.M., Ssomad, M.A.H.A., Rizuwan, Y.M., Asimi, M.N.N. and Abdullah, A.B. (2011) Development of automatic alkaloid removal system for *Dioscorea hispida. Frontiers in Science* 1, 16–20.
Humbert, H. (1946–1984) Flore de Madagascar et des Comores. Museum National D'Histoire Naturelle; Paris http://www.biodiversitylibrary.org/ (accessed 28 March 2018).
IFAD (2016) Gabon - International Fund for Agricultural Development. https://www.ifad.org/en/web/operations/country/id/gabon (accessed 18 October 2019).
Igwilo, N. and Okoli, O.O. (1988) Evaluation of yam cultivars for seed yam production, using the minisett technique. *Field Crops Research* 19, 81–89.
IITA (2009) Annual report, International Institute of Tropical Agriculture, Ibadan, Nigeria.
IITA (2013) Growing seed yams in the air. In IITA weekly Bulletin No 2175 of 3–7 June 2013, 3–4.
Ike, P.C. and Odjuvwederdie, E.I. (2006) Determination of yam production and economic efficiency among small-holder farmers in South Eastern Nigeria. *Journal of Central European Agriculture* 7, 337–340.
Ikeemobi, C.O., Chelvarajan, R.L. and Ukoija, A.I. (1986) Biochemical aspects of wound healing in yams. Proceedings of the third triennial symposium of the International Society for Tropical Root Crops - Africa branch held in Owerri, Nigeria, 17–23 August 1986 page 157 (abstract).
Ikotun, T. (1983) Postharvest microbial rot of yam tubers in Nigeria. *Fitopatología Brasüdra* 8, 1–7.
Ile, E.I., Craufurd, P.Q., Battey, N.H. and Asiedu, R. (2006) Phases of Dormancy in Yam Tubers (*Dioscorea rotundata*). *Annals of Botany* 97, 497–504.
Imakawa, S. (1967) Studies on the browning of Chinese yam. *Memoir of the Faculty of Agriculture, Hokkaido University, Japan* 6, 181.
Imeh, J., Onimisi, M.Y. and Jonah, S.A. (2012) Effect of irradiation on sprouting of water yam (*Dioscorea alata*) using different doses of gamma radiation. *American Journal of Chemistry* 2, 137–141.
Infonet (2019) Yam. https://www.infonet-biovision.org/PlantHealth/Crops/Yam (accessed 3 November 2019).

IPGRI/IITA (1997) Descriptors for yam (*Dioscorea* species). International Institute of Tropical Agriculture, Ibadan, Nigeria/International Plant Genetic Resources Institute, Rome, Italy.

Itefa Degefa Alemu and Baressa Anbessa (2017) Traditional agronomic practices of yam (*Dioscorea* species) in Abaya Woreda, Southern Ethiopia. *International Journal of Chemical and Natural Sciences* 5, 399–503.

Itodo, I.N. and Daudu, J.O. (2007) Performance evaluation of a yam (*Dioscorea* species) harvester. *Agricultural Mechanization in Asia Africa and Latin America* 38, 79–83.

Itodo, I.N. and Daudu, J.O. (2013). Performance evaluation of a tractor mounted yam [Dioscorea spp.] harvester fitted with a collection unit. Conference: 2013 Kansas City, Missouri, July 21–24, 2013. DOI:10.13031/AIM.20131592710

Jaiaree, N., Itharat, A. and Kumapava, K. (2010) Cytotoxic saponin against lung cancer cells from *Dioscorea birmanica* Prain and Burkill. *Journal of the Medical Association of Thailand* 93, 192–197.

Jaiarree, N., Itharat A. and Poomirat, S. (2018) Cytotoxic activity of *Dioscorea birmanica* Prain and Burkill extract and its cytotoxic compound against breast and ovarian cancer cells. *Thammasat Medical Journal* 18, 545–552.

Jairo, S.M., Jorge, F.F., Fernando, F. and Jorge, H. (2018) Lintnerized yam starches (*Dioscorea alata* cv. Diamante 22) for food industry. *Advance Journal of Food Science and Technology* 16, 234–243.

Jaleel, C.A., Gopi, R., Manivannan, P., Kishorekumar, A., Gomathinayagam, M. and Panneerselvam, R. (2007) Changes in biochemical constituents and induction of early sprouting by triadimefon treatment in white yam (*Dioscorea rotundata* Poir.) tubers during storage. *Journal of Zheijang University Science B* 8, 283–288.

Jarmai, S. and Montford, L.C. (1968) Yam flour for the production of fufu. *Ghana Journal of Agricultural Science* 1, 161–163.

Jasons (2019) Festivals of New Caledonia. https://www.jasons.com/new-caledonia/festivals-of-new-caledonia/ (accessed 25 October 2019)

Jeannoda, V.H., Razanamparany, J.L., Rajaonah, M.T., Monneuse, M-O., Hladik, A. and Hladik C.M. (2007) Les ignames (*Dioscorea* species) de Madagascar: espèces endémiques et formes introduites; diversité, perception, valeurnutritionnelle et systèmes de gestion durable. *Revue d'Ecologie, Terre et Vie, Société Nationale de Protection de la Nature* 62, 191–207.

Jeffers, J.P.W. (1977) Mechanization of yam and sweet potato production in Barbados. Proceedings of the 4th Symposium of the International Society for Tropical Root Crops (Colombia, 1976), IDRC-080e (Cock, J., MacIntyre, R. and Graham, M. eds). Ottawa, Canada: International Development Research Centre 275–277.

Jiménez-Montero, M. and Aguilar, A. (2016) Evaluación morfoagronómica de la papa de aire (*Dioscorea bulbifera* L.) en Panamá. *Cultivos Tropicales* 37, 14–21.

Jin, S-S., Wang, Y., Li, X-J., Li, X., Zhao, W-S., Zhou, B., Zhao, C-C. and Gao, W-Y. (2017) Phytochemical and chemotaxonomic studies on *Dioscorea collettii*. *Biochemical Systematics and Ecology* 71, 10–15.

Jin, Y. and Sun, L. (2009) Development of composite functional beverage with chinese yam and red dates [J]. *Hubei Agricultural Sciences* 7th edition.

Joachim, R., Farant, M., Amory, R. and Osseux, J. (2003) *L'igname. Manuel du Planteur*. Chambres d'Agriculture de Guadeloupe et de Martinique, French West Indies.

Joob, B. and Wiwanitkit, V. (2014) Asiatic bitter yam intoxication. *Asian Pacific Journal of Tropical Biomedicine* 4, S42.

Jova, M.C., Kosky, R.G., Morales, S.R., Torres, J.L., Cabrera, A.R., Pérez, M.B., Pino, A.S., Vega, V.M. and Rodriguez, G.R. (2008) Multiplicacionin vitrode segmentos nodales del clon de name Blanco de Guinea (*Dioscorea cayenensis–D. rotundata*) en sistemas de cultivo semi automatizado. *Biotechnologie, Agronomie, Société et Environnement* 10, 97–103.

Jova, M.C., Kosky, R.G. and Cuellar, E.E. (2011) Effect of liquid media culture systems on yam plant growth (*Dioscorea alata* L. 'Pacala Duclos'). *Biotechnologie, Agronomie, Société et Environnement* 15, 515–521.

Joy Online (2005) Kwame Nkrumah University of Science and Technology, lecturers develop fufu machine, https://www.modernghana.com/news/68826/1/knust-lecturers-develop-fufu-machine.html (accessed 1 May 2018).

JSTOR (2018) Global Plants https://plants.jstor.org/compilation/Dioscorea.aculeata (accessed 12 May 2018).

JSTOR (2019) Digital library. https://www.jstor.org (accessed 24 December 2019).

Kadiri, M., Ojewumi, A. W., Aniola, T. O. and Musa, A. H. (2014) Ethno-botanical survey of plants commonly used for ceremonial activities among Yoruba tribe of south West Nigeria. *Direct Research Journal of Health Pharmacology* 2, 1–5.

Kagoda, F., Coyne, D.L., Asiedu, J., Wanyera, N., Mudiope, J., Dusabe, J. and Nabulime, C. (2010) Yam germplasm conservation and screening in Uganda. In: Tenywa, J.S., Adipala, E., Nampala, P., Tusiime, G., Okori, P. and Kyamuhangire, W. (eds) 7th African Crop Science Conference Proceedings, Entebbe, Uganda 7, 245–249.

Kaladhar, D.S.V.G.K., Narsinga Rao, V. Sreenu Barla and Harasreeramulu, S. (2008) Comparative antimicrobial studies of *Dioscorea hamiltonii* Hook.f. tubers with *Azadirachta indica* stem. *Journal of Pharmaceutical Science and Technology* 2, 284–287.

Kang, B.T. and Wilson, J.E. (1981) Effect of mound size and fertilizer on white guinea yam (*Dioscorea rotondata*) in Southern Nigeria. *Plant and Soil* 61, 319–327.

Kanika Patel, Manoj Gadewar, Vijay Tahilyani and Dinesh Kumar Patel (2012) A review on pharmacological and analytical aspects of diosgenin: a concise report. *Natural Products and Bioprospecting* 2, 46–52.

Karam, L.B., Ferrero, C., Martino, M.N., Zaritzky, N.E. and Grossmann, M.V.E. (2006) Thermal, microstructural and textural characterisation of gelatinised corn, cassava and yam starch blends. *Institute of Food Science and Technology* 41, 805–812.

Karnick, C.R. (1969) *Dioscorea* (yams) – the food of the slaves, with potentials for newer drugs: a review. *Pharmaceutical Biology* 9, 1372–1391.

Karthikeyan, S., Jain, S.K., Nayar, M.P. and Sanjappa, M. (1989) *Florae Indicae Enumeratio*: *Monocotyledonae*. Fl. India, Ser. 4. Botanical Survey of India, Calcutta, pp. 73–76.

Kawakam, K. (1970) Yam culture in Japan. Tropical root and tuber crops tomorrow. Proceedings of the 2nd International Symposium on Tropical Root and Tuber Crops 1, 102.

Kawasaki, T., Komori, T. and Setoguchi, S. (1968) Furanoid nordoterpenes from dioscoreaceae plants. *Chemical and Pharmaceutical Bulletin* 16, 2430–2435.

Kay, D.E. (revised by Gooding E.G.B.) (1987) *Root Crops. Crop and Product Digest No.2*, Second Edition. London: Tropical Development and Research Institute

Kayode, G.O. (1985) Effects of NPK fertilizer on tuber yield, starch content and dry matter accumulation of white Guinea yam (*Dioscorea rotundata*) in a forest alfisol of South Western Nigeria. *Experimental Agriculture*, 21, 389–393.

Kelmanson, J.E., Jager, A.K. and van Staden, J. (2000) Zulu medicinal plants with antibacterial activity. *Journal of Ethnopharmacology* 69, 241–246.

Kennerley, J.A. and Wilkin, P. (2017) *Dioscorea pteropoda*. *The IUCN Red List of Threatened Species* 2017: e.T66770271A66770273. http://dx.doi.org/10.2305/IUCN.UK.2017-3.RLTS.T66770271A66770273.en (accessed 16 September 2019).

Kenyon, L., Shoyinka, S.A., Hughes, J.A. and Odu, B.O. (2001) An overview of viruses infecting *Dioscorea* yam in sub-Saharan Africa. *Plant Virology in Sub-Saharan Africa*. pp. 432–438.

Kenyon L., Lebas BS. and Seal SE. (2008) Yams (*Dioscorea* species) from the South Pacific Islands contain many novel *Badnaviruses*: implications for international movement of yam germplasm. *Archives of Virology* 153, 877–889.

Kermarrec, A., Degras, L. and Anais, A. (1981) The yam nematode *Scutellonema bradys* in the Caribbean: distribution and international quarantine. *Agronomie Tropicale* 36, 364–368.

Kew (2014) Catalogue of Life: 2014 Annual Checklist. http://www.catalogueoflife.org/annual-checklist/2014/details/species/id/9743532 (accessed 20 November 2017).

Kew (2020) http://apps.kew.org/herbcat/getImage.do?imageBarcode=K001150180 (accessed 1 June 2020).

Kew Science (2019) Plants of the World. http://powo.science.kew.org/taxon/urn:lsid:ipni.org:names:317936-1 (accessed 9 August 2019).

Kewscience (2019) Plants of the world online. *D. wallichii*, http://powo.science.kew.org/taxon/urn:lsid:ipni.org:names:318787-1 (accessed 9 July 2019).

Khare, C.P. (2010) *Encyclopedia of Indian Medicinal Plants*. Springer Verlag.

Kim, H.Y., Kim, S.S., Bae, S.H., Bae, E.H., Ma, S.K. and Kim, S.W. (2014) Acute interstitial nephritis induced by *Dioscorea quinqueloba*. *BMC Nephrology* 3, 15 143.

Kim, J.G., Shin, Y-K. and Kim, K-Y. (2018) Promotion of keratinocyte proliferation by tracheloside through ERK1/2 stimulation. *Evidence-Based Complementary and Alternative Medicine*. Published online 2018 Jul 26. doi: 10.1155/2018/4580627.

Kim, J.Y., Lee, Y.C., Kim, Y.C., Kim, Y.B. and Choi, H.W. (2016) Research on processing properties of yam (*Dioscorea batatas*) by drying methods and milling methods. *Food Engineering Progress* 20, 307–313.

Kim, K.H., Choi, S.U., Choi, S.Z., Son, M.W. and Lee, K.R. (2011a) Withanolides from the rhizomes of *Dioscorea japonica* and their cytotoxicity. *Journal of Agricultural and Food Chemistry* 59, 6980–6984.

Kim, K.H., Kim, M.A., Moon, E., Kim, S.Y., Choi, S.Z., Son, M.W. and Lee, K.R (2011b) Furostanol saponins from the rhizomes of *Dioscorea japonica* and their effects on NGF induction. *Bioorganic & Medicinal Chemistry Letters* 21, 2075–2078.

Kim, S.K., Lee, S.C., Lee, B.H., Choi, H.J., Kim, K.U. and Lee, I.J. (2003a) Bulbil formation and yield responses of Chinese yam to application of gibberellic acid, mepiquat chloride and trinexapac-ethyl. *Journal of Agronomy and Crop Science* 189, 255–260.

Kim, S.K., Lee, S.C., Shin, D.H., Jang, S.W., Nam, J.W. and Lee, I.J. (2003b) Quantification of endogenous gibberellins in leaves and tubers of Chinese yam *Dioscorea opposite* Thunb. cv. Tsukune during tuber enlargement. *Plant Growth Regulation* 39, 125–130.

Kim, W.K., Hwang, Y-S. and Yu, S.H. (2008) Two species of *Penicillium* associated with blue mold of yam in Korea. *Mycobiology* 36, 217–221.

Kirizawa, M., Xifreda, C.C., Couto, R. and Araújo, D. (2015) Dioscoreaceae in Lista de Espécies da Flora do Brasil. Jardim Botânico do Rio de Janeiro. Disponível em: http://floradobrasil.jbrj.gov.br/jabot/floradobrasil/FB52722 (accessed 16 December 2019)

Kirizawa, M., Xifreda, C.C. and Silva, J.H. (2016) Diversidade florística de Dioscoreaceae na Reserva Biológica do Alto da Serra de Paranapiacaba, Santo André, São Paulo, Brasil. *Hoehnea* 43, 99–117.

Kleinkopf, G.E., Brandt, T.L., Frazier, M.J. and Moller, G. (1997) CIPC residues on stored Russet Burbank potatoes. 1. Maximum label application. *American Potato Journal* 74, 107–117.

Klu, G.Y.P., Asare, E.K., Blay, E.T. and Ng, S.Y.C. (2005) Effect of medium type and incubation duration on improved *in vitro* tuberization in three *Dioscorea rotundata* Poir. cultivars. *Plant Genetic Resources Newsletter* 144, 24–29.

Knoth, J. (1993) Traditional storage of yams and cassava and its improvement. Deutshe Gesellschaft für Technische Zusammenarbeit (GTZ), Eschborn, Germany.

Koli, S.E. (1973) The response of yam (*Dioscorea rotundata*) to fertilizer application in Northern Ghana. *Journal of Agricultural Science* 80, 245–249.

Korada, R.R., Naskar, S.K. and Edison, S. (2010) Insect pests and their management in yam production and storage: a world review. *International Journal of Pest Management* 56, 337–349.

Košanin, N. (1914) *Dioscorea balcanica. Österreichische Botanische Zeitschrift.* 64, 38–39.

Košanin, N. (1929) Nachträge zur *Dioscorea balcanica* Kos. *Bulletin de L'institute et du Jardin Botaniques de L'université de Beograd* 1, 199–203.

Kouadio, N.C., Patrice, Y.A.Y.D., Gnanwa, M.J., Konan, H.K., Kouadio, J.P.E.N. and Kouame, L.P. (2017) Effects of post-harvest storage and boiling time on some biochemical parameters of two yams species (*Dioscorea cayenensis-rotundata* variety 'Tambi' and *Dioscorea alata* variety 'Azaguié') tubers. *Journal of Basic and Applied Research* 3, 16–21.

Kouakou Dje, M., Dabonne, S. Tagro Guehi, S. and Patrice Kouame, L. (2010) Monitoring of some biochemical parameters of two yam species (*Dioscorea* spp.) tubers parts during post-harvest storage. *Advance Journal of Food Science and Technology* 2, 178–183.

Kovács, A., Forgo, P., Zupkó, I., Réthy, B., Falkay, G., Szabó, P. and Hohmann, J. (2007) Phenanthrenes and a dihydrophenanthrene from *Tamus communis* and their cytotoxic activity. *Phytochemistry* 68, 687–691.

Kumar, P.L., Bhattacharjee, R., Lopez-Montes, A., Abberton, M.T., Perkee, E.and Richardson, O. (2013) YAMS 2013: First Global Conference on Yams. In: Ibadan, Nigeria: International Institute of Tropical Agriculture. https://dioscorea.org/where.html (accessed 18 May 2018).

Kumar, S., Das, G., Han-Seung Shin and Patra, J.K. (2017) *Dioscorea* species (a wild edible tuber): a study on its ethnopharmacological potential and traditional use by the local people of Similipal Biosphere Reserve, India. *Frontiers in Pharmacology* 8, 52.

Kumoro, A.C. and Hartati, I. (2015) Microwave assisted extraction of dioscorin from Gadung (*Dioscorea Hispida* Dennst) tuber flour. *Procedia Chemistry* 14, 47–55.

Kwon, C-S., Sohn, H.Y., Kim, S.H., Kim, J.H., Son, K.H., Lee, J.S., Lim, J.K. and Kim, J-S. (2003) Anti-obesity effect of *Dioscorea nipponica* Makino with lipase-inhibitory activity in rodents. *Bioscience, Biotechnology and Biochemistry* 67, 1451–1456.

Kwoseh, C.K., Plowright, R.A., Bridge, J. and Asiedu, R. (2007) Search for *Scutellonema bradys* resistance in yams (*Dioscorea* species). *Journal of Science and Technology* 27, 33–42.

Labuschagne, N., Baimey, H., Kossi, B. and Coyne, D. (2006) Effect of fertiliser application on yam nematode (*Scutellonema bradys*) multiplication and consequent damage to yam (*Dioscorea* species) under field and storage conditions in Benin. *International Journal of Pest Management* 52, 63–70.

Lacointe, A. and Zinsou, C. (1987) Effet de la date de plantation sur la croissance et le développement de plantules d'igname (*Dioscorea alata* L.) produites par culture *in vitro*. *Agronomie* 7, 475–481.

Lallu, N. (1997) Low temperature breakdown in kiwifruit. *Acta Horticulturae* 444, 579–585.

Lancaster, P.A. and Coursey, D.G. (1984) FAO Agricultural Services Bulletin 59. Food and Agriculture Organization of the United Nations Rome, 1984.

Lankester, E. (1832) *Vegetable Substances Used for the Food of Man*. Great Britain, Charles Knight, https://books.google.co.uk/books?id=1aAUAAAAYAAJandpg=PA167andlpg=PA167anddq=Dioscorea+sativaandsource=blandots=gdekdC-N1mandsig=PmYD01V6CM7aL2Nhwl1_yEchT-0andhl=enandsa=Xandved=2ahUKEwjh-rT2jNPcAhUIKMAKHQy6CWYQ6AEwDXoECAAQAQ#v=onepageandq=Dioscorea%20sativaandf=false (accessed 3 October 2019).

Lauzer, D., Laublin, G., Vincent, G. and Cappadocia, M. (1992) *In vitro* propagation and cytology of wild yams, *Dioscorea abyssinica* Hoch. and *D. mangenotiana* Miège. *Plant Cell, Tissue and Organ Culture* 28, 215–223.

Lebot, V. (2009) *Tropical root and tuber crops: cassava, sweet potato, yams and aroids*. CAB International, Wallingford, UK.

Lebot, V. (2019) *Tropical root and tuber crops: cassava, sweet potato, yams and aroids*. 2nd edition. CAB International, Wallingford, UK.

Lebot, V., Malapa, R. and Kuttolamadathil, A. (2016) The Pacific yam (*Dioscorea nummularia* Lam.), an under-exploited tuber crop from Melanesia. *Genetic Resources and Crop Evolution*, 64, 217–235.

Lebot, V., Malapa, R. and Abraham, K. (2017) The Pacific yam (*Dioscorea nummularia* Lam.), an under-exploited tuber crop from Melanesia. *Genetic Resources and Crop Evolution* 64, 217–235.

Lebot, V., Faloye, B., Okon, E. and Gueye, B. (2019) Simultaneous quantification of allantoin and steroidal saponins in yam (*Dioscorea* spp.) powders. *Journal of Applied Research on Medicinal and Aromatic Plants* 13, 100–108.

Lee, M.H., Lin, Y.S., Lin, Y.H., Hsu, F.L. and Hou, W.C. (2003) The mucilage of yam (*Dioscorea batatas* Decne) tuber exhibited antiotensin converting enzyme inhibitory activities. *Botanical Bulletin Academia Sinica Taipei* 44, 267–273.

Lee, N.N., Kim, J.A., Kim, Y.W. and Kim, T.D. (2018) Microtuberization and acclimatization in the *Dioscorea cayenensis* Thunb. by the carbon source. *Journal of Plant Biotechnology* 45, 125–130.

Leng, M., I. Gouado, and R. Ndjouenkeu. (2011) Blanching and drying behaviour of *Dioscorea schimperiana* and impact on cellular changes and on calcium, ascorbic acid and β-carotene contents. *American Journal of Food Technology* 6, 362–373.

Lentza-Rizos, C. and Balokas, A. (2001) Residue levels of chlorpropham in individual tubers and composite samples of postharvest-treated potatoes. *Journal of Agricultural and Food Chemistry* 49, 710–714.

Leonard, T.M. (2005) *Encyclopaedia of the Developing World*. Psychology Press, Hove.

Levya, J.F. and Guttierrez, E. (1937) Toxicological studies of Nami, *Dioscorea hispida*. *Journal of the Philippine Medical Association* 17, 349.

Li, B., Xu, P., Wu, S., Jiang, Z., Huang, Z., Li, Q.,and Chen, D. (2018) Diosgenin attenuates lipopolysaccharide-induced Parkinson's disease by inhibiting the TLR/NF-κB pathway. *Journal of Alzheimer's Disease* 64, 943–955.

Li, H., Huang, W., Wen, Y., Gong, G., Zhao, Q. and Yu, G. (2010) Anti-thrombotic activity and chemical characterization of steroidal saponins from *Dioscorea zingiberensis* C.H. Wright. *Fitoterapia* 81, 1147–1156.

Li, L., Meng, X., Fan, K. Wang, Y., Xiao, Y and Wany, Y. (2016) Advanced studies on *Rhizoma Dioscoreae Collettii*. *Chinese Journal of Integrative Medicine* 1-6. https://doi.org/10.1007/s11655-016-2263-1 (accessed 29 July 2019)

Li, M., Li, J., Liu, W., Liu, L., Lu, J., Niu, J., Liu, X. and Yang, Q. (2014) A protocol for *in vitro* production of microtubers in Chinese yam (*Dioscorea opposita*). *Bioscience, Biotechnology, and Biochemistry* 78, 1005–1009.

Li, Q-M., Li, Y., Zou, J-H., Guo, S-Y., Wang, F., Yu, P. and Su, X-J. (2020) Influence of adding Chinese yam (*Dioscorea opposita* Thunb.) flour on dough rheology, gluten structure, baking performance, and antioxidant properties of bread. *Foods* 9, 256.

Likhitwitayawuid, K., Angerhofer, C.K., Chai, H, Pezzuto, J.M., Cordell, G.A. and Ruangrungsi, N. (1993) Cytotoxic and antimalarial alkaloids from the bulbs of *Crinum amabile*. *Journal of Natural Products* 56, 1331–1338.

Lin, D., Li, C., Peng, Y., Gao, H. and Zheng, J. (2014) Cytochrome P450–Mediated Metabolic Activation of Diosbulbin B. *Drug Metabolism and Disposition* 42, 1727–1736.

Lin, J.Y., Lu, S., Liou, Y.L. and Liou, H.L. (2006) Antioxidant and hypolipidaemic effects of a novel yam-boxthorn noodle in an *in vivo* murine model. *Food Chemistry* 94, 377–384.

Lin, Y. P., Lee, T.Y., Tsen, J.H. and King, A.E. (2007) Dehydration of yam slices using FIR-assisted freeze drying. *Journal of Food Engineering* 79, 1295–1301.

Liu, H.W., Hu, K., Zhao, Q.C., Cui, C.B., Kobayashi, H. and Yao, X.S. (2002) Bioactive saponins from *Dioscorea futschauensis*. *Pharmazie* 57, 570–572.

Loko, Y.L., Dansi, A., Tamo, M., Bokonon-Ganta, A.H. Assogba, P., Dansi, M., Vodouhè, R., Akoegninou, A. and Sanni, A. (2013) Storage insects on yam chips and their traditional management in Northern Benin. *The Scientific World Journal* 2013(6), 484–536.

Loko, Y.L., Bhattacharjee, R., Agre, A.P., Dossou-Aminon, I., Orobiyi, A., Djedatin, G.L. and Dansi, A. (2016) Genetic diversity and relationship of Guinea yam (*Dioscorea cayenensis* Lam.–*D. rotundata* Poir. complex) germplasm in Benin (west Africa) using microsatellite markers. *Genetic Resources and Crop Evolution* 64, 1205–1219.

López, C.Q., Corral, P., Lorrain-Lorrette, B., Martinez-Swatson, K., Michoux, F. and Simonsen, H.T. (2018) Use of a temporary immersion bioreactor system for the sustainable production of thapsigargin in shoot cultures of *Thapsia garganica*. *Plant Methods* 14, 1–17.

Low, T. (1988) *Wild Food Plants of Australia*. Angus and Roberstson, North Ryde, NSW, Australia.

Lubag, A.J.M., Laurena, A.C. and Tecson-Mendoza, E.M. (2008) Antioxidants of purple and white greater yam (*Dioscorea alata* L.) varieties: identification of anthocyanidin alatanin with high antioxidant potential. *Philippine Journal of Science* 137, 61–67.

Lugo, W.I., Lugo, H.M., Gonzalez, A., Rafols, N. and Almodovar, C. (1993) Tillage and fertilizer rate effects on yam yields (*Dioscorea alata* L.). *Journal of Agriculture of the University of Puerto Rico* 77, 153–159.

Lyonga, S.N. (1985) Studies on the fertilization of yams and on yam tuber storage in Cameroon. Proceedings of the Sixth International Symposium on Tropical Root Crops. Lima, Peru.

Lyonga, S.N., Fayemi, A.A. and Agboola, A.A. (1973) Agronomic studies on edible yam in the grassland plateau region of the United Republic of Cameroon. In: Leakey, C.L.A. (ed.) Proceeding of Third Symposium of the International Society of Tropical Root Crops, pp. 340–346.

Ma, F., Zhang, Y., Yao, Y., Wen, Y., Hu, W., Zhang, J., Liu, X., Bell, A.E. and Tikkanen-Kaukanen, C. (2017) Chemical components and emulsification properties of mucilage from *Dioscorea opposita* Thunb. *Food Chemistry* 228, 315–322.

Ma, J., Kang, S.Y., Meng, X., Kang, A.N, Park, J.H., Park, Y-K. and Jung, H.W. (2018) Effects of rhizome extract of *Dioscorea batatas* and its active compound, allantoin, on the regulation of myoblast differentiation and mitochondrial biogenesis in C_2C_{12} myotubes. *Molecules* 23, E2023.

Maalekuu, B.K., Saajah, J.K. and Addae, A.K. (2014) Effect of three storage methods on the quality and shelf-life of White Yam (*Dioscorea rotundata*) cultivars Pona and Tela. *Journal of Agricultural Science* 6, 221–230.

Magwé-Tindo, J., Wieringa, J.J., Sonké, B., Zapfack, L., Vigouroux. Y., Couvreur, T.L.P. and Scarcelli, N. (2018) Guinea yam (*Dioscorea* species) wild relatives identified using whole plastome phylogenetic analyses. *Taxon* 67, 905–915.

Mahmud, A. and Idris, N.M. (2017) Assessment of post-harvest losses of yam (*Dioscorea* species) in selected districts of Karu local government area, Nasarawa state. *FUW Trends in Science and Technology Journal* 2, 213–219.

Maiden, J.H. (1889) *The useful native plants of Australia: including Tasmania*. Turner and Henderson, Sydney, Australia.

Maina, M. (2008) Yam (*Dioscorea* species). How will this crop be revived to enhance food security in East Africa? In: Agricultural Biosciences Vol. I. Proceedings of the 1st International e-Conference on Agricultural Biosciences 2008. www.e-conference.elewa.org/agriculture (accessed 3 November 2019).

Makame, M. (1994) Performance of water yam (*Dioscorea alata*) under by various agronomic practices. *Acta Horticulturae* 380, 519–519.

Malapa, R., Arnau, G., Noyer, J.L. and Lebot, V. (2005) Genetic diversity of the greater yam (*Dioscorea alata* L.) and relatedness to *D. nummularia* Lam. and *D. transversa* Br. as revealed with AFLP markers. *Genetic Resources and Crop Evolution* 52, 919–929.

Malapa, R., Noyer, J.L., Marchand, J.L. and Lebot, V. (2006) Genetic relationship between *Dioscorea alala* L. and *D. nummularia* Lam. as revealed by AFLP markers. In: Motley, T.J., Zerega, N. and Cross, H. (eds) *Darwin's harvest: new approaches to the origins, evolution, and conservation of crops*. Columbia University Press, New York, NY, pp. 239–265.

Maneenoon, K., Sirirugsa, P. and Sridith, K. (2008) Ethnobotany of *Dioscorea* L. (Dioscoreaceae), a major food plant of the Sakai Tribe at Banthad Range, Peninsular Thailand. *Ethnobotany Research and Applications*, 6, 385–394.

Mansfeld's World Database of Agriculture and Horticultural Crops (2019) http://mansfeld.ipk-gatersleben.de/apex/f?p=185:3:0 (accessed 29 August 2019).

Mantell, S.H. and Hugo, S.A. (1989) Effect of photoperiod, minimal medium strength, inorganic ammonium, sucrose and cytokinin on root, shoot and microtuber development in shoot cultures of *Dioscorea alata* L. and *D. bulbifera* L. yams. *Plant Cell, Tissue and Organ Culture* 16, 23–37.

Mantell, S.H., Haque, S.Q. and Whitehall, A.P. (1978) Clonal multiplication of *Dioscorea alata* L. and *Dioscorea rotundata* Poir. yams by tissue culture. *Journal of Horticultural Science* 53, 95–98.

Marchant, N.G., Wheeler, J.R., Rye, B.L., Bennett, E.M., Lander, N.S. and Macfarlane, T.D. (1987) *Flora of the Perth Region*. Western Australian Herbarium.

Marcos, J., Lacointe, A., Tournebize, R., Bonhomme, R. and Sierra, J. (2009) Water yam (*Dioscorea alata* L.) development as affected by photoperiod and temperature: experiment and modeling. *Field Crops Research* 111, 262–268.

Markson, A.A., Amadioha, A.C., Omosun, G., Madunagu, B.E., Udo, S.E. and Umana, E.J. (2012) Control of *Botryodiplodia theobromae* causing Tissue Rot of White Yam (*Dioscorea rotundata* Poir.). *Scholarly Journal of Agricultural Science* 2, 1–7.

MARNDR (2017) Agricultural production data 2014-2016. Ministère de l'Agriculture, des Ressources Naturelles et du Développement Rural, Haiti.

Marsden, W. (1812) *A Dictionary of the Malayan Language*. Cox & Baylis, London.

Martin, F. and Rubeste, R. (1976) The polyphenol of *Dioscorea alata* yam tubers associated with oxidative browning. *Journal of Agricultural and Food Chemistry* 14, 67–70.

Martin, F.W. (1969) The species of *Dioscorea* containing sapogenin. *Economic Botany* 23, 373–379.

Martin, F.W. (1973) A collection of West African yams. International Society for Tropical Root Crops 3rd symposium 23–27, http://www.istrc.org/images/Documents/Symposiums/Third/3rd_symposium_proceedings_10.pdf (accessed 6 November 2019).

Martin, F.W. (1974a) *Tropical yams and their potential.* Part 1. *Dioscorea esculenta*, United States Department of Agriculture, *Agriculture Handbook*, 457.

Martin, F.W. (1974b) *Tropical Yams and their Potential. Series* – Part 1. *Dioscorea esculentum*. USDA Agriculture Handbook No. 466.

Martin, F.W. (1974c) *Tropical Yams and their Potential.* Series – Part 2. *Dioscorea bulbifera*. USDA Agriculture Handbook No. 466.

Martin, F.W. (1976) *Tropical Yams and their Potential.* Part 3. *Dioscorea alata*, United States Department of Agriculture, Agriculture Handbook No. 495.

Martin, F.W. (1977) Prolonging the storage life of yams by removal of shoots and by treatment with gibberillic acid and waxes. *Tropical Root and Tuber Crops Newsletter* 10, 39–44.

Martin, F.W. (Ed). (1984) *Hand Book of Tropical Food Crops*. USA: CRC Press, Inc. Boca Raton, Florida.

Martin, F.W. and Degras, L. (1974) *Tropical yams and their potential.* Part 2. *Dioscorea bulbifera*, USDA Agriculture Handbook No. 466.

Martin, F.W. and Degras, L. (1978) *Tropical yams and their potential.* Part 5. *Dioscorea trifida*. United States Department of Agriculture, Agriculture Handbook, 522.

Martin, F.W. and Gaskins, M.J. (1968) Cultivation of the sapogenin-bearing *Dioscorea* species. *US Department of Agriculture, Production Research Report* No.1 03, 3–5, 13–14.

Martin, F.W. and Ortiz, S. (1962) Origin and anatomy of tubers of *Dioscorea floribunda* and *D. spiculiflora*. *Botanical Gazette* 124, 416.

Martin, F.W. and Ortiz, S. (1963) Chromosome numbers and behaviours in some species of *Dioscorea*. *Cytologia* 28, 96–101.

Martin, F.W. and Ruberté R. (1972) Yams (*Dioscorea* spp.) for production of chips and French fries. *Journal of Agriculture of University of Puerto Rico* 56, 228–234.

Martin, F.W. and Sadik, S. (1977) *Tropical yams and their potential.* Part 4. *Dioscorea rotundata* and *Dioscorea cayenensis*. United States Department of Agriculture, *Agriculture Handbook* 502.

Martin, F.W., Cabanillas, E. and Gaskins, M.H. (1966) Economics of the saponin-bearing yam as a crop plant in Puerto Rico. *The Journal of Agriculture and the University of Puerto Rico* 1, 55, 57, 60.

Martin, F.W., Cabanillas, E. and Guadalupe, R. (1975) Selected varieties of *Dioscorea alata* L., the Asian Greater yam. *Journal of Agriculture, University of Puerto Rico* 59, 165–171.

Maundu, P.M., Ngugi, G.W. and Kabuye, C.H.S. (1999) *Traditional Food Plants of Kenya*. National Museum of Kenya, Report.

McAnuff-Harding, M.A., Omoruyi, F.O. and Asemota, H.N. (2006) Intestinal disaccharidases and some renal enzymes in streptozotocin-induced diabetic rats fed sapogenin extract from bitter yam (*Dioscorea polygonoides*). *Life Sciences* 78, 2595–2600.

McDonald, F.D., Alleyne, A.T., O'Garro, L.W. and Delauney, A.J. (1998) Yam anthracnose in the English-speaking islands of the Eastern Caribbean: successes and research advances in disease management. *Tropical Agriculture* 75, 53–57.

McKoy, M., Thomas, P., Asemota, H., Omoruyi, F. and Simon, O. (2014) Effects of Jamaican bitter yam (*Dioscorea polygonoides*) and diosgenin on blood and fecal cholesterol in rats. *Journal of Medicinal Food* 17, 1183–1188.

Melteras, M.V. (2005) Soil fertility as a factor affecting the production of yams (*Dioscorea* species) in Vanuatu. MPhil. thesis, School of Land and Food Sciences, University of Queensland, Brisbane, Australia.

Mengesha, W.A., Demissew, S., Fay, M.F. Smith, R.J., Nordal, I. and Wilkin, P. (2013) Genetic diversity and population structure of Guinea yams and their wild relatives in South and South West Ethiopia as revealed by microsatellite markers. *Genetic Resources and Crop Evolution* 60, 529–541.

Mercantilia (1989) *Guide to food transport: fruit and vegetables*. Mercantilia Publishers, Copenhagen, Denmark.

Meyer, J.Y. (2007) Rapport de mission sur l'Ile d'Uvea (Wallis and Futuna) du 6 au 17 Novembre 2007: Inventaire preliminaire de la flore vasculaire secondaire. Papeete, Tahiti: Ministère de l'Education, l'Enseignement Supérieur et la Recherche.

Michereff, S.J., Silveira, N.S.S., Reis, A. and Mariano, R.L.R. (1994) Epiphytic bacteria antagonistic to Curvularia leaf spot of yam. *Microbial Ecology* 28, 101–110.

Miège, J. (1958) Deux ignames nouvelles d'Afrique occidentale à tubercules vivaces. *Bulletin de l'Institut Fondamental d'Afrique Noire, série A* 20, 39–55.

Mignouna, H.D., Abang, M.M., Green, K.R. and Asiedu, R. (2001a) Inheritance of resistance in water yam (*Dioscorea alata*) to anthracnose (*Colletotrichum gloeosporioides*). *Theoretical and Applied Genetics* 103, 52–55.

Mignouna, H.D., Hjukeng, P., Abang, M.M. and Asiedu, R. (2001b) Inheritance of resistance to Yam Mosaic Virus, genus Potyvirus, in white yam (*Dioscorea rotundata*). *Theoretical and Applied Genetics* 103, 1196–1200.

Mignouna, H.D., Abang, M.M. and Asiedu, E. (2007) Yams. In: Kole, C. (ed.) *Genome Mapping and Molecular Breeding in Plants*. Vol. 3. Pulses, Sugar and Tuber Crops. Springer-Verlag, Berlin, Heidelberg, Germany, 271–296.

Mignouna, D.B., Akinola, A.A., Abdoulaye, T. and Maroya, N. (2016) Economic impacts of yam productivity research in West Africa: a case of YIIFSWA Project. Invited paper presented at the 5th International Conference of the African Association of Agricultural Economists, September 23–26, 2016, Addis Ababa, Ethiopia, 1–19.

Mignucci, J., Torres-Lopez, R., Hepperly, P.R. and Ramos-Busigo, D. (1985) Field diseases of tropical yams (*Dioscorea* spp.) and their control in Puerto Rico. VI of the International Society for Tropical Root Crops, Gosier (Guadeloupe), 1–6 July 1985, 285–290.

Milne-Redhead, E. (1975) Dioscoreaceae *Flora of Tropical East Africa*. Royal Botanic Gardens, Kew.

Ministry of Agriculture and Fisheries (2013) Jamaica - Demand study for yellow yam. http://www.ja-mis.com/Companionsite/Repository_Doc/Yellow%20Yam%20Research%20Paper.pdf (accessed 17 October 2019).

Miyazawa, M., Shimamura, H., Sei-ichi Nakamura and Hiromu Kameoka (1996) Antimutagenic Activity of (+)-β-Eudesmol and Paeonol from *Dioscorea japonica*. *Journal of Agricultural and Food Chemistry* 44, 1647–1650.

Miyoshi, N., Nagasawa, T., Mabuchi, R., Yasui, Y., Wakabayashi, K., Tanaka, T. and Ohshima, H. (2011) Chemoprevention of azoxymethane/dextran sodium sulfate-induced mouse colon carcinogenesis by freeze-dried yams Sanyaku and its constituent diosgenin. *Cancer Prevention Research* 4, 924–934.

Moerman, D.E. (1998) *Native American ethnobotany*. Portland, Oregan, Timber Press.

Mohan, V.R. and Kalidass, C. (2010) Nutritional and antinutritional evaluation of some unconventional wild edible plants. *Tropical and Subtropical Agroecosystems* 12, 495–506.

Mollica, J.Q., Cara, D.C., D'Auriol, M., Oliveira, V.B., Cesar, I.C. and Brandão, M.G.L. (2013) Anti-inflammatory activity of American yam *Dioscorea trifida* L.f. in food allergy induced by ovalbumin in mice. *Journal of Functional Foods* 5, 1975–1984.

Monteiro, D.A. and Peressin, V.A., (2002) Cultura do inhame. In: Cereda, M.P. (ed.) *Agricultura: tuberosas amiláceas Latino Americanas*. Fundação Cargill, São Paulo, pp. 511–518.

Montes, E., Torres, R., Andrade, R., Perez, O., Marimon, J. and Meza, I. (2009) Models of desorption isotherms of yam (*Dioscorea rotundata*). *Dyna* 76, 146.

Morales, M. and Bosch, S.M. (2015) Secret of long life lies underground. *New Phytologist*, 205, 463–467.

Morat, P. and Veillon, J.M. (1985) Contributions à la conaissance de la végétation et de la flore de Wallis et Futuna. *Bulletin du Muséum National d'Histoire Naturelle. Section B, Adansonia* 7, 259–329.

Moriya, C., Hosoya, T., Agawa, S., Sugiyama, Y., Kozone, I., Shin-ya, K., Terahara, N. and Kumazawa, S. (2015) New acylated anthocyanins from purple yam and their antioxidant activity. *Food and Nutrition Science* 79, 1484–1492.

Mosissa, D. and Abraha, B. (2018) Tuber wild plants of Shinasha people as food and medicine in Bullen district, North West of Ethiopia. *Biodiversity International Journal* 2, 53–58.

Mothogoane, M.S. (2013) National Herbarium, Pretoria. http://pza.sanbi.org/dioscorea-dregeana (accessed 7 August 2019.

Moula, N., Detiffe, N., Farnir F., Antoine-Moussiaux, N. and Leroy, P. (2012) Aviculture familiale au Bas-Congo, République Démocratique du Congo (RDC) *Livestock Research for Rural Development* 24, Article #74. http://www.lrrd.org/lrrd24/5/moul24074.htm (accessed 21 February 2020).

Mudiope, J., Speijer, P.R., Maslen, N.R. and Adipala, E. (1998) Evaluation of yam host-plant response to root knot nematodes in Uganda. *African Plant Protection* 4, 119–122.

Mudiope, J., Coyne, D.L., Adipala, E. and Talwana, H.A.L. (2012) Damage to yam (*Dioscorea* spp.) by root-knot nematode (*Meloidogyne* spp.) under field and storage conditions in Uganda. *Nematropica* 42, 137–145.

Mulholland, D.A., Crouch, N., Decker, B. and Smith, M.T. (2002) The isolation of the Amaryllidaceae alkaloid crinamine from *Dioscorea dregeana* (Dioscoreaceae). *Biochemical Systematics and Ecology* 30, 183–185.

Murray, A.P., Faraoni, M.B., Castro, M.J., Alza, N.P. and Cavallaro, V. (2013) Natural AChE inhibitors from plants and their contribution to Alzheimer's disease therapy. *Current Neuropharmacology* 11, 388–413.

Murray, R.D.H., Jorge, Z., Khan, N.H., Shahjahan, M. and Quaisuddin, M. (1984) Diosbulbin D and 8-epidiosbulbin E acetate, norclerodane diterpenoids from *Dioscorea bulbifera* tubers. *Phytochemistry* 23, 623–625.

Muthamia, Z.K., Nyende, A.B., Mamati, E.G., Ferguson, M.E. and Wasilwa, J. (2014) Determination of ploidy among yam (*Dioscorea* species) landraces in Kenya by flow cytometry. *African Journal of Biotechnology* 13, 394–402.

Muthukumarasamy, M. and Panneerselvam, R. (2000) Carbohydrate metabolism in yam tubers during storage and sprouting. *Tropical Science* 40, 63–66.

Mwirigi, P.N., Kahangi, E. M., Nyende, A. B. and Mamati, E.G. (2009) Morphological variability within the Kenyan yam (*Dioscorea* species). *Journal of Applied Biosciences* 16, 894–901.

Na Nyum Lee, Ji Ah Kim, Yong Wook Kim, and Tae Dong Kim (2018b) Microtuberization and acclimatization in the *Dioscorea cayenensis* Thunb. by the carbon source. *Journal of Plant Biotechnology* 45, 125–130.

Nadia, L., Wirakartakusumah, M.A., Andarwulan, N., Purnomo, E.H., Koaze, H. and Noda, T. (2014) Characterization of physicochemical and functional properties of starch from five yam (*Dioscorea alata*) cultivars in Indonesia. *International Journal of Chemical Engineering and Applications* 5, 489–496.

Nagai, T. and Nagashima, T. (2006) Functional properties of dioscorin, a soluble viscous protein from Japanese yam (*Dioscorea opposita* Thunb.) tuber mucilage Tororo. *Zeitschrift für Naturforschung C* 61, 792–798.

Nath, R.C. and Mukherjee, B. (2000) *Dioscorea floribunda,* a potential source of nematicides of plant origin *Nematologia Mediterranea* 28, 145–149.

Nations Encyclopaedia (2008) https://www.nationsencyclopedia.com/Africa/Guinea-AGRICULTURE.html (accessed 24 October 2019).

Ndegwe, N.A., Ikpe, F.N., Gbosi, S.D. and Jaja, E.T. (1990) Effect of staking method on yield and its components in sole-cropped white Guinea yam (*Dioscorea rotundata* Poir.) in a high-rainfall area of Nigeria. *Tropical Agriculture* 67, 29–32.

New Guyana Marketing Corporation (2004) Postharvest Care and Market Preparation National Agricultural Research Institute Technical Bulletin No. 32.

News 784 (2018) St Vincent, 18 December 2018.

Ngo-Ngwe, M.F.S., Joly, S., Bourge, M., Brown, S. and Omokolo, D.N. (2014) Nuclear DNA content analysis of four cultivated species of yams (*Dioscorea* species) from Cameroon. *Journal of Plant Breeding and Genetics* 2, 87–95.

Ngo-Ngwe, M.F.S., Omokolo, D.N. and Joly, S. (2015) Evolution and phylogenetic diversity of yam species (*Dioscorea* species): implication for conservation and agricultural practices. *PLoS ONE* 10, e0145364, https://doi.org/10.1371/journal.pone.0145364 (accessed 6 November 2019).

Nimenibo-Uadia, R. (2003) Effect of aqueous extract of *Canavalia ensiformis* seeds on hyperlipidaemia and hyperketonaemia in alloxan-induced diabetic rats. *Biochemistry* 15, 7–15.

Nimenibo-Uadia, R. and Oriakhi, A. (2017) Proximate, mineral and phytochemical composition of *Dioscorea dumetorium* Pax. *Journal of Applied Sciences and Environmental Management* 21, 771–774.

Niño, J., Díez, M.J. and Nuez, F. (2006) Molecular variability of a collection of *Dioscorea polygonoides* (Willd.) Humb. and Bonpl. (wild yam) of the Gran Caldas region (Colombia) using AFLPs. *The Journal of Horticultural Science and Biotechnology* 81, 803–808.

Njogu, M.K., Geofrey, G.K. and Daniel, P.M. (2015) Comparative effects of foliar application of gibberellic acid and benzylaminopurine on seed potato tuber sprouting and yield of resultant plants. *American Journal of Agriculture and Forestry* 3, 192–201.

Njoh Ellong, E., Billard, C., Pétro, D., Adenet, S. and Rochefort, K. (2015) Physicochemical, nutritional and sensorial qualities of Boutou yam (*Dioscorea alata*) varieties. *Journal of Experimental Biology and Agricultural Sciences* 3, 138–150.

Nkere, C.K., Ihezie, C.I. and Mbanaso, E.N.A. (2011) Micropropagation of yam (*Dioscorea rotundata)*: assessment of performance in cassava starch-gelled medium. *Biotechnology* 10, 551–553.

Nnodu, E.C. and Nwankiti, A.O. (1986) Chemical control of post-harvest deterioration of yam tubers. *Fitopatologia Brasileira* 1, 865–871.

Nonglapat Sasiwatpaisit, Worathat Thitikornpong, Chanida Palanuvej and Nijsiri Ruangrungsi (2014) Dioscorine content in *Dioscorea hispida* dried tubers in Thailand by TLC-densitometry and TLC image analysis. *Journal of Chemical and Pharmaceutical Research* 6, 803–806.

Noon, R.A. (1978) Storage and market diseases of yam. *Tropical Science* 20, 177–188.

Noon, R.A. and Colhoun, J. (1979) Market and storage diseases of yams imported into the UK. *Phytopathologische Zeitschrift* 94, 289–302.

Noon, R.A. and Colhoun, J. (1981) Observations on the nature of chilling injury in yams. *Phytopathologische Zeitschrift* 100, 10–18.

Norman, P.E., Whyte, J.B.A. Samura, A.E., Massaquoi, A., Sesay, L., Dixon, A.G.O., Fomba, S.N., Benya, M.T. and Sowa, M.M. (2015) Effect of staking and non-staking systems on disease severity, yield and quality attributes of yams (*Dioscorea alata*). *Journal of Agriculture and Ecology Research International* 2, 219–229.

Nwankiti, A.O. and Arene, O.B. (1978) Diseases of yam in Nigeria. *Pest Articles and News Summaries* 24, 486–494.

Nweke, F.I. (2016) *Yam in West Africa: food, money and more*. Michigan State University Press, East Lansing.

Nyadanu, D., Dapaah, H. and Agyekum, A.D. (2014) Resistance to post-harvest microbial rot in yam: integration of genotype and storage methods. *African Crop Science Journal* 22, 89–95.

Nystrom, L.W. Shrum Jr., J.E. and Dawson, R.F. (1973) A mechanical harvester for diosgenin-producing yams. 3rd International Symposium on Tropical Root Crops, Ibadan, Nigeria 272–276.

Nzeako, S.O., Barisuka, G.A., Torle, M.E., Imafidor, H.O. and Eze, C.N. (2015) Plant Parasitic Nematodes Status of Seed Yams in Khana L.G.A, Rivers State, Nigeria. *International Journal of Scientific Research in Environmental Sciences* 3, 266–273.

Oben, J.A., Egbe, A.E., Chuyong, G.B. and Tabot, P.T. (2016) Diversity of yam (*Dioscorea* species) populations in South Western Region of Cameroon. *American Journal of Life Sciences* 4, 187–194.

Obi, S. (1981) Pectinasic activity of aerobic bacteria associated with soft rot of yams *Dioscorea rotundata*. *Applied and Environmental Microbiology* 81, 41–43.

Obidiegwu, J.E. and Akpabio E.M. (2017) The geography of yam cultivation in southern Nigeria: Exploring its social meanings and cultural functions. *Journal of Ethnic Foods* 4, 28–35.

Obilo, O.P., Oguamanam, K.N., Ogbedeh, K.O., Onyia, V.N. and Ofor, M.O. (2005) The use of plant extracts in the control of *Aspergillus niger* in the rot of yam (*Dioscorea* species) during storage. *International Journal of Agriculture and Rural Development* 6, 74–80.

Ocitti p'Obwoya, C.N. (2001) Yams (Dioscorea spp.). In: Mukibi, J.K. (ed.) *Agriculture in Uganda Vol II Crops*. Fountain Publishers/ CTA/ NARO 2, 229–232.

Oduro, I., Opata, D.D., Asiedu-Larbi, J. and Ellis, W.O. (2009) The suitability of water yam (*Dioscorea alata*) for couscous production. *Tropical Roots and Tubers in a Changing Climate: A Convenient Opportunity for the World. Fifteenth Triennial Symposium of the International Society for Tropical Root Crops, Lima, Peru, 2-6 November 2009*, 82–88. International Society for Tropical Root Crops-Peru Branch.

Ogali, E.L., Opadokun, J.S. and Okobi, A.O. (1991) Effect of lime and local gin on post-harvest rot of yams. *Tropical Science* 31, 365–370.

Ogaraku, A.O. and Usman H.O. (2008) Storage rot of some yams (*Dioscorea* spp) in Keffi and environs, Nasarawa State, Nigeria. *Polymers for Advanced Technologies* 4, 22–27.

Ogbo Frank, C. and Agu Kingsley, C. (2014a) Role of fungal rots in post-harvest storage losses in some Nigerian varieties of *Dioscorea* species. *British Microbiology Research Journal* 4, 343–350.

Ogbo Frank, C. and Agu Kingsley, C. (2014b) Proximate composition, physiological changes during storage, and shelf life of some Nigerian varieties of yams (*Dioscorea* species). *Journal of Scientific Research and Reports* 3, 553–562.

Ogbuagu, D.H. and Ayoade, A.A. (2012) Presence and levels of common polynuclear aromatic hydrocarbons (PAHs) in staple foods of Nigerians. *Food and Public Health* 2, 50–54.

Ogundana S.K. (1983) Life cycle of *Botryodiplodia theobromae*, a soft rot pathogen of yam. *Phytopathologische Zeitschrift* 106, 204–213.

Ogundana S.K. and Dennis C. (1981) Assessment of fungicides for the prevention of storage rot of yam tubers. *Pesticide Science* 12, 491–494.

Ogundana, S.K., Naqvi, S.H.Z. and Ekundayo, J.A. (1970) Fungi associated with soft rot of yams (*Dioscorea* species) in storage in Nigeria. *Transactions of the British Mycological Society* 54, 445–451.

Ogunleye, A.O. and Ayansola, O.T. (2014) Studies of some isolated rot-causing mycoflora of yams (*Dioscorea* spp.). *American Journal of Microbiology and Biotechnology* 1, 9–20.

Ohwi, J. (1965) Flora of Japan. Meyer, F.G. and Walker, E.H. (eds) *Flora of Japan*, Smithsonian Institution, Washington, D.C., https://library.si.edu/digital-library/book/floraofjapaninen00oiji (accessed 19 December 2019).

Ojo, O.I. and Ojo, J.O. (2009) A comparative study of effects of storage on basic nutritional composition of two major edible *Dioscorea dumentorum* varieties. *Global Journal of Pure and Applied Sciences* 15, 353–355.

Okafor, N. (1966) Microbial rotting of stored yams (*Dioscorea* species) in Nigeria. *Experimental Agriculture*, 2, 179–182.

Okagami, N. and Kawai, M. (1977) Gibberellin-induced inhibition or promotion in seed germination of *D. tokoro* and *D. tenuipes* in relation to light quality. *Plant Physiology* 60, 360–362.

Okagami, N. and Tanno, N. (1991) Dormancy in *Dioscorea*: comparison of dormant characteristics in bulbils of Northern species (*D. opposita*) and Southern species (*D. bulbifera* var. *vera*). *Journal of Plant Physiology*, 138, 559–565.

Okeke, S.E. (2004) The taxonomic position of members of the so-called *Dioscorea cayenensis-rotundata* complex (Dioscoreaceae). *Nigerian Journal of Botany* 17, 95–103.

Okigbo, R.N. (2002) Mycoflora of tuber surface of white yam (*Dioscorea rotundata* Poir.) and postharvest control of pathogens with *Bacillus subtilis*. *Mycopathologia* 156, 81–85.

Okigbo, R.N. (2003) Fungi associated with peels of post harvest yams in storage. *Global Journal of Pure and Applied Sciences* 9, 19–23.

Okigbo, R.N. (2005) Biological control of postharvest fungal rot of yam (*Dioscorea* species) with *Bacillus subtilis*. *Mycopathologia* 159, 307–314.

Okigbo, R.N. and Emeka, A.N. (2010) Biological control of rot-inducing fungi of water yam (*Dioscorea alata*) with *Trichoderma harzianum, Pseudomonas syringae* and *Pseudomonas chlororaphis*. *Journal of Stored Products and Postharvest Research* 1, 18–23.

Okigbo, R.N. and Ikediugwu, F.E.O. (2000) Untersuchungen zur biologischen Bekämpfung der Lagerfäule von Yamswurzeln (*Dioscorea* species) mit *Trichoderma viride*. *Journal of Phytopathology* 148, 351–355.

Okigbo, R.N. and Ikediugwu, F.E.O. (2001) Biological control of tuber surface mycoflora of yam (*Dioscorea rotundata*). *Tropical Science* 41, 85–89.

Okigbo, R.N. and Ikediugwu, F.E.O. (2008) Studies on biological control of postharvest rot in yams (*Dioscorea* spp.) using *Trichoderma viride*. *Journal of Phytopathology* 148, 351–355.

Okoli, O.O. (1975) Yam production from seeds: prospects and problems. Paper presented at the 11th Conference of the Agricultural Society of Nigeria, Enugu, Nigeria, July 1–15, 1975.

Okpogba, T.C., Sobowale, A.A. and Gbadamos, I.T. (2019) Control of some Penicillium and Aspergillus rots of *Dioscorea alata* Poir. and *Dioscorea rotundata* L. using extracts of *Xylopia aethiopica* (Dunal.) Linn. and *Syzygium aromaticum* (Linn.) Merr. *African Journal of Plant Science* 13, 113–124.

Oli, M.T., Becker, H and Maass, B.L. (2008) Diversity, distribution and management of yam landraces (*Dioscorea* spp.) in Southern Ethiopia. *Genetic Resources and Crop Evolution* 55, 115–131.

Olivier, K.A., Konan, K.N., Anike, F.N., Agbo, G.N. and Dodo, H.W. (2012) *In vitro* induction of minitubers in yam (*Dioscorea cayenensis-D. rotundata* complex). *Plant Cell, Tissue and Organ Culture* 109, 179–189.

Olorunda, A.O., McKelvie, A.D. and Macklon, A.E.S. (1974) Effects of temperature and chlorpropham on the storage of the yam. *Journal of the Science of Food and Agriculture* 25, 1233–1238.

Olu, M., Ogunmoyela, O.A.B., Adekoyeni, O.O., Jimoh, O., Oluwajoba, S.O. and Sobanwa, M.O. (2012) Rheological and functional properties of soy-poundo yam flour. *International Journal of Food Science and Nutrition Engineering* 2, 101–107.

Onayemi, O. (1983) Observation on the dehydration characteristics of different varieties of yam and cocoyam. Abstract 6th symposium of the Int. Soc. For Tropical Root crops, Peru.

Ondo Ovono, P., Kevers, C. and Dommes, J. (2010) Effets des conditions de conservation sur la germination des microtubercules d'igname (complexe *Dioscorea cayenensis–D. rotundata*). *Comptes Rendus Biologies* 333, 28–34.

Ondo Ovono, P., Kevers, C. and Dommes, J. (2015) Behaviour of various types of seeds of two species of yams tuber (*Dioscorea cayenensis* Lam. and *Dioscorea rotundata* Poir.) in Gabon. *International Research Journal of Agricultural Science and Soil Science* 5, 58–66.

Ondo Ovono, P., Kevers, C. and Dommes, J. (2016) Effects of planting methods and tuber weights on growth and yield of yam cultivars (*Dioscorea rotundata* Poir.) in Gabon. *International Research Journal of Agricultural Science and Soil Science* 6, 32–42.

Onwueme, I.C. (1973) The sprouting process in yam, *Dioscorea* species tuber pieces. *Journal of Agricultural Science* 8, 11375–11379.

Onwueme, I.C. (1977a) Field comparison of West African planting and harvesting practices in yam (*Dioscorea rotundata*): pre-sprouting, dry season planting, and double-harvesting. *Journal of Agricultural Science* 84, 503–505.

Onwueme, I.C. (1977b) Field comparison of West African planting and harvesting practices in yam (*Dioscorea rotundata*): pre-sprouting, dry season planting, and double-harvesting. *Journal of Agricultural Science* 88, 311–318.

Onwueme, I.C. (1977c) Effect of varying the time of the first harvest, and of late planting, on double-harvest yield of yam (*Dioscorea rotundata*) in field plots. *Journal of Agricultural Science* 88, 737–741.

Onwueme, I.C. (1978) *The Tropical Tuber Crops: Yams, Cassava, Sweet potato, Cocoyams*. John Wiley and Sons, Ltd, Chichester, UK.

Onwueme, I.C. (2016) *Plant Resources of South-East Asia*. http://uses.plantnet-project.org/en/Dioscorea_hispida_(PROSEA) (accessed November 2017).

Onwueme, I.C. and Charles, W.B. (1994) *Tropical Root and Tuber Crops: Production, Perspectives and Future Prospects. FAO Plant Production and Protection*, Paper No. 126, FAO, Rome.

Onwueme, I.C. and Haverkort, A.J. (1991) Modelling growth and productivity of yams (Dioscorea spp): prospects and problems. *Agricultural Systems* 36, 351–367.

Onyeka, T.J., Petro, D., Ano, G., Etienne, S. and Rubens, S. (2006a) Resistance inwater yam (*Dioscorea alata*) cultivars in the French West Indies to anthracnose disease based on tissue culture derived whole plant assay. *Plant Pathology* 55, 671–678.

Onyeka, T.J., Petro, D., Etienne, S., Jacqua, G. and Ano, G. (2006b) Optimizing controlled environment assessment of levels of resistance to yam anthracnose disease using tissue culture-derived whole plants. *Journal of Phytopathology* 154, 286–292.

Opara, L.U. (1999) Yam Storage. In: Bakker-Arkema, F.W. (ed) CIGR *Handbook of Agricultural Engineering 171 Volume IV Agro Processing*, The American Society of Agricultural Engineers, St. Joseph, 182–214.

Opara, L.U. (2003) Yams: Post-harvest operation - post-harvest compendium, AGST/FAO, http://www.fao.org/fileadmin/user_upload/inpho/docs/Post_Harvest_Compendium_-_Yams.pdf (accessed 25 October 2019)

Opata, D.D., Asiedu–Larbi. J., Ellis, W.O. and Oduro, I. (2009) Securing livelihoods through yams. In: Nkamleu, B., Annang, D. and Bacco, N.M. (eds) Proceedings of a technical workshop on progress in yam research for development in West and Central Africa held in Accra, Ghana, 11–13 September 2007. IFAD TAG 704, IITA, Nigeria, pp. 165–178.

Orhevba, B.A. and Osunde, Z.D. (2006) Effect of storage conditions and CIPC treatments on sprouting and weight loss of stored yam tubers. Proceedings of the Seventh International Conference and 28th Annual General Meeting of the Nigerian Institution of Agricultural Engineers held at Zaria, Kaduna State 28, 352–360.

Orkwor, G.C. and Asadu, C.L.A. (1998) Agronomy. In: Orkwor, G.C., Asiedu, R. and Ekanayake, I.J. (eds) *Food Yams: Advances in Research*. International Institute of Tropical Agriculture, Ibadan, Nigeria, 105–141.

Orkwor, G.C., Asiedu, R.and Ekanayake, I.J. (1998) Food Yams: Advances in Research, IITA, Ibadan and NRCRI, Umudike, Nigeria.

Osagie, A.U., (1992) The yam tuber in storage. Postharvest Research Unit, University of Benin, Benin, Nigeria.

Osai, E.O. and Ikotun, T. (1994) Microbial rot of yam minisetts. *Fitopatologia Brasileira* 19, 408–412.

Osai, E.O., Ikotun, T., Asiedu, R. and Hahn, S.K. (1996) Susceptibility of yam cultivars to minisett rot pathogens. *Tropical Science* 36, 201–205.

Osei-Adu, J., Acheampong, P.P., Amengor. N.E. and Sagoe, R. (2016) Input supply structure for yam production in Ghana. *Journal of Economics and Sustainable Development* 7, 72–84.

Osunde, Z.D. (2008) Minimizing postharvest losses in yam (*Dioscorea* spp.): treatments and techniques. In: Robertson, G.L. and Lupien, J.R. (eds) *Using Food Science and Technology to Improve Nutrition and Promote National Development*. International Union of Food Science and Technology 1–12.

Osunde, Z.D. and. Orhevba, B.A. (2011) Effects of storage conditions and pre-storage treatment on sprouting and weight loss of stored yam tubers. *Journal of Stored Products and Postharvest Research* 2, 29–36.

Ota, A.B. (2007) Primitive tribal groups (PTGs) of Orissa: an overview. *Adivasi* 47, 41–47.

Ota, A.B., Mohanty, S.C. and Kodamasingh, A. (2013). *Ho. Bhubaneswar.* Scheduled Castes & Scheduled Tribes Research and Training Institute.

Otegbayo, B., Aina, J., Asiedu, R. and Bokanga, M. (2005) Microstructure of boiled yam (*Dioscorea* spp.) and its implication for assessment of textural quality. *Journal of Texture Studies* 36, 324–332.

Otoo, E., Kleih, U., Boadu, P. and Amponsah, S.K. (2013) Ghana: In-depth value chain analysis - Final Report: Yam Improvement for Income and Food Security in West Africa Project, p. 75.

Otoo, E., Okonkwo, C.C. and Asiedu, R. (2006a) Cultivar evaluation and megaenvironment investigation of *Dioscorea cayenensis* cultivars in Ghana based on the GGE biplot analysis. *Journal of Food, Agriculture and Environment* 4, 162–166.

Otoo, E., Okonkwo, C.C. and Asiedu, R. (2006b) Stability studies of hybrid yam (*Dioscorea rotundata* Poir.) genotypes in Ghana. *Journal of Food, Agriculture and Environment* 4, 234–238.

Otoo, J.A., Osiru, D.S.O., Ng, S.Y. and Hahn, S.K. (1987) *Improved technology for seed yam production.* International Institute of Tropical Agriculture, Ibadan, Nigeria.

Otto, E., Anchirina, V.M., Ennin, S.A. and Asiedu, R. (2008) Sustainable yam production in Ghana the non-staking option. *Journal of Food, Agriculture and Environment* 6, 391–396.

Otusanya, M.O. and Jeger, M.J. (1996) Effect of *Aspergillus niger* on shoot emergence and vine development in field-sown yams (*Dioscorea* species) and rot development under long-term storage conditions. *International Biodeterioration and Biodegradation* 38, 89–100.

Ou-yang, S., Jiang, T., Zhu, L. and Yi, T. (2018) *Dioscorea nipponica* Makino: a systematic review on its ethnobotany, phytochemical and pharmacological profiles. *Chemistry Central Journal* 12, 57.

Oviedo Prieto, R., Herrera Oliver, P., Caluff, M.G, *et al.*, (2012) National list of invasive and potentially invasive plants in the Republic of Cuba – 2011. (Lista nacional de especies de plantas invasoras y potencialmente invasoras en la República de Cuba – 2011) Bissea: Boletin sobre Conservacion de Plantas del Jardin Botanico Nacional de Cuba, 6 (Special Issue 1), 22–96.

Ovono, P.O., Kevers, C and Dommes, J, (2010) Effects of storage conditions on sprouting of microtubers of yam (*Dioscorea cayenensis-D. rotundata* complex). *Comptes Rendus Biologies* 333, 28–34.

Oyama, M., Tokiwano, T., Kawaii, S., Yoshida, Y., Mizuno, K., Oh, K. and Yoshizawa, Y. (2017) Protodioscin, isolated from the rhizome of *Dioscorea tokoro* collected in northern Japan is the major antiproliferative compound to HL-60 leukemic cells. *Current Bioactive Compounds* 13, 170–174.

Oyeleke, S.B., Auta, S.H. and Egwim, E.C. (2010) Production and characterization of amylase produced by *Bacillus megaterium* isolated from a local yam peel dumpsite in Minna, Niger State. *Journal of Microbiology and Antimicrobials* 2, 88–92.

Padhan, B. and Panda, D. (2018) Variation of photosynthetic characteristics and yield in wild and cultivated species of yams (*Dioscorea* species) from Koraput, India. *Photosynthetica* 56, 1010–1018.

Pagare, R., Arisdason, W., Sellappan, K. and Janarthanam, M.K. (2015) Lectotypification of *Dioscorea vexans* (Dioscoreaceae), an endemic and rare species from Andaman and Nicobar Islands, India. *Phytotaxa* 233, 298.

Pakakrong Thongdeeying, Arunporn Itharat, Kaoru Umehar and Srisopa Ruangnoo (2016) A novel steroid and cytotoxic constituents from *Dioscorea membranacea* Pierre against hepatocellular carcinoma and cholangiocarcinoma cells. *Journal of Ethnopharmacology* 194, 91–97.

Panneerselvam, R., Abdul Jaleel, C., Somasundaram, R., Sridharan, R. and Gomathinayagam, M. (2007) Carbohydrate metabolism in *Dioscorea esculenta* (Lour.) Burk. tubers and *Curcuma longa* L. rhizomes during two phases of dormancy. *Colloids and Surfaces B: Biointerfaces* 59, 59–66.

Park, Y.M. and Kwon, K.Y. (1999) Prevention of the incidence of skin blackening by postharvest curing procedures and related anatomical changes in 'Niitaka' pears. *Journal of the Korean Society for Horticultural Science* 40, 65–69.

Park, P., Ishii, H., Adachi, Y., Kanematsu, S., Ieki, H. and Umemoto, S. (2000) Infection behavior of *Venturia nashicola*, the cause of scab on Asian pears. *Phytopathology* 90, 1209–1216.

Passam, H.C. (1982) Dormancy of yams in relation to storage. In: Miège, J. and Lyonga, S.N. (eds) *Yams.* Oxford University Press, Oxford, pp. 285–293.

Passam, H.C. (1999) The curing of yam (*Dioscorea* spp.) tubers for improved storage. In: Gerasopoulos, D. (ed.) *Post-harvest losses of perishable horticultural products in the Mediterranean region*. Chania: CIHEAM, 57–62.

Passam, H.C., Hughes, P.A. and Read, S.J. (1977) Wound repair in yam tubers: the effects of inhibitors of respiration and protein synthesis on repair physiology. *New Phytologist* 79, 593–603.

Passam, H.C., Read, S.J. and Rickard, J.E. (1978) The respiration of yam tubers and its contribution to storage losses. *Tropical Agriculture* 55, 207–213.

Paterne, E.A.S.A., Antoine, A., Estelle, L.Y.L., Clément, A. and Dansi A. (2019) Yam (*Dioscorea rotundata* Poir and *D. cayenensis* Lam complex) in the traditional agriculture of Benin: present-day cultivar diversity and farmers' perception on their tolerance to tuber dry rot caused by the nematode *Scutellonema bradys*. *International Journal of Current Microbiology and Applied Sciences* 8, 1119–1138.

Pedralli, G.G. (2002) Levantamento florístico das dioscoreaceae (r. Br.) Lindley da cadeia do Espinhaço, Minas Gerais e Bahia, Brasil. *Boletim de Botânica da Universidade de São Paulo* 20, 63–119.

Pedralli, G., Carmo, C.A.S., Cereda, M. and Puiatti, M. (2002) Uso de nomes populares para as espécies de Araceae e Dioscoreaceae no Brasil. *Horticultura Brasileira* 20, 530–532.

Penet, L., Barthe, E., Alleyne, A. and Blazy, J. M. (2016) Disease risk perception and diversity of management strategies by farmers: the case of anthracnose caused by *Colletotrichum gloeosporioides* on water yams (*Dioscorea alata*) in Guadeloupe. *Crop Protection* 88, 7–17.

Peng, B., Zhang, Y., Sun, X., Li, M., Xue, J. and Hang, Y. (2016) Genetic relationship and identification of *Dioscorea polystachya* cultivars accessed by ISAP and SCAR markers. *Archives of Biological Sciences*. 69, 277–284.

Pengelly, A. and Bennett, K. (2011) Appalachian plant monographs: *Dioscorea villosa* L., Wild Yam. http://www.frostburg.edu/aces/appalachian-plants/ (accessed 21 December 2019).

Pérez, E., Gibert, O., Rolland-Sabaté, A., Segovia,X., Sánchez, T., Reynes, M. and Dufour, D. (2010) Physicochemical, functional, and macromolecular properties of waxy yam starches discovered from 'mapuey' (*Dioscorea trifida*) genotypes in the Venezuelan Amazon. *Journal of Agricultural and Food Chemistry* 59, 263–273.

Pérez, E.E., Faks, J., Cira, L.E., Schroeder, M. and Diez, N. (2014) Estimacón de la dioscorina extraída de variedades de ñames (género *Dioscorea*) cultivados en el Amazonas Venezolano. *Acta Horticulturae* 1016, 53–60.

Perlas, F.B., Ruiz, B.B. and Pante, R.E. (2010) Field evaluation of selected yam (*Dioscorea alata*) accessions in acid soils and saline-prone areas 19th World Congress of Soil Science, Soil Solutions for a Changing World 1–6 August 2010, Brisbane, Australia. 18–21.

PFAF (2018) Plants For A Future. https://pfaf.org/user/Default.aspx (accessed 25 May 2021)

PFAF ((nd)) *D. japonica* https://pfaf.org/User/Plant.aspx?LatinName=Dioscorea+japonica (accessed 21 May 2021).

PFAF (2019) *kamoonensis*: https://www.pfaf.org/USER/Plant.aspx?LatinName=Dioscorea+kamoonensis; *villosa*: https://pfaf.org/user/Plant.aspx?LatinName=Dioscorea+villosa (accessed 10 September 2019).

Phillips, S., Briddon, R.W., Brunt, A.A. and Hull, R. (1999) The partial characterization of a badnavirus infecting the greater asiatic or water yam (*Dioscorea alata*). *Journal of Phytopathology* 147, 265–269.

Piening, L.J. (1962) A check list of fungi recorded from Ghana. Ghana, Ministry of Agriculture. Bulletin no. 2.

PIER ((nd)) *Pacific Island Ecosystems at Risk*. http://www.hear.org/pier/species/dioscorea_sansibarensis.htm (accessed November 2017).

Pike, S. (2016) Yam investment profile http://www.jamaicatradeandinvest.org/sites/default/files/SandorPike_Yam_15Nov2016.pdf (accessed December 2017).

Pinto Da Luz, C.F., de Oliveira Barbuglio-Santiago, L., Simões, A.R.G., Da Silva, J.H., Dos Santos, V.L. and Kirizawa, M. (2020) Pollen morphology of *Dioscorea* (Dioscoreaceae) from the Atlantic Forest in south-east Brazil (São Paulo) with a contribution to the systematics of Neotropical species, *Grana* 59.

Pinzón-Rico, Y.A. and Raz, L. (2017) Commercialization of Andean wild yam species (*Dioscorea* L.) for Medicinal use in Bogotá, D.C., Colombia. *Economic Botany* 71, 45–57.

Plant List D. *Burkillii* (2019) http://www.plantlist.org/about/#status (accessed 28 November 2019).

Plants of the World Online (2019) *Dioscorea abysmophila* – Plants of the World Online – Kew Science (accessed 24 November 2019).

Plumbley, R.A. and Rees, D.P. (1983) An infestation by *Araecerus fasciculatus* (Degeer) (Coleoptera: Anthribidae) and *Decadarchis mmuscula* (Walsingham) (Lepidoptera: Tineidae) on stored fresh yam tubers in South-East Nigeria. *Journal of Stored Products Research* 19, 93–95.

Plumbley, R.A., Hernandez Montes, A. and Thompson, A.K. (1984) Benomyl tolerance in a strain of *Penicillium sclerotigenum* infecting yams and the use of imazalil as a means of control. *Tropical Agriculture* 61, 182–185.

Plumbley, R.A., Cox, J., Kilminster, K., Thompson, A.K. and Donegan, L. (1985) The effect of imazalil in the control of decay in yellow yam caused by *Penicillium sclerotigenum*. *Annals of Applied Biology* 106, 277–284.

Polycarp, D. (2017) Textural characteristics of seven different yams (*Dioscorea* species) grown and consumed in Ghana. *International Journal of Scientific and Technology Research* 6, 281–284.

Polycarp, D., Afoakwa, E.O., Budu, A.S. and Otoo, E. (2012) Characterization of chemical composition and anti-nutritional factors. in seven species within the Ghanaian yam. *International Food Research Journal* 19, 985–992.

Poornima, G.N. and Ravishankar, R.V. (2009) Evaluation of phytonutrients and vitamin contents in a wild yam, *Dioscorea belophylla* (Prain) Haines. *African Journal of Biotechnology* 8, 971–973.

Prain, D. and Burkill, I.H. (1927) The genus *Dioscorea* in Siam. *Bulletin of Miscellaneous Information, Royal Botanic Gardens, Kew* 1927, (6) 225–247.

Prain, D. and Burkill, I.H. (1936) An account of the genus *Dioscorea* in the East, Part 1. The species which twine to the left. *Annals of the Royal Botanic Garden (Calcutta)* 14, 1–210.

Prain, D. and Burkill, I.H. (1938) An account of the genus *Dioscorea* in the East, Part 2. The species which twine to the right. *Annals of the Royal Botanic Garden (Calcutta)* 14, 211–528.

Preston, W.H., Haun, J.R., Garvin, J.W. and Daum, R.J. (1964) Several aspects of growth, development and sapogenin yield of tubers of *Dioscorea spiculiflora*. *Economic Botany* 18, 323–328.

Price, E.J., Wilkin, P., Sarasan, V. and Frasera, P.D. (2016) Metabolite profiling of *Dioscorea* (yam) species reveals underutilised biodiversity and renewable sources for high-value compounds. *Scientific Reports* 6, 29.

Price, E.J., Bhattacharjee, R., Lopez-Montes, A. and Fraser, P.D. (2017) Metabolite profiling of yam (*Dioscorea* species) accessions for use in crop improvement programmes. *Metabolomics* 13, 144.

Priestley, J.H. and Woffenden, M. (1923) The healing of wounds in potato tubers and their propagation by cut sets. *Annals of Applied Biology* 10, 96–115.

PROSEA orbiculata (2019) *Dioscorea orbiculata* Hook.f. http://powo.science.kew.org/taxon/urn:lsid:ipni.org:names:318350-1 (accessed 28 March 2019).

PROTA (2017) PROTA4U web database. Grubben, G.J.H. and Denton, O.A. (eds) Wageningen, Netherlands: Plant Resources of Tropical Africa. http://www.prota4u.org/search.asp (accessed 21 October 2019).

PROTA4U (2018) https://www.prota4u.org/database/protav8.asp?andg=peandp=Dioscorea+cayenensis+Lam (accessed 24 August 2018).

Purdue University (2019) https://www.purdue.edu/hla/sites/famine-foods/famine_food/dioscorea-daunea/ (accessed 6 August 2019).

Purdue University Famine Foods (2018) https://www.purdue.edu/hla/sites/famine-foods/famine_food/dioscorea (accessed 8 May 2018).

Purseglove, J.W. (1972) *Tropical Crops Monocotyledons*. Longman, London.

Purwandari, U., Suprianto, G., Milanita, D., Supriyanto and Burhan. (2011) Development of functional bread using yam (*Dioscorea* sp.) flour. International Scientific Conference on Nutraceuticals and Functional Food. Kosice, Slovakia: Pavol Jozef Safarik University.

Quansah, J., Saalia,, F., Abbey, L. and Annor, G. (2010) Performance of yam as an alternative to frozen potato French. *Natural Sciences* 8, 70–78.

Quattrocch, U. (2016) *World Dictionary of Medicinal and Poisonous Plants: Common Names*. CRC Press, Boca Raton.

Quenerhervé, P. (1998). Les nematodes de l'igname. In: Berthaud, J., Bricas, N. and Marchand J.L. (eds) *L'igname, plante séculaire et culture d'avenir*. CIRAD, Montpellier, France, pp. 193–204.

Rached, L.B., De Vizcarrondo, C.A., Rincón, A.M. and Padilla, F. (2006) Evaluation of the flour and starch from white and purple varieties of mapuey (*Dioscorea trifida*). *Archivos Latinoamericanos de Nutrición* 56, 375–383.

Raí, S.K., Thakur, R.N., Singh, P. and Sastry, K.S.M. (1986) Tuber rot - a new disease of *Dioscorea composita* and its control. *Current Science* 55, 411–412.

Rajyalakshmi, P. and Geervani, P. (1994) Nutritive value of the foods cultivated and consumed by the tribals of South India. *Plant Foods for Human Nutrition* 46, 53–61.

Rakotobe, L., Berka, M., Huet, H., Djediat, C., Jeannoda, V., Bodo, B., Mambu, L., Crespeau, F. and Edery, M. (2010) Effects of Madagascar yam extracts, *Dioscorea antaly*, on embryo-larval development of medaka fish, *Oryzias latipes*. *Toxicon* 55, 87–91.

Rakotodratsimba (2008) quoted by Andriamparany, J.N., Brinkmann, K., Wiehle, M., Jeannoda, V. and Buerkert, A. (2015) Modelling the distribution of four *Dioscorea* species on the Mahafaly Plateau of south-western Madagascar using biotic and abiotic variables. *Agriculture, Ecosystems and Environment* 212, 38–48.

Ramanujam, T. and Nair, S.G. (1982) Control of sprouting in edible yams *Dioscorea* spp. *Journal of Root Crops* 8, 49–54.

Ramos-Escudero, F., Santos-Buelga, C., Pérez-Alonso, J.J., Yáñez, J.A. and Dueñas, M. (2010) HPLC-DAD-ESI/MS identification of anthocyanins in *Dioscorea trifida* L. yam tubers (Purple Sachapapa). *European Food Research and Technology* 230, 745–752.

Ramser, J., Weising, K., Terauchi, R., Kahl, G., Lopez-Peralta, C. and Terhalle, W. (1997) Molecular marker-based taxonomy and phylogeny of Guinea yam (*Dioscorea rotundata-D. cayenensis*). *Genome* 40, 903–915.

Randriamboavonjy, T., Wilkin, P. Rajaovelona, L.R. and Cable, S. (2013) Sustainable use and conservation of yams in Ambositra-Vondrozo forest corridor, Madagascar. *Scripta Botanica Belgica* 50, 341–348.

Rankine, L.B. (1973) Comparative performance: large-scale production verses small-scale production of yams and sweet potatoes in the Caribbean. International Society for Tropical Root Crops 3rd symposium 187–191.

Rao, A.N and Tan, A.S. (1976) Shoot apex and bulbil development in *Dioscorea sansibarensis* Pax. *Botanical Journal of the Linnaean Society* 72, 285–298.

Rao, P.S. and Beri, R.M. (1952) Tubers of *Dioscorea hispida* Dennst. *The Indian Forester* 78, 146–152.

Rašper, V. and Coursey, D.G. (1967a) Anthocyanins of *Dioscorea alata* L. *Experientia* 23, 611–612.

Rašper, V. and Coursey, D.G. (1967b) Properties of starches of some West African yams. *Journal of the Science of Food and Agriculture* 18, 240–244.

Rätsch, C. and Müller-Ebeling, C. (2003) *The encyclopedia of aphrodisiacs psychoactive substances for use in sexual practices*. Toronto, Canada, Park Street Press.

Ravindran, G. and Wanasundera, J.P.D. (1993) Chemical changes in yam tubers (*Dioscorea alata* and *D. esculenta*) during storage. *Tropical Science* 33, 57–62.

Raz, L. and Agudelo Zamora, H. (2020) Catálogo de Plantas y Líquenes de Colombia. Version 1.1. Universidad Nacional de Colombia. Checklist dataset. https://doi.org/10.15472/7avdhn (accessed via GBIF.org on 27 November 2020).

Raz, L. and Pérez-Camacho, J. (2016) A new species of *Dioscorea* (Dioscoreaceae) from Central Cuba. *Brittonia* 69, 109–113.

Raz, L., Pérez, D., Pinzón Rico, Y.A. and Rojas, M. (2013) *Renewed pharmacological interest in* Dioscorea*: a case study from Colombia*, https://www.researchgate.net/publication/274083015_RENEWED_PHARMACOLOGICAL_INTEREST_IN_DIOSCOREA_A_CASE_STUDY_FROM_COLOMBIA (accessed 4 December 2017).

Red List (2019) The International Union for Conservation Nature's of Threatened Species Red List https://www.iucnredlist.org/ (accessed 18 September 2019).

Reddy, P.P. (2015) *Plant protection in tropical root and tuber crops*. Springer Nature, Switzerland.

Rees, D. and Hammond, L. (2002) Biology of plant commodities. In: Golob, P., Farrell, G. and Orchard, J.E,). *Crop Post-Harvest: Science and Technology*, Vol. 1: *Principles and Practice*. Blackwell Science Ltd, Oxford.

Reid, D.P. (1988) *Chinesische Naturheilkunde*. Orac, Wien.

Retamales, J., Cooper, T. and Montealegre, J. (1997) Effects of cooling and curing regime on ethylene production and storage behaviour of kiwifruit. *Acta Horticulturae* 444, 567–571.

Rickard, J.E. and Coursey, D.G. (1981) Cassava storage. Part 1: Storage of fresh cassava roots. *Tropical Science* 23, 1–32.

Rodriguez-Montero, W. (1997) *Crop Physiology of the Greater Yam* (Dioscorea alata). Ulrich E. Grauer, Stuttgart, Germany.

Rojas, R., Alba, J., Magaña-Plaza, I., Cruz, F. and Ramos-Valdivia, A.C. (1999) Stimulated production of diosgenin in *Dioscorea galeottiana* cell suspension cultures by abiotic and biotic factors. *Biotechnology Letters* 21, 907–911.

Rokem, J.S., Tal, B. and Goldberg, I. (1985) Methods for increasing diosgenin production by *Dioscorea* cells in suspension cultures. *Journal of Natural Products* 48, 210–220.

Rouanet G. (1967) Experiments on yams in Guadeloupe. Proceedings of the International Symposium on Tropical Root Crops. Tai, E.A., Charles, W.B., Haynes, P.H., Iton, E.F. and Leslie, K.A. (eds), Vol. 1, Section 111, pp. 152-158. St. Augustine, Trinidad: University of the West Indies.

Rout, S.D. and Panda, S.K. (2010) Ethnomedicinal plant resources of Mayurbhanj district, Orissa. *Indian Journal of Traditional Knowledge* 9, 68–72.

Royal Botanic Garden Edinburgh (2019) https://chileanendemics.rbge.org.uk/taxa/dioscorea-variifolia-bertero-ex-colla (accessed 5 September 2019).

Rubatzky, V.E. and Yamaguchi, M. (1997) Yams, *Dioscorea*. In: *World Vegetables, Principles, Production and Nutritive Values*, 2nd edition. Boston, Springer, pp.162–182.

Ruffo, C.K., Birnie, A. and Tengnas, B. (2002) *Edible wild plants of Tanzania*. Regional Land Management Unit, Nairobi.

Sadik, S. and O.U. Okereke (1975) A new approach to improvement of yam *Dioscorea rotundata*. *Nature* 254, 134–135.

Sagwansupyakom, C. and Chantraprasong, C. (1984) Yam of Thailand species: importance and utilization. Proceedings 1st Regional Workshop, held in Bangkok, Thailand, 5–8 June 1984, 213–215. http://www.istrc.org/images/Documents/Symposiums/Fifth/5th_symposium_proceedings_0026_213.pdf (accessed 13 April 2018).

Sahoré, D.A., Amani, N.G. and Nemlin, G.J. (2005) The properties of starches from some Côte d'Ivoire wild yam (*Dioscorea*) species. *Tropical Science* 45, 122–125.

Saikia, B. and Sarma, J. (2015) *Dioscorea daunea* Prain and Burkill (Dioscoreaceae): a new report to the Flora of India. *Pleione* 9, 231–234.

Salda, V.B., Ramsden, L., Sun, M. and Corke, H. (1998) Genetic variation in physical properties of flour from selected Asian yams (*Dioscorea* species). *Tropical Agriculture* 75, 212–216.

Sankara Rao, K. (2019) Flora of Peninsular India http://flora-peninsula-indica.ces.iisc.ac.in/index.php (accessed 1 September 2019).

Sasidharan, N. and Pal Fellow, B.P. (2019) *Dioscorea tomentosa J. Koenig ex Spreng*. India Biodiversity Portal, Species https://indiabiodiversity.org/biodiv/species/show/32176 (accessed 25 September 2019).

Sastrapradja, S. (1982) *Dioscorea alata*: its variation and importance in Java, Indonesia. In: Miège, J. and Lyonga, S.N. (eds) *Yams*. Oxford University Press, Oxford, pp. 45–49.

Sato, A. (2012) Export strategies of Japanese agricultural products: a focus on production management in Japanese agricultural co-operatives. *ICCS Journal of Modern Chinese Studies* 5, 62–71.

Sato, H. (2006) A brief report on a large mountain-top community of *Dioscorea praehensilis* in the tropical rainforest of Southeastern Cameroon. *African Study Monographs*, Supplement 33, 21–28.

Satoh, I. and Tanabe, K. (1971) Studies on the use of maleic hydrazide in the cultivation of Chinese yams (*Dioscorea batatas*). *Bulletin of the Faculty of Agriculture, Tottori University* 23, 47–52.

Sautour, M., Mitaine-Offer, A.C., Miyamoto, T., Dongmo, A. and Lacaille-Dubois, M.A. (2004) Antifungal steroid saponins from *Dioscorea cayenensis*. *Planta Medica* 70, 90–92.

Sautour, M., Miyamoto, T., Lacaille-Dubois, M.A. (2006) Steroidal saponins and flavan-3-ol glycosides from *Dioscorea villosa*. *Biochemical Systematics and Ecology* 34, 60–63.

Sautour, M., Mitaine-Offer, A-C. and Lacaille-Dubois, M-A. (2007) The *Dioscorea* genus: a review of bioactive steroid saponins. *Journal of Natural Medicines* 61, 91–101.

Scarcelli, N., P. Cubry, R. Akakpo, A-C. Thuillet, J. Obidiegwu, M.N. Baco, E. Otoo, B. Sonké, A. Dansi, G. Djedatin, C. Mariac, M. Couderc, S. Causse, K. Alix, H. Chaïr, O. François and Y. Vigouroux (2019) Yam genomics supports West Africa as a major cradle of crop domestication. *Science Advances* 5, eaaw1947.

Schales, F.D. and Isenberg, F.M. (1963) The effect of curing and storage on chemical composition and taste acceptability of winter squash. *Proceedings of the American Society for Horticultural Science* 83, 667–674.

Schmidt, R.J. and Moult, S.P. (1983) The dermatitic properties of black bryony (*Thamus communis*). *Contact Dermatitis* 9, 390–396.

Schols, P., Furness, C.A., Wilkin, P., Huysmans, S. and Smets, E. (2001) Morphology of pollen and orbicules in some *Dioscorea* (Dioscoreaceae) species and its systematic implications. *Botanical Journal of the Linnean Society* 136, 295–311.

Schols, P., Furness, C.A., Wilkin, P., Smets, E., Cielen, V. and Huysmans, S. (2003) Pollen morphology of *Dioscorea* (Dioscoreaceae) and its relation to systematics. *Botanical Journal of the Linnean Society* 143, 375–390.

SeaLand (1991) *Shipping guide to perishables*. SeaLand Services Inc. New Jersey, USA.

Segarra-Moragues, J.G., Palop-Esteban, M., González-Candelas, F. and Catalán, P. (2005) On the verge of extinction: genetics of the critically endangered Iberian plant species, *Borderea chouardii* (Dioscoreaceae) and implications for conservation management. *Molecular Ecology* 14, 969–982.

Seguchi, M., Ozawa, M., Nakamura, C., and Tabara, A. (2012) Development of gluten-free bread baked with yam flour. *Food Science and Technology Research* 18, 543–548.

Selina Wamucii (2019) Tanzania yam. https://www.selinawamucii.com/produce/tubers/tanzania-yam/ (accessed 25 October 2019).

Selina Wamucii (2020) https://www.selinawamucii.com/produce/tubers/rwanda-yams/ (accessed 25 February 2020).

Sengupta, J., Mitra, G.C. and Sharma, A.K. (1984) Organogenesis and tuberization in cultures of *Dioscorea floribunda*. *Plant Cell, Tissue and Organ Culture* 3, 325–331.

Shah, H.J. and Lele, S.S. (2012) *In vitro* propagation of *Dioscorea alata* var. *purpurae*. *Applied Biochemistry and Biotechnology* 167, 1811–1817.

Shajeela, P.S., Mohan, V.R., Jesudas, L.L. and Tresina Soris, P. (2011) Nutritional and antinutritional evaluation of wild yam (*Dioscorea* species) *Tropical and Subtropical Agroecosystems* 14, 723–730.

Sharma, S., Gupta R., Deswal R. (2017) *Dioscorea alata* tuber proteome analysis shows over thirty dioscorin isoforms and novel tuber proteins. *Plant Physiology and Biochemistry* 114, 128–137.

Sheikh, N., Kumar, Y., Misra, A.K. and Pfoze, L. (2013) Phytochemical screening to validate the ethnobotanical importance of root tubers of *Dioscorea* species of Meghalaya, North East India. *Journal of Medicinal Plants Studies* 1, 62–69.

Shi, W., Zhang, C., Zhao, D., Wang, L., Li, P. and Li, H. (2018) Discovery of hepatotoxic equivalent combinatorial markers from *Dioscorea bulbifera* tuber by fingerprint-toxicity relationship modelling. *Scientific Reports* 8, 462.

Shin, J.H., Kang, D.K. and Sohn, J.K. (2013) Production of yam mosaic virus (YMV)-free *Dioscorea opposita* plants by cryotherapy of shoot-tips. *Cryo Letters* 34, 149–157.

Shiwachi, H., Ayankanmi, T. and Asiedu, R. (2005a) Effect of photoperiod on the development of inflorescences in white Guinea yam (*Dioscorea rotundata*). *Tropical Science* 45, 126–130.

Shiwachi, H., Kikuno, H. and Asiedu, R. (2005b) Mini tuber production using yam (*Dioscorea rotundata*) vines. *Tropical Science* 45, 163–169.

Shriram, V., Jahagirdar, S., Latha, C., Kumar, V., Puranik, V., Rojatkar, S., Dhakephalkar, P. and Shitole, M. (2008) A potential plasmid-curing agent, 8-epidiosbulbin E acetate, from *Dioscorea bulbifera* L. against multidrug-resistant bacteria. *International Journal of Antimicrobial Agents* 32, 405–410.

Shuka, L., Matevski, V. and Caković, D. (2018) *Dioscorea balcanica*. The IUCN Red List of Threatened Species 2018: e.T13133568A18612550. http://dx.doi.org/10.2305/IUCN.UK.2018-1.RLTS.T13133568A18612550.en (accessed 25 December 2019).

Siadjeu, C., Pucker, B., Viehöver, P., Albach, D.C. and Weisshaar, B. (2020) High contiguity *de novo* genome sequence assembly of trifoliate yam (*Dioscorea dumetorum*) Using Long Read Sequencing. *Genes (Basel)* 11, 274–283.

Sibanda, S., Sychawska, B. and Sibanda, S. (2000) A comparative study of wild yam starch from *Dioscorea schimperiana*. *Journal of Applied Science in Southern Africa* 6, 79–87.

Simoons, F.J. (1990) *Food in China: a cultural and historical inquiry*. CRC Press.

Siqueira, M.V.B.M. (2011) Yam: a neglected and underutilized crop in Brazil. *Horticultura Brasileira* 29, 16–20.

Snowdon, A.L. (1992) *A colour atlas of post harvest diseases and disorders of fruits and vegetables. Volume 2 Vegetables*. Wolfe Scientific Ltd.

Solano, X. (1996) Field practices to improve *Dioscorea trifida* L. planting material production. *Journal of Clinical Biochemistry and Nutrition* 21, 215–217.

Solomon Islands (2014) Yam and pana bulking sites show positive results. https://www.pacific.undp.org/content/pacific/en/home/presscenter/pressreleases/2014/08/28/yam-and-pana-bulking-sites-show-positive-results.html (accessed 25 October 2019).

Somnuk Promdang, Rongrong Homhual, Monthar Wongmaneeroj, Rattana Agarum and Surak Jamjumrus (2017) Nutritional Constituents of Man Lueat (*Dioscorea alata*). *Journal of Science and Technology* 6, 1–13.

Somorin, Y.M., Bankole, M.O., Omemu, A. M. and Atanda, O.O. (2011) Impact of milling on the microbiological quality of yam flours in Southwestern Nigeria. *Research Journal of Microbiology* 6, 480.

Sorh, S., Koné, F.M.T., Binaté, S., Dabonné, S. and Kouamé, L.P. (2015) Nutritional composition and enzyme activities changes occurring in water yam (*Dioscorea alata*) cultivar 'Brazo' during the postharvest storage. *International Journal of Food and Nutritional Sciences* 4, 6–12.

Soto Gomez, M., L. Pokorny, M. B. Kantar, F. Forest, I. J. Leitch, B. Gravendeel, P. Wilkin, S. W. Graham, and J. Viruel. (2019) A customized nuclear target enrichment approach for developing a phylogenomic baseline for *Dioscorea* yams (Dioscoreaceae). *Applications in Plant Sciences* 7, e11254.

Sotomayor-Ramírez, D., González-Vélez, A. and Román-Paoli E. (2003) Yam (*Dioscorea* species) response to fertilization in soils of the semiarid southern coast of Puerto Rico. *The Journal of Agriculture of the University of Puerto Rico* 52, 91–103.

Souleymane, C., Maniga, W., Aimée, A.T., Jean, N. G. and Patrice, K. (2018) Physicochemical properties of pretreated flours and organoleptic characteristics of couscous of four varieties of yams (*Dioscorea cayenensis-rotundata*) grown in Côte d'Ivoire. *International Journal of Pure Applied* Biosciences 6, 747–761.

Sparshott, E.N. (1935) Observations on the formation, development, and structure of the tuber of *Testudinaria elephantipes*, and on the origin of the vegetative shoot. *The Journal of the Linnean Society, Botany*. 49, 593–610.

SPYN, (2003) *Yam: Cultivar Selection for Disease Resistance and Commercial Potential in the Pacific Islands*. South Pacific Yam Network EU-INCO-DC Project Final Report, Cirad, Montpellier, France (accessed 26 April 2018).

Srivastava, A., Dagbenonbakin, G.D. and Gaiser, T. (2010) Effect of fertilization on yam (*Dioscorea rotundata*) biomass production. *Journal of Plant Nutrition* 33, 1056–1065.

St John, H. (1954) The Hawaiian variety of *Dioscorea pentaphylla*, an edible yam. Hawaiian Plant Studies 22. *The Journal of the Polynesian Society* 63, 27–34.

Steingroewer, J., Bley, T., Georgiev, V., Ivanov, I., Lenk, F., Marchev, A. and Pavlov, A. (2013) Bioprocessing of differentiated plant *in vitro* systems. *Engineering in Life Sciences* 13, 26–38.

Stennett, D., Oladeinde, F., Wheatley, A., Bryant, J., Dilworth, L. and Asemota, H. (2014) Effects of *Dioscorea polygonoides* (Jamaican Bitter Yam) supplementation in normocholesterolemic and genetically modified hypercholesterolemic mice species. *Journal of Food Biochemistry* 38, 28–37.

Stephens, J.M. (2009) Cushcush - *Dioscorea trifida* L. HS590. Florida Cooperative Extension Service, University of Florida, Institute of Food and Agricultural Sciences.

Su, P. (1987) Pollen morphology of *Dioscorea* in China. *Acta Phytotaxonomica Sinensia* 25, 357–365.

Sulit, J. I. (1967) Processing and utilization of name (*Dioscorea hispida* Dennst.) tubers. *Araneta Journal of Agriculture* 14, 203–221.

Sumalee Panthong, Srisopa Ruangnoo, Pakakrong Thongdeeying, Busarawan Sriwanthana, and Arunporn Itharat (2014) Immunomodulatory activity of *Dioscorea membranacea* Pierre rhizomes and of its main active constituent Dioscorealide B. *BMC Complementary and Alternative Medicine* 14, 403.

Sun, J.H., Li, Z-C., Jewett, D.K., Britton, K.O., Ye, W.H. and Ge, X-J. (2005) Genetic diversity of *Pueraria lobata* (kudzu) and closely related taxa as revealed by inter-simple sequence repeat analysis. *Weed Research* 45, 255–260.

Sun, X-Q., Zhu, Y-J., Guo, J-L., Peng, B., Bai, M-M., Hang, Y-Y. (2012) DNA barcoding the *Dioscorea* in China, a vital group in the evolution of monocotyledon: use of *mat*K gene for species discrimination. *PLoS ONE* 7, e32057. https://doi.org/10.1371/journal.pone.0032057 (accessed 11 October 2019).

Supinya Tewtrakul and Arunporn Itharat (2006) Anti-allergic substances from the rhizomes of *Dioscorea membranacea*. *Bioorganic and Medicinal Chemistry* 14, 8707–8711.

Susan John, K., George, J., Shanida Beegum, S.U. and Shivay, Y.S. (2016) Soil fertility and nutrient management in tropical tuber crops-An overview. *Indian Journal of Agronomy* 61, 263–273.

Swain, M.R., Ray, R.C. and Nautiyal, C.S. (2008) Biocontrol efficacy of *Bacillus subtilis* strains isolated from cow dung against postharvest yam (*Dioscorea rotundata* L.) pathogens. *Current Microbiology* 57, 407–411.

Swarnkar, S. and Katewa, S.S. (2008) Ethnobotanical observation on tuberous plants from tribal area of Rajasthan (India). *Ethnobotanical Leaflets* 12, 647–666.

Swinkels, J.J.M. (1985) Composition and properties of commercial native starches. *Starch - Stärke* 37, 1–5.

Sykes, W.R. (2003) Dioscoreaceae, new for the adventive flora of New Zealand. *New Zealand Journal of Botany* 41, 727–730.

Tamiru, M., Becker, H.C. and Maass, B.L. (2008) Diversity, distribution and management of yam landraces (*Dioscorea* spp.) in Southern Ethiopia. *Genetic Resources and Crop Evolution* 55, 115–131.

Tamiru, M., Natsume, S., Takagi, H., White, B., Yaegashi, H., Shimizu, M., Yoshida, K., Uemura, A., Oikawa, K., Abe, A., et al. (2017) Genome sequencing of the staple food crop white Guinea yam enables the development of a molecular marker for sex determination. *BMC Biology* 15, 86.

Tang, W. and Eisenbrand, G. (1992) *Chinese Drugs of Plant Origin: Chemistry, Pharmacology, and Use in Traditional and Modern Medicine*. Springer, pp. 459–474.

Tapera, M. and Machacha, S. (2017) Matrix solid phase dispersion extraction and screening of phytochemicals from *Dioscorea steriscus* tubers of Mashonaland Central Province, Zimbabwe. *Der Chemica Sinica* 8, 117–122.

Tariq, M.A. (1999) Effect of curing parameters on the quality and storage life of damaged citrus fruits. PhD thesis, Cranfield University, UK.

Taylor, S.J. and Bancroft, R.D. (2002) Storage of horticultural produce. In: Golob, P., Farrell, G. and Orchard, J.E. *Crop post-harvest science and technology Volume 1 principles and practice*. Blackwell Science, Oxford.

Tchabi, A., Coyne, D., Hountondji, F., Lawouin, L., Wiemken, A. and Oeh, F. (2010) Efficacy of indigenous arbuscular mycorrhizal fungi for promoting white yam (*Dioscorea rotundata*) growth in West Africa. *Applied Soil Ecology* 45, 92–100.

Techen, N., Parveen, I. and Khan, I.A. (2017) A single molecular marker to distinguish between species of Dioscorea. *Genome* 60, 201–207, https://doi.org/10.1139/gen-2015-0105.

Technical Bulletin No. 23 (2004) Postharvest handling: ginger postharvest care and market preparation. www.agrinetguyana.org.gy/moa_mfcl (accessed 4 April 2012).

Teponno, R.B., Tapondjou, A.L., Abou-Mansour, E., Stoeckli-Evans, H., Tane, P. and Barboni, L. (2008) Bafoudiosbulbins F and G, further clerodane diterpenoids from *Dioscorea bulbifera* L. var *sativa* and revised structure of Bafoudiosbulbin B. *Phytochemistry* 69, 2374–2379.

Terauchi, R., Terachi, T. and Tsunewaki, K. (1991) Intraspecific variation of chloroplast DNA in *Dioscorea bulbifera* L. *Theoretical and Applied Genetics* 81, 461–470.

Tetteh, J.P. and Saakwa, C. (1994) Prospects and constraints to yam production in Ghana. *Acta Horticulturae* 380, 355–359.

Thapyai, C., Wilkin, P. and Chayamarit, K. (2004) A rare endemic Thai yam rediscovered: *Dioscorea inopinata* Prain and Burkill (Dioscoreaceae) and its affinities. *Thai Forestry Bulletin (Botanical)* 32, 151–158.

Thapyai, C., Wilkin, P. and Chayamarit, K. (2005a) *Dioscorea craibiana* Prain and Burkill: an endemic Thai yam rediscovered, with new data on its distribution and the morphology of the male plant. *Thai Forest Bulletin (Botany)* 33, 185–192.

Thapyai, C., Wilkin, P. and Chayamarit, K. (2005b) *Dioscorea scortechinii* Prain and Burkill, a new record for Thailand and its infraspecific variation. *Thai Forest Bulletin (Botany)*, (33), 206–212. //www.tci-thaijo.org/index.php/ThaiForestBulletin/article/view/24318 (accessed 18 February 2018).

Thapyai, C., Wilkin, P. and Chayamarit, K. (2005c). The rediscovery of *Dioscorea rockii* Prain and Burkill, endemic to Northern Thailand. *Thai Forest Bulletin (Botany)* 33, 220–227.

Thapyai, C., Wilkin, P. and Chayamarit, K. (2005d). The *Dioscorea* species of Doi Chiang Dao with particular reference to *Dioscorea collettii* Hook.f. (Dioscoreaceae), a new record for Northern Thailand. *Thai Forest Bulletin (Botany)* 33, 213–219.

The Plant List (2013) The Plant List: a working list of all plant species. Available at: 862 http://www.theplantlist.org/ (accessed 26 February 2017).

The Plant List (2018) http://www.theplantlist.org/browse/A/Dioscoreaceae/, http://www.theplantlist.org/tpl/record/kew-240126 (accessed, 7 June 2018).

Thereis Choo (2009) Waging war against *Dioscorea*. *Gardenwise*. https://www.nparks.gov.sg/-/media/nparks-real-content/biodiversity/wildlife-in-sg/invasive-alien-species/2009jangwwagingwaragainstdioscorea.pdf?la=en&hash=25F718959FE0F354F2D2D5AD9B3B2A053960C212 (accessed 21 February 2020).

Thompson, A.K. (1967) The effects of sub-toxic concentrations of plant growth substances on the potato crop. PhD thesis, University of Leeds.

Thompson, A.K. (1972a) Storage and transport of fruit and vegetables in the West Indies. In: *Proceedings of the Seminar/Workshop on Horticultural Development in the Caribbean, Maturin, Venezuela*, 170–176.

Thompson, A.K. (1972b) Report on an assignment on secondment to the Jamaican Government as food storage advisor October 1970-October 1972. *Tropical Products Institute Report*, R278. 36 pp.

Thompson, A.K. (2003) *Fruit and vegetables*. Second edition. Blackwell Publishing, Oxford.

Thompson, A.K. (2015) *Fruit and vegetables. Harvesting, handling and storage*, Third edition, Volume 2, *Vegetables, mushrooms, herbs*. Wiley-Blackwell Publishing, Oxford.

Thompson, A.K., Been, B.O. and Perkins, C. (1973a) Reduction of wastage in stored yams. *Proceedings of the Third Symposium of the International Society for Tropical Root Crops, International Institute for Tropical Agriculture, Nigeria* 443–449.

Thompson, A.K., Been, B.O. and Perkins, C. (1973b) Nematodes in stored yams. *Experimental Agriculture* 9, 281–286.

Thompson, A.K., Been, B.O. and Perkins, C. (1973c) Reduction of wastage in stored yams. Proceedings of the Third Symposium of the International Society for Tropical Root Crops, International Institute for Tropical Agriculture, Nigeria, 443–449.

Thompson, A.K., Been, B.O. and Perkins, C. (1977) Fungicidal treatments of stored yams. *Tropical Agriculture* 54, 179–183.

Thompson, A.K., Booth, R.H. and Proctor, F.J. (1972) Onion storage in the tropics. *Tropical Science* 14, 19–34.

Thompson, A.K., Prange, R.K., Bancroft, R.D. and Tongchai Puttongsiri (2018) *Controlled atmosphere storage of fruits and vegetables*, 3rd edition. CAB International, Wallingford, UK.

Tiama, D., Zoundjihekpon, J., Bationo Kando, P., Sawadogo, N., Nebie, B., Nanema, R. K., Traore, E.R., Tao, I., Mahamadou, S. and Zongo, J.D. (2016) Les ignames 'Yuya' de la province de Passore au Burkina Faso. *International Journal of Innovation and Applied Studies* 14, 1075–1085.

Tindall, H.D. (1983) *Vegetables in the tropics*. Macmillan Press, London.

TNAU (2019) Tamil Nadu Agricultural University, India.

Tobih, F.O. (2014) Methods of insecticide application to control yam tuber beetles (*Heteroligus* spp; coleoptera: dynastidae) in yam monocrop system in delta state, Nigeria. *Agriculture and Biology Journal of North America* 5, 259–264.

Tohda, C., Yang, X., Matsui, M., Inada, Y., Kadomoto, E., Nakada, S., Watari, H. and Shibahara, N. (2017) Diosgenin-rich yam extract enhances cognitive function: a placebo-controlled, randomized, double-blind, crossover study of healthy adults. *Nutrients* 9, 1160.

Toktok, N. (2006) Yam Staking. National Agricultural Research Institute, Wet-lowlands mainland programme. Papua New Guinea.

Tomlins, K.I., Ndunguru, G.T., Rwiza, E. and Westby, A. (2000) Postharvest handling, transport and quality of sweet potato in Tanzania. *Journal of Horticultural Science and Biotechnology* 75, 586–590.

Tonga (2019) Kumara and giant yams sustaining life and livelihoods in Tonga https://food.caritas.org/sweet-potatoes-and-giant-yams-in-tonga/ (accessed 25 October 2019).

Tongchai Puttongsiri and Thompson, A.K. (2019) Controlled atmospheres for fruit and vegetable storage and ripening. In: Cachon, R., Voilley, A. and Girardon, P. (eds) *Gases in Agro-Food Processes*. Elsevier, USA.

Tongco, M.D., Cayetano, M., Restum, E., Restum, J., Alejo, J. and McClatchey, M. (2014) Nutritional analysis of wild yam (*Dioscorea divaricata* Blanco), a culturally important species to the Magbukun Ayta of Kanawan, Morong, Bataan, Philippines. https://ethnobiology.org/nutritional-analysis-wild-yam-dioscorea-divaricata-blanco-culturally-important-species-magbukun-ayta (accessed 14 January 2020).

Tonini, G., Barberini, K., Bassi, F. and Proni, R. (1999) Effects of new curing and controlled atmosphere storage technology on Botrytis rots and flesh firmness in kiwifruit. *Acta Horticulturae* 498, 285–291.

Torres, S. (2001) Preliminary diagnostic of Haitian yellow yam (*Dioscorea cayenensis-rotundata*) and Malanga (*Xanthosoma* species). Hillside Agricultural Program Technical Study, USAID.

Tortoe, C., Akonor, P.T., Nketia, S., Owusu, M., Glover-Amengor, M., Hagan, L. and Otoo, E. (2014a) Evaluation of the sensory characteristics and consumer acceptability of deep-fried yam (*Dioscorea rotundata*) crisps in the Accra Metropolitan Area. *Journal of Food and Nutrition Sciences* 2, 19–23.

Tortoe, C., Dowuona, S., Dziedzoave, N.T. and Rees, D. (2014b) Effect of curing treatments on seven key farmers' yams (*Dioscorea* species) in Ghana. *Agricultural Sciences* 5, 1119–1128.

Tostain, S., Agbangla, C., Scarcelli, N., Mariac, C., Daïnou, O., Berthaud, J. and Pham, J-L. (2007) Genetic diversity analysis of yam cultivars (*Dioscorea rotundata* Poir.) in Benin using SSR markers. *Plant Genetic Resources: Characterization and Utilization* 5, 71–81.

Tridge (2019) Yam, Costa Rica. https://www.tridge.com/intelligences/yam/CR (accessed 26 December 2019).

Tropicos (2021) http://legacy.tropicos.org/Home.aspx (accessed 30 March 2021).

Tschannen, A.T. (2003) Controlling post-harvest losses of yam (*Dioscorea* spp.) by application of gibberellic acid. Dissertation N° 14942, Swiss Federal Institute of Technology, Zurich.

Tsen, C.C. and Tang, R.T. (1971) K-State process for making high-protein breads. *Baker's Digest* 45, 26–27, 30–32.

Tu, M. (2002) *Dioscorea oppositifolia*. In: Randall, J. and Martin, T. (eds) *Global Invasive Species Team*. The Nature Conservancy, Davis, CA, https://wiki.bugwood.org/Archive:TNC/Dioscorea_oppositifolia (accessed 31 March 2021).

Ubuebo, A.E. (1971) The effect of external supply of growth substances on axillary proliferation and development in *Dioscorea bulbifera*. *Annals of Botany* 35, 159, 24.

Uguru, G.C., Akinyanju, J.A. and Sani, A. (1997) The use of yam peel for growth of locally isolated *Aspergillus niger* and amylase production. *Enzyme and Microbial Technology* 21, 48–51.

Ukom, A.N., Ojimelukwe, P.C., Ezeama, C.F., Ortiz, D.O. and Aragon, I.I. (2014) Proximate composition and carotenoid profile of yams (*Dioscorea* species) and cocoyam (*Xanthosoma maffa* Scoth) root tubers from Nigeria. *American Journal of Food and Nutrition* 4, 1–10.

UNDP (2012) *Chad MDG acceleration framework, food and nutrition security. http://www.undp.org › Chad ENG - april 4 PRINT* (accessed 24 May 2021).

University of Hawaii Manoa Library (2018) Traditional Pacific Island Crops: Yam. https://guides.library.manoa.hawaii.edu/paccrops/yam (accessed 7 August 2018).

University of Massachusetts (2013) Garlic harvest, curing, and storage. https://ag.umass.edu/vegetable/fact-sheets/garlic-harvest-curing-storage (accessed 14 November 2019).

Uphof, J.C. (1959) *Useful Tropical Plants*. Germany, Weinheim. http://tropical.theferns.info/viewtropical.php?id=Dioscorea+trifoliata (accessed 9 July 2019).

USDA (2019) Germplasm Resources Information Network (GRIN-Taxonomy). National Germplasm Resources Laboratory, Beltsville, Maryland. https://npgsweb.ars-grin.gov/gringlobal/taxonomydetail.aspx?id=418158 (accessed 6 September 2019).

USDA-NRCS (2012) The PLANTS Database. http://plants.usda.gov. National Plant Data Center, Baton Rouge, LA 70874-4490 USA.

Useful Plants of Boyacá Project (2019) https://www.kew.org/science/our-science/projects/boyaca-useful-plants (accessed 19 July 2019).

USNAS (1981) *Post-harvest food losses in developing countries*. United States National Academy of Science, Washington, D.C.

Valéria de Souza, A., Bertoni, B.W., França, S.C. and Soares Pereira, A.M. (2011) Micropropagação de *Dioscorea multiflora* Grised. *Ciência e Agrotecnologia* 35, 92–98.

Van Wijmeersch, P. (1986) Root crop production in Tonga. Field Document 13. Food and Agriculture Organization of the United Nations in association with the South Pacific Commission, Suva, Fiji.

Van Wijmeersch, P. and Bule, F. (1988) Research and development of root crops in Vanuatu. Food and Agriculture Organization of the United Nations, TCP/VAN/4403 and 6652, Santo

Vanisree, M., Lee, C-Y., Lo, S-F., Nalawade, S.M., Lin, C.Y. and Tsay, H-S. (2004) Studies on the production of some important secondary metabolites by plant tissue cultures. *Botanical Bulletin Academia Sinica Taipei* 45, 1–22.

Vasudevan, K. and Jos, J.S. (1992) Gamma ray effects on yams (*Dioscorea alata* L and *D. esculenta* Lour Burk). *Journal of Root Crops* 18, 94–98.

Veasey, E.A., Bressan, E.A., Siqueira, M.V.B.M., Borges, A., Queiroz-Silva, Jr., Pereira, K.J.C., Recchia, G.H. and Ming, L.C. (2012) Genetic characterization of cassava (*Manihot esculenta* Crantz) and yam (*Dioscorea trifida* L.) landraces in swidden agriculture systems in Brazil. In: Gepts, P., Famula, T., Bettinger, R., Qualset, C., McGuire, P., Brush, S. and Damania, A. (eds) *Biodiversity in agriculture: domestication, evolution and sustainability*. Cambridge University Press, Cambridge, pp. 344–360.

Vernier, P., Dossou, R.A. and Letourmy, P. (1999) La fabrication d'igname à partir de *Dioscorea alata*: influence de la variété et du type de chips sur le séchage, la conservation et les qualités organoleptiques. *African Journal of Root and Tuber Crops* 3, 62–67.

Vernier, P., Dossou, R. A. and Letourmy, P. (2000) Influence de la fertilisation chimique sur les qualités organoleptiques de l'igname. *Cahiers Agricultures* 9, 131–134.

Vernier P., Goergen G., Dossou R. A., Letourmy P. and Chaume J. (2005) Utilization of biological insecticides for the protection of stored yam chips. *Outlook on Agriculture* 34, 173–179.

Viana, A.M. and Felippe, G.M. (1990) Effects of storage on germination of *Dioscorea composita* (Dioscoreaceae) seeds. *Economic Botany* 44, 311–317.

Vidal, P.J. (1960) Useful plants of Laos. https://www.persee.fr/doc/jatba_0021-7662_1960_num_7_9_2624 (accessed 3 April 2018).

Viruel, J., Segarra-Moragues, J.G., Raz, L., Forest, F., Wilkin, P. and Sanmartin, I. (2016) Late Cretaceous-Early Eocene origin of yams (*Dioscorea*, Dioscoreaceae) in the Laurasian Palearctic and their subsequent Oligocene-Miocene diversification. *Journal of Biogeography* 43, 750–762.

Von Teichman, I., van der Schijff, H.P. and Robbertse, P.J. (1975) The genus *Dioscorea* L. in South Africa. *Boissiera* 24, 215–224.

Wagner, W.L., Herbst, D.R. and Sohmer, S.H. (1999) *Manual of the flowering plants of Hawai'i*. Revised edition: Volume 1. Bishop Museum Special Publication 97. Honolulu, HI: University of Hawai'i Press; Bishop Museum Press.

Waizel, J. (2009) The traditional use of the species of the genus *Dioscorea*. *Revista de Fitoterapia* 9, 53–67.

Walck, J.L., Cofer, M.S. and Hidayati, S.N. (2010) Understanding the germination of bulbils from an ecological perspective: a case study on Chinese yam (*Dioscorea polystachya*). *Annals of Botany* 106, 945–955.

Walter, W.M. and Schadel, W.E. (1983) Structure and composition of normal skin (periderm) and wound tissue from cured sweetpotatoes. *Journal of the American Society for Horticultural Science* 108, 909–914.

Wanasundera, J.P. and Ravindran, G. (1994) Nutritional assessment of yam (*Dioscorea alata*) tubers. *Plant Foods Human Nutrition* 46, 33–39.

Wang, L., Meng, Q., Wang, C., Liu, Q., Peng, J., Huo, X., Sun, H., Ma, X. and Liu, K. (2013) Dioscin restores the activity of the anticancer agent adriamycin in multidrug-resistant human leukemia K562/adriamycin cells by down-regulating MDR1 via a mechanism involving NF-κB signaling inhibition. *Journal of Natural Products* 76, 909–914.

Wang, W., Zhao, Y., Jing, W., Zhang, J., Xiao, H., Zha, Q. and Liu, A. (2015) Ultrahigh-performance liquid chromatography-ion trap mass spectrometry characterization of the steroidal saponins of *Dioscorea panthaica* Prain et Burkill and its application for accelerating the isolation and structural elucidation of steroidal saponins. *Steroids* 95, 51–65.

Wang, S., Liu, H., Gao, W., Chen, H., Yu, J. and Xiao, P. (2006a) Characterization of new starches separated from different Chinese yam (*Dioscorea opposita* Thunb.) cultivars. *Food Chemistry* 99, 30–37.

Wang, S., Liu, H., Gao, W., Chen, H., Yu, J. and Xiao, P. (2006b) Studies on the physicochemical, morphological, thermal and crystalline properties of starches separated from different *Dioscorea opposita* cultivars. *Food Chemistry* 99, 38–44.

Wang, S.L., Cai, B., Cui, C.B., Liu, H.W., Wu, C.F. and Yao, X.S. (2004) Diosgenin-3-O-alpha-L-rhamnopyranosyl-(1 --> 4)-beta-D-glucopyranoside obtained as a new anticancer agent from *Dioscorea futschauensis* induces apoptosis on human colon carcinoma HCT-15 cells via mitochondria-controlled apoptotic pathway. *Journal of Asian Natural Products Research* 6, 115–125.

Wang, Y., Liu, H., Bao, J., Hong, Y., Yang, Z. and Zhang, C. (2008) The saccharification-membrane retrieval-hydrolysis (SMRH) process: a novel approach for cleaner production of diosgenin derived from *Dioscorea zingiberensis*. *Journal of Cleaner Production* 16, 1133–1137.

Washaya, S., Mupangwa, J.F. and Muranda, E. (2016) Nutritional value and utilization of yams (*Dioscorea steriscus*) by residents of Bindura town high density suburbs, *Zimbabwe. Advances in Agriculture*, article id 5934738, http://dx.doi.org/10.1155/2016/5934738 (accessed 21 November 2017).

Watt, M.P. (2012) The status of temporary immersion system (TIS) technology for plant micropropagation. *African Journal of Biotechnology* 11, 14025–14035.

Weber, O., Wilkin, P. and Rakotonasolo, F. (2005) A new species of edible yam (*Dioscorea* L.) from Western Madagascar. *Kew Bulletin* 60, 283–291.

Webster, J., Beck, W. and Ternai, B. (1984) Toxicity and bitterness in Australian *Dioscorea bulbifera* L. and *Dioscorea hispida* Dennst. from Thailand. *Journal of Agricultural and Food Chemistry* 32, 1087–1090.

Wickham, L.D. (1988) Extension of dormancy in cush-cush yams (*Dioscorea trifida*) by treatment with gibberellic acid. *Tropical Science* 28, 75–77.

Wickham, L.D., Wilson, L.A, and Passam, H.C. (1981) Tuber germination and early growth in four edible *Dioscorea* species. *Annals of Botany* 47, 87–95.

Wickham, L.D., Passam, H.C. and Wilson, L.A. (1984a) Tuber development, storage and germination in yams (*Dioscorea* species) in response to pre-harvest application of plant growth regulators. *Journal of Agricultural Science* 102, 437–442.

Wickham, L.D., Passam, H.C. and Wilson, L.A. (1984b) Dormancy responses to post-harvest application of growth regulators in *Dioscorea* species: I. Responses of bulbils, tubers and tuber pieces of three *Dioscorea* species. *Journal of Agricultural Science* 102, 427–432.

Wickham, L.D., Wilson, L.A, and Passam, H.C. (1982) The origin, development and germination of bulbils in two *Dioscorea* species. *Annals of Botany* 50, 621–627.

Wikiwand (2018) *Dioscorea*. http://www.wikiwand.com/en/Dioscorea (accessed 2 August 2018).

Wilkin, P. (1999) A Revision of the compound-leaved yams (*Dioscorea*, Dioscoreaceae) of Africa. *Kew Bulletin* 54, 19–39.

Wilkin, P. and Muasya, A.M. (2015) Clarifying the *Dioscorea buchananii* Benth. species complex: a new potentially extinct subspecies for South Africa. *PhytoKeys* 48, 51–72.

Wilkin, P. and Thapyai, C. (2009) Dioscoreaceae. In: Santsuk, T. and Larsen, K. (eds) *Flora of Thailand* 10, 1–140.

Wilkin, P. and Timberlake, J. (2009); Dioscoreaceae. In: Timberlake, J. (ed.) *Flora Zambesiaca* 12, 109–140. Royal Botanic Gardens, Kew.

Wilkin, P., Rajaonah, M.T., Jeannoda, V.H., Hladik, A., Jeannoda, V.L. and Hladik, C.M. (2008) An endangered new species of edible yam (*Dioscorea*, Dioscoreaceae) from Western Madagascar and its conservation. *Kew Bulletin* 63, 113–120.

Wilkin, P., Hladik, A., Jeannoda, V. and Weber, O. (2009) The threatened edible yams of the *Dioscorea sambiranensis* R. Knuth species complex (Dioscoreaceae): a new species and subspecies. *Adansonia* 31, 249–266, https://www.researchgate.net/publication/232669003_The_threatened_edible_yams_of_the_Dioscorea_sambiranensis_RKnuth_species_complex_Dioscoreaceae_A_new_species_and_subspecies (accessed March 21 2018).

Wilkin, P., Kennerley, J.A., Rajaonah, M.T. et al. (2017) A new species of critically endangered edible yam endemic to northern Madagascar, *Dioscorea irodensis* (Dioscoreaceae) and its conservation. *Kew Bull* 72, 15. https://doi.org/10.1007/s12225-017-9677-6

Wilson, J.E. (1980) Careful storage of yams. Commonwealth Secretariat, London.

Wilson, J.E. (1982) Progress in the breeding of yam, *Dioscorea* species. In: Miège, J. and Lyonga, S.N. (eds) *Yams*. Oxford University Press, Oxford, pp. 17–22.

Wilson, J.E. and Hamilton, L.S. (1988) A practical guide to identifying yams the main species of *Dioscorea* in the pacific islands. USP Institute for Research, Extension and Training in Agriculture (IRETA). USP Alafua Campus P.O. Private Bag Apia, Western Samoa.

Wilson, L.A., Wickham, L.D. and Ferguson, T.U. (1998) Alternative manifestations in origin, form and function of the primary nodal complex of yams (*Dioscorea* species): A review. *Tropical Agriculture* 75, 77–83.

Winch, J.E., Newhook, F.J., Jackson, G.V.H. and Cole, J.S. (1984) Studies of *Colletotrichum gloeosporioides* disease on yam (*Dioscorea alata*) in the Solomon Islands. *Plant Pathology* 33, 467–477.

Wireko-Manu, F., Oduro, I., Ellis, W.O., Robert, A. and Maziya-Dixon, B. (2013) Food quality changes in water yam (*Dioscorea alata*) during growth and storage. *Asian Journal of Agriculture and Food Sciences* 1, 66–72.

Wireko-Manu, F.D., Ellis, W.O., Oduro, I., Asiedu, R. and Maziya-Dixon, B. (2014) Prediction of the suitability of Water Yam (*Dioscorea alata*) for amala product using pasting and sensory characteristics. *Journal of Food Processing and Preservation* 38, 1339–1345.

Woldetsadik, S.K. and Workneh, T.S. (2010) Effects of nitrogen levels, harvesting time and curing on quality of shallot bulb. *African Journal of Agricultural Research* 5, 3342–3353.

Wu, Z-G., Li, X.X, Lin, X.C., Jiang, W., Tao, Z.M., Mantri, N., Fan, C.Y. and Bao, X.Q. (2014) Genetic diversity analysis of yams (*Dioscorea* species) cultivated in China using ISSR and SRAP markers. *Genetic Resources and Crop Evolution* 61, 639–650.

Wu, Z-G., Jiang, W., Wei, Y-H. and Tao, Y. (2015) New variety breeding of *Dioscorea alata*, cultivar 'Wenshanyao No.1'. *China Journal of Chinese Materia Medica* 40, 1705–1709.

Xia, W., Zhang, B., Xing, D., Li, Y., Wu, W., Xiao, Y., Sun, J., Dou, Y., Tang, W., Zhang, J., Huang, X., Xu, Y., Xie, J., Wang, J. and Huang, D. (2019) Development of high-resolution DNA barcodes for *Dioscorea* species discrimination and phylogenetic analysis. *Ecology and Evolution* 9, 10843–10853.

Xiao, H.W., Yao, X.D., Lin, H., Yang, W.X., Meng, J.S. and Gao, Z.J. (2012) Effect of SSB (superheated steam blanching) time and drying temperature on hot air impingement drying kinetics and quality attributes of yam slices. *Journal of Food Process Engineering* 35, 370–390.

Xifreda, C.C. (1989) Estudios en Dioscoreaceae V. Rehabilitacion de *Dioscorea monadelpha* y identificacion de *Dioscorea subhastata*. *Darwiniana* 29, 401–404.

Xifreda, C.C. and Kirizawa, M. (2003) Observaciones morfológicas, nomenclaturales y taxonómicas en *Dioscorea monadelpha* y *D. subhastata* (Dioscoreaceae). *Darwiniana* 41: 207–215.

Xifreda, C.C. and Seo, M.N. (2009) Flora del Valle de Lerma. Herbario Mcns, Facultad de Ciencias Naturales, Universidad Nacional De Salta. http://eprints.natura.unsa.edu.ar/408/1/DIOSCOREACEAE.pdf (accessed 9 June 2018).

Xu, G.J. and Xu, L.S. (1997) *Species systematization and quality evaluation of commonly used Chinese traditional drugs*. Vol. II, Fujian Science and Technology Press, Fuzhou.

Yan, H., Yang, L and, Li, Y. (2011) Improved growth and quality of *Dioscorea fordii* Prain et Burk and *Dioscorea alata* plantlets using temporary immersion system. *African Journal of Biotechnology* 10, 19444–19448.

Yang, M. and Chen, Y. (1983) Steroidal sapogenins in *Dioscorea collettii*. *Planta Medica* 49, 38–42.

Yang, M.H., Chin, Y-W. and Kim, J. (2014) Phenolic compounds with pancreatic lipase inhibitory activity from Korean yam (*Dioscorea opposita*). *Journal of Enzyme Inhibition and Medicinal Chemistry* 29, 1–6.

Yang, S.L. and Liu, X.K. (2005) Three new diaryl propanes from *Dioscorea composita*. *Chinese Chemical Letters* 16, 57–60.

Yang, X. and Tohda, C. (2018) Diosgenin restores Aβ-induced axonal degeneration by reducing the expression of heat shock cognate 70 (HSC70) *Scientific Reports* 8, 11707.

Yi, T., Fan, L-L., Chen, H-L., Zhu, G-Y., Suen, H-M., Tang, Y-N., Zhu, L., Chu, C., Zhao, Z-Z. and Chen, H-B. (2014) Comparative analysis of diosgenin in *Dioscorea* species and related medicinal plants by UPLC-DAD-MS. *BMC Biochemistry* 15, 19.

Yin, J., Kouda, K., Tezuka, Y., Tran, Q.L., Miyahara, T., Chen, Y. and Kadota, S. (2003) Steroidal glycosides from the rhizomes of *Dioscorea spongiosa*. *Journal of Natural Products* 66, 646–650.

Yoshida, Y., Seino, T. and Kanahama, K. (1999) Combination effects of temperature and photoperiod on the development of flowering spikes and tubers in Chinese yams (*Dioscorea opposita* cv. Ichoimo). *Journal of the Japanese Society of Horticultural Science* 68, 603–607.

Yoshida, Y., Takahashi, H., Kanda, H. and Kanahama, K. (2000) Interactive effects between photoperiod and seed tuber weight on the development of flowering spikes and tubers in Chinese yams (*Dioscorea opposita* cv. Ichoimo). *Journal of the Japanese Society of Horticultural Science* 69, 472–476.

Yoshida, Y., Takahashi, H., Kanda, H. and Kanahama, K. (2001) Interactive effects of photoperiods and plant growth regulators on the development of flowering spikes and tubers in Chinese yams (*Dioscorea opposita* cv. Ichoimo). *Journal of the Japanese Society of Horticultural Science* 70, 304–309.

Yuan, S.L. (2008) Research advances on chemical compositions and bioactivities of *Dioscorea opposite* Thunb. (Chinese yam). *Food Research and Development* 29, 176–179.

Yusuf, C. and Okusanya, B.A.O. (2008) Fungi associated with the storage rot of yam *(Discorea rotundata* Poir.) in Yola, Adamawa State. *Journal of Sustainable Development in Agriculture and Environment* 3, 99–103.

Yusuf, M., Begum, J., Hoque. M.N. and Chowdhury, J.U. (2009) *Medicinal plants of Bangladesh.* BCSIR Chittagong.

Zhang, Y.L., Li, C.S., Zhang, B., Qi, J.S., Xu, S.T. and Ma, L.G. (2018) First report of root rot of Chinese yam caused by *Pythium ultimum* var. *ultimum* in China. *Plant Disease* 102, 687.

Zhang, Z., Gao, W., Wang, R. and Huang, L. (2014) Changes in main nutrients and medicinal composition of Chinese yam (*Dioscorea opposita*) tubers during storage. *Journal of Food Science and Technology* 51, 2535–2543.

Zhao, G., Kan, J., Li, Z. and Chen, Z. (2005) Structural features and immunological activity of a polysaccharide from *Dioscorea opposita* Thunb roots. *Carbohydrate Polymers* 61, 125–131.

Zheng, Y., Xia, B., Hang Y., Wang X., Zhou Y. and Wu B. (2007) Molecular authentication and relationships of *Dioscorea polystachya* Turcz. and its related species. *Journal of Nanjing Agricultural University* 2, 55–59.

Zohouri, G.P. (1998) Les maladies de l'igname (*Dioscorea* species) In: Berthaud, J., Bricas, N. and Marchand, J.L. (eds) *L'igname, plante séculaire et culture d'avenir.* Proceedings of International Symposium. Editions CIRAD, Montpellier, France, pp. 223–230.

Zulu, D., Ellis, R.H. and Culham, A. (2019) Collection, consumption, and sale of Lusala (*Dioscorea hirtiflora*) – a Wild Yam – by rural households in Southern Province, Zambia. *Economic Botany* 73, 47–63.

Index

Note: Page numbers in **bold** type refer to **figures**; Page numbers in *italic* type refer to *tables*